# Dioxygen Binding and Sensing Proteins

A Tribute to Beatrice and
Jonathan Wittenberg

# PROTEIN REVIEWS

Recent Volumes in this Series

# Dioxygen Binding and Sensing Proteins

## A Tribute to Beatrice and Jonathan Wittenberg

**Edited by**
**Martino Bolognesi**
*Department of Biomolecular Sciences and Biotechnology,*
*University of Milan, Milan, Italy*

**Guido di Prisco**
*Institute of Protein Biochemistry, CNR, Naples, Italy*

**Cinzia Verde**
*Institute of Protein Biochemistry, CNR, Naples, Italy*

 Springer

Martino Bolognesi
Department of Biomolecular Sciences and Biotechnology
University of Milan, Milan, Italy

Guido di Prisco
Institute of Protein Biochemistry, CNR, Naples, Italy

Cinzia Verde
Institute of Protein Biochemistry, CNR, Naples, Italy

Library of Congress Control Number: 2008930799

ISBN: 978-88-470-0806-9 Springer Milan Berlin Heidelberg New York
e-ISBN: 978-88-470-0807-6

Springer is a part of Springer Science+Business Media
springer.com
© Springer-Verlag Italia 2008

Printed on acid-free paper

Typesetting: Graphostudio, Milan, Italy
Printing and binding: Arti Grafiche Nidasio, Assago (MI), Italy

*Printed in Italy*
Springer-Verlag Italia S.r.l., Via Decembrio 28, I-20137 Milan

# Contents

# Contributors

**Paolo Ascenzi**
National Institute for Infectious Diseases, I.R.C.C.S. "Lazzaro Spallanzani",
Rome, Italy
*and*
Interdepartmental Laboratory for Electron Microscopy, University "Roma Tre",
Rome, Italy

**Abdu I. Alayash**
Laboratory of Biochemistry and Vascular Biology, Center for Biologics
Evaluation and Research, Food and Drug Administration, Bethesda, MD, USA

**Cyril Appleby**
CSIRO Division of Plant Industry, Canberra, Australia

**Xavier Bailly**
Station Biologique de Roscoff, Roscoff, France

**Michael Berenbrink**
Integrative Biology Research Division, School of Biological Sciences,
University of Liverpool, Liverpool, UK

**Martino Bolognesi**
Department of Biomolecular Sciences and Biotechnology, University of Milan,
Milan, Italy

**Celia Bonaventura**
Nicholas School of the Environment and Earth Sciences, Duke University
Marine Laboratory, Beaufort, NC, USA

**Alberto Boffi**
Department Biochemical Sciences "A. Rossi Fanelli", University of Rome
"La Sapienza", Rome, Italy

**Tom Brittain**
Metalloprotein Structure & Function Laboratory, School of Biological Sciences,
University of Auckland, Auckland, New Zealand

**Maurizio Brunori**
Department of Biochemical Sciences, University of Rome "La Sapienza",
Rome, Italy

**Thorsten Burmester**
Institute of Zoology, Molecular Animal Physiology, University of Mainz, Mainz,
Germany

**Emilia Chiancone**
Department of Biochemical Sciences "A. Rossi-Fanelli", University of Rome
"La Sapienza", Rome, Italy

**Gianni Colotti**
Institute of Molecular Biology and Pathology, CNR, Rome, Italy

**Alvin L. Crumbliss**
Department of Chemistry, Duke University, Durham, NC, USA

**David Dantsker**
Department of Physiology and Biophysics, Albert Einstein College of Medicine,
Bronx, New York, NY, USA

**Sylvia Dewilde**
Department of Biomedical Sciences, Antwerp University, Belgium

**Guido di Prisco**
Institute of Protein Biochemistry, CNR, Naples, Italy

**Gabriella Fanali**
Department of Structural and Functional Biology, and Center of Neuroscience,
University of Insubria, Busto Arsizio (VA), Italy

**Mauro Fasano**
Department of Structural and Functional Biology, and Center of Neuroscience,
University of Insubria, Busto Arsizio (VA), Italy

**Riccardo Fesce**
Department of Structural and Functional Biology, and Center of Neuroscience,
University of Insubria, Busto Arsizio (VA), Italy

**Joel Friedman**
Department of Physiology and Biophysics, Albert Einstein College of Medicine, Bronx, New York, NY, USA

**Eva Geuens**
Department of Biomedical Sciences, Antwerp University, Belgium

**Michel Guertin**
Department of Biochemistry, Faculty of Sciences and Engineering, Laval University, Quebec, Canada

**Thomas Hankeln**
Institute of Molecular Genetic, Johannes Gutenberg University Mainz, Mainz, Germany

**Robert Henkens**
Nicholas School of the Environment and Earth Sciences, Duke University Marine Laboratory, Beaufort, NC, USA

**David Hoogewijs**
Department of Biology and Center for Molecular Phylogeny, Ghent University, Ghent, Belgium

**Kiyohiro Imai**
Yokohama City University, Tsurumi, Yokohama, Japan

**Juliette Lecomte**
Department of Chemistry, The Pennsylvania State University, University Park, PA, USA

**Michael C. Marden**
Inserm U779, University of Paris, Le Kremlin-Bicetre, France

**Mario Milani**
CNR-INFM c/o Department of Biomolecular Sciences and Biotechnology, University of Milan, Milan, Italy

**Lelio Mazzarella**
Department of Chemistry, University of Naples "Federico II", Naples, Italy

**Luc Moens**
Department of Biomedical Sciences, Antwerp University, Belgium

**Kiyoshi Nagai**
Laboratory of Molecular Biology, Medical Research Council Centre, Cambridge, UK

**Marco Nardini**
Department of Biomolecular Sciences and Biotechnology, University of Milan,
Milan, Italy

**Robert W. Noble**
Department of Medicine, University at Buffalo School of Medicine, VA Medical
Center, Buffalo, TX, USA

**John Olson**
Department of Biochemistry and Cell Biology, Houston, TX, USA
*and*
Keck Center for Structural and Computational Biology, Rice University,
Houston, TX, USA

**Sam-Yong Park**
Yokohama City University, Tsurumi, Yokohama, Japan

**Alessandra Pesce**
Department of Physics, CNR-INFM and Center for Excellence in Biomedical
Research, University of Genoa, Genoa, Italy

**Robert K. Poole**
Department of Molecular Biology and Biotechnology, The University of
Sheffield, Sheffield, UK

**Camille R. Roche**
Department of Physiology and Biophysics, Albert Einstein College of Medicine,
Bronx, New York, NY, USA

**Uri Samuni**
Department of Chemistry, Queens College of the City University of New York,
Queens, New York, NY, USA

**Jeremy Tame**
Protein Design Laboratory, Yokohama City University, Yokohama, Japan

**Nora B. Terwilliger**
Oregon Institute of Marine Biology, University of Oregon, Charleston, OR,
USA

**Lesley Tilleman**
Department of Biomedical Sciences, Antwerp University, Belgium

**Ken van Holde**
Department of Biochemistry and Biophysics, Oregon State University, Corvallis,
OR, USA

# 1
# Introduction: A Dedication to Beatrice and Jonathan Wittenberg

Kensal E. van Holde

## Abstract

This introduction to the volume presents a dedication to Beatrice and Jonathan Wittenberg, and summarizes their many joint contributions to our understanding of oxygen transport and utilization.

It is with appreciation and admiration that we dedicate this volume to Beatrice and Jonathan Wittenberg, who together have played such a leading role in the study of oxygen transport. Each has published nearly one hundred papers, most dealing with oxygen transport and transport proteins. Most remarkable are the facts that over fifty of these papers have been joint efforts and that this collaboration has extended over nearly half a century. Surely the Wittenbergs represent one of the most effective and productive of scientific collaborations. In what follows I shall summarise briefly the fruits of their joint work, citing only what seem to be the most important publications.

The Wittenbergs married in 1954, and a few years later their first joint paper appeared (Wittenberg and Wittenberg 1961). This initial work dealt with the transport of oxygen and other gases into the swim bladders of fishes. The problem is still of importance, especially in connection to the Root Effect, a topic of special attention at recent symposia.

This work was followed by a pioneering exploration of the nerve haemoglobins of a number of invertebrate species (Wittenberg, Briehl and Wittenberg 1965). This established that these were monomeric, myoglobin-like proteins that functioned in oxygen transport. It predated, by nearly 40 years, the recent excitement about such proteins. A parallel study, begun in 1973, and continued over the next decade, explored in depth the structure and function of leghaemoglobin, the haemoglobin of symbiotic bacteria in the root nodules of leguminous plants (see, for example, Appleby et al. 1983).

During this same period, the Wittenbergs embarked on what was to prove one of their outstanding achievements – the definitive study of the role of myoglobin in muscle physiology. The record begins in 1975 with a publication in *The Journal of Biological Chemistry*: "Role of myoglobin in the oxygen supply to red skeletal muscle" (Wittenberg, Wittenberg and Caldwell 1975). In a long series of papers, extending to the present, every aspect of the problem of how oxygen is supplied to muscle has been explored. The studies range from fundamental physiological studies (i.e. Wittenberg and Wittenberg 1987) to a relevant physical analysis of myoglobin self-diffusion (Riveros-Moreno and Wittenberg 1972). The whole subject of myoglobin function has been elegantly summarised in a recent review (Wittenberg and Wittenberg 2003).

Despite the fact that they have contributed to the field for nearly half a century, the Wittenbergs continue to explore new areas. In recent years (see Wittenberg et al. 2002), they have become interested in the truncated haemoglobins, and have contributed in a major way to the understanding of their oxygen-binding function.

The examples cited above do not nearly convey the breadth of Jonathan and Beatrice's contributions to the oxygen transport field. All of their work has been characterised by a unique combination of physiological insight and biophysical rigour. The quality of their work reflects, I think, the special individual talents of this remarkable team. All of us who study oxygen transport and all who value good science are in their debt.

## *References*

Appleby, C. A., Bradbury, C. I., Morris, R. I., Wittenberg, B. A., Wittenberg, J. B., and Wright, P. E. 1983. Leghemoglobin kinetics, nuclear magnetic resonance, and optical studies of pH dependence of oxygen and carbon monoxide binding. J. Biol. Chem. 258:2254–2259.

Riveros-Moreno, V., and Wittenberg, J. B. 1972. The self-diffusion coefficient of myoglobin and hemoglobin in concentrated solutions. J. Biol. Chem. 247:895–901.

Wittenberg, B. A., and Wittenberg, J. B. 1987. Myoglobin mediates oxygen delivery to mitochondria of isolated cardiac myocytes. Proc. Natl. Acad. Sci. U.S.A. 84:7403–7407.

Wittenberg, B. A., Briehl, R. W., and Wittenberg, J. B. 1965. Haemoglobins of invertebrate tissues. Nerve haemoglobins of *Aphrodite*, *Aplisia*, and *Halosydna*. Biochem. J. 96:363–371.

Wittenberg, B. A., Wittenberg, J. B., and Caldwell, P. R. 1975. Role of myoglobin in the oxygen supply to red skeletal muscle. J. Biol. Chem. 250:9038–9043.

Wittenberg, J. B., and Wittenberg B. A. 1961. Secretion of oxygen into the swim-bladder of fish. II. J. Gen. Physiol. 44:527–542.

Wittenberg, J. B., Bolognesi, M., Wittenberg, B. A., and Guertiu, M. 2002. Truncated haemoglobins: a new family of haemoglobins widely distributed in bacteria, multicellular eukaryotes and plants. J. Biol. Chem. 277:871–874.

Wittenberg, J. B., and Wittenberg, B. A. 2003. Myoglobin function reassessed. J. Exp. Biol. 206:2011–2020.

# 2
# Reflections on Beatrice and Jonathan Wittenberg

Michel Guertin

Dear Beatrice and Jonathan,

Our relationship as friends and colleagues began when we met at a Gordon conference in 1994. It marked the beginning of animated, challenging and respectful scientific discussions that have greatly benefited my development as a scientist. The work sessions in our respective laboratories, often very intensive, have also been privileged moments for me and my graduate students. Jonathan, your unique ability to gather and make people work together, and Beatrice, your talent to get people organised, focused and efficient, together synergistically created the energy, direction and momentum needed to launch the truncated haemoglobin field. Thanks to both of you for having communicated to me your passion for the fascinating world of globins. Thanks also for having introduced me to Martino Bolognesi, Joel Friedman, Jack Peisach and Denis Rousseau from whom I have learned much about the biophysics of globins. Jonathan and Beatrice you have been wonderful teachers.

With time we became friends and this has been an even more enriching experience. I am one of those who have had the chance to enjoy your wonderful hospitality and generosity. I have always found it wonderful and inspiring to be around you not only for the science but also for joyous delights such as exchanging views with you and your friends on diverse topics such art and music, or travelling through the rich fabric of time that is revealed from your many wonderful memories.

So as a friend and as a scientist, I am truly happy that your colleagues are able to recognise and honour your achievements and contributions through this impressive volume.

Michel

# 3
# The Wittenbergs:
# A Personal Appreciation

Cyril A. Appleby

## Abstract

In this chapter the author presents his appreciation of the interacting personalities and scientific skills of Beatrice and Jonathan Wittenberg that have made them a pre-eminent partnership in haemoglobin research. His impressions are based on 40 years of observation and occasional joint experimentation in respect of their work defining the role of haemoglobins in the facilitated diffusion of oxygen within animal and plant tissues.

Forty years ago a brief confrontation with Jonathan Wittenberg at a haemoprotein symposium in Japan, followed soon after by exciting discussion with him and wife Beatrice (Bea) at their New York home, led to the considerable enrichment of my scientific life. For this and an enduring friendship I honour them. The ensuing years of personal and laboratory interaction, correspondence and three-person international phone calls have brought me to an understanding of how these two, with their different personalities and talents, achieved such remarkable synergy.

Whereas Professor van Holde's Dedication presents a succinct account of the Wittenbergs' overall career, what I offer are anecdotes related to our joint experimental investigation of the facilitated diffusion of oxygen by plant kingdom haemoglobins, our many conversations about this phenomenon in mammalian myocytes and skeletal muscle, and the pleasure of watching them in action.

In the late 1950s I became interested in relationships between the nitrogen fixing *Rhizobium* bacteria of legume root nodules and an associated monomeric haemoglobin, the so-called leghaemoglobin (Lb) (see Appleby 1984, 1992). My early direct-equilibrium measurement of oxygen affinity for the principal components of soybean leghaemoglobin suggested that they were half saturated at ~70 nM dissolved gas (Appleby 1962). I proposed oxygen transport at low free concentration as a specific natural function for leghaemoglobin, citing pioneering work on haemoglobin by Scholander (1960), but then busied myself in searching for a high-affinity rhizobial oxidase rather than contemplation of oxygen transport mechanisms. That is, until August 1967, when I found myself face to face with an intense American, Jonathan Wittenberg, at a haemoprotein conference in Osaka and the conversation began something like this:

"So you're Appleby; you must be god-damned stupid to suppose that your leghaemoglobin could have anything to with facilitated diffusion of oxygen." "Why?" "Because everyone knows high oxygen-affinity haemoglobins such as the one from *Ascaris* have such slow off-rate constants that they could not support oxygen flux." "They don't know it for leghaemoglobin. The only reported values of its high oxygen affinity, mine and an earlier one of Keilin's, were based on hazardous equilibrium procedures. Why don't I send pure leghaemoglobin for your measurement of oxygenation kinetics?" "Why don't we do it together? Come to talk with Beatrice and me on your way back to Australia to set something up." And thus our interaction began. I realised later that Jonathan's initial irritation might have been due to the fact that in Appleby (1962) I cited Scholander (1960) instead of Wittenberg (1959) as the first to propose an oxygen transport function for myoglobin.

I had a 1960s-style round-world airline ticket without limit on stopovers, so a month later, in England with several saved days before other long-arranged USA lab visits, I hesitatingly phoned Jonathan: "This is Cyril Appleby, calling from London." "Fine, I'll meet you. What flight will you be on? You will stay with us of course?" A typical Jonathan response, and a first indication of the generous hospitality and friendship to be offered by him and Bea. Even so, two mornings later, when sitting in a dingy BOAC plane en route to Kennedy Airport, I wondered what I might be letting myself in for. Perhaps a double dose of Jonathan? But on meeting Bea that evening, serene and gloriously pregnant with daughter Rebecca – to be born a month later – and observing wonderful interaction between her and Jonathan, I sensed the prospect of an exhilarating collaboration.

To make the Australian feel at home they had bought a saddle of lamb, and after this was barbecued and eaten we talked and talked, with help from a duty-free bottle of whisky. They led me through the basics of facilitated diffusion, exemplified by Jonathan's impressive model system (Wittenberg 1966), where haemoglobin or myoglobin in solution in Millipore membranes augments the rate of steady-state diffusion of oxygen through buffer. They emphasised that this occurs only when there is a downward gradient of oxygenation between loading and unloading boundaries. If haemoglobin becomes completely oxygenated, facilitated flux disappears. They kept at me until I understood and accepted this basic point.

Then Bea struck: "Cyril, our aim is to prove that myoglobin-facilitated diffusion of oxygen is important in working heart muscle as well as in Jonathan's model system. But so far we have no way of measuring work done by muscle fibres, or energy conservation by myocyte mitochondria, during monitored oxygenation of myocyte myoglobin. A few months ago your Canberra CSIRO colleagues Bergersen and Turner (1967) showed for the first time that *(Brady)Rhizobium japonicum* bacteroids isolated anaerobically from soybean root nodules retained nitrogen-fixing activity. What is the possibility that in the presence of part-oxygenated leghaemoglobin all of us could demonstrate an increased efficiency of bacteroid respiration and consequent nitrogen fixation?" Jonathan, Bea and I agreed that a necessary preliminary to such investigation would be the examination of leghaemoglobin oxygenation kinetics.

I returned to Canberra, improved the procedure for making stable, ligand-free soybean leghaemoglobin, and in early 1971 brought a supply of the purified principal subcomponents Lb$a$ and Lb$c$ to the Wittenberg laboratory at the Albert Einstein College of Medicine, New York for ligand kinetics and structural studies. Just before leaving Australia I met the distinguished biophysicist J. T. Edsall, at the Australian National University on a pre-retirement sabbatical. After telling him of my work and plans he said: "Ah yes, Jonathan is a nice boy but he does have some strange ideas." Jonathan was much pleased to hear this assessment of him as a challenger of the system, made by his revered Harvard professor. I appreciated but did not need the gentle warning. I had seen already how Bea could follow Jonathan's glittering, imaginative expositions with plans for real-life collaborative experiments. Of her determination to isolate undamaged, functional cardiac myocytes for use in copybook facilitated diffusion experiments involving endogenous myoglobin, more later.

The Wittenbergs' stopped-flow apparatus was a 1960s commercial (Durrum-Gibson) unit based on Quentin Gibson's original but in which, thankfully, a temperamental hydraulic lever-operated syringe drive had been replaced by a button-actuated pneumatic drive. A memory oscilloscope with attached Polaroid camera enabled one to record optical transmittance changes at chosen monochromator wavelength as reactions proceeded. With touch-typist dexterity Jonathan would set valves for sample syringe fill, drive or flush operation, and button pushing as required. With equal dexterity Bea would photograph the oscilloscope trace and check the quality of a Polaroid print before she and Jonathan could decide to repeat the previous run or move to new wavelength, sample or ligand conditions. Meanwhile her technician would begin the digitising of photographed traces and use of a bulky 1960s Wang electronic calculator to calculate reaction rate constants.

My own involvement was more leisured. Before leaving Australia I had recorded absolute and difference spectra of ferric and ferrous Lb$a$ and Lb$c$ and relevant ligand complexes. These were to help Jonathan and Bea choose, for kinetics studies, monochromator wavelengths that might be isosbestic for unwanted side reactions. Knowing that oxyferrous and deoxyferrous leghaemoglobin were very susceptible to degradation catalysed by dithionite breakdown products or heavy-metal ions, I was pleased to be tutored by Jonathan in the preparation of day-long stable stock solutions of sodium dithionite by gently sifting fresh, reagent-grade powder into ice-cold 1 mM NaOH being sparged with pure argon. As well, I was pleased to see on his lab shelves bottles of reagent-grade, metal-free buffer constituents and large carboys filled with the best double glass-distilled water, labelled "DANGER: NOR-METHANOL" to keep them from marauding physiologists. This commitment to quality (evident also in Bea's adjacent myocyte-investigation lab) meant that I could have reliable solutions of leghaemoglobin and ligands ready for their day's anticipated work or make others at short notice.

The outcome of our foray into leghaemoglobin oxygenation kinetics (Wittenberg, Appleby and Wittenberg 1972) was exciting and satisfying. The oxygen Off rate at 20°C was calculated as 4.4 s$^{-1}$ for Lb$a$ and 4.9 s$^{-1}$ for Lb$c$; fast enough to satisfy Jonathan that leghaemoglobin could be involved in the phenomenon of facilitated diffusion. The oxygen On rate for Lb$a$ was 118 $\mu$M$^{-1}$ s$^{-1}$ and for Lb$c$ was 97 $\mu$M$^{-1}$ s$^{-1}$, the fastest yet recorded for a haemoglobin. Oxygen dissociation constants were calculated thereby as 37 nM for Lb$a$ and 54 nM for Lb$c$, in reasonable agreement with those measured by the earlier (Appleby 1962) equilibrium procedure.

With my 1971 New York semester still less than half gone, and no crystal structure for guidance, Bea, Jonathan and I had then a wonderful time arguing and beginning experiments to help work out what might allow leghaemoglobin to have such extraordinary oxygen affinity yet fast turnover. It was known from Nils Ellfolk's work in Helsinki and mine in Canberra that ferric leghaemoglobin, unlike ferric myoglobin, could combine with acetate or straight-chain homologues at high affinity to form 100% high-spin spectral species. I supposed, in wishful-thinking mode, that the distal pocket of leghaemoglobin might be open or flexible enough to allow the binding of acetate carboxylate to heme iron with the hydrocarbon moiety, at least for longer homologues, poking out of the pocket. Jonathan was fascinated by the alternative that acetate or un ionised acetic acid might, by binding to a remote site on the protein surface, cause a linked allosteric change in heme configuration. Bea remained calm and carefully neutral until one day, with Jonathan and I shouting at each other, she cracked and said: "Right, you two. Stay for no longer than ten minutes in this instrument room, then come out as friends." Jonathan and I glared at each other for nine and a half minutes until he burst out laughing and said: "Bea wins; she always does." We emerged to hugs all round and I realised that this must be the way she dealt with Jonathan and two strong-willed sons. I was flattered to have been accorded honorary family status; a situation now enjoyed for 37 years.

Our stopped-flow kinetic studies showed that the acetate complex of ferric leghaemoglobin could not be reduced until acetate was dissociated and that fluoride, a known ferric-heme ligand, bound to leghaemoglobin in direct competition with acetate. These observations and later EPR analysis (see below) of leghaemoglobin acetate suggested, and others' crystal structures proved, the direct ligation of acetate anion to heme Fe and allowed the assumption that bulkier homologues such as $n$-valerate could bind the same way. This concept of a flexible, commodious distal pocket was reinforced by our identification of an exogenous ligand, found in soybean nodule extracts and able to bind to ferric or ferrous leghaemoglobin with high affinity, as nicotinic acid (pyridine 3-carboxylate) anion. Our kinetics and EPR analyses suggested, and later X-ray analysis by University of Sydney colleagues proved, the direct ligation of the nicotinate ring N to heme Fe. Moreover, the X-ray structure showed the presence of a hydrogen bond between nicotinate carboxylate and $N^{\delta 1}$ of the displaced distal histidine (Ellis et al. 1997).

Whereas deoxyferrous leghaemoglobin, like deoxyferrous myoglobin, has an undoubted five-coordinate structure, Ellfolk had shown unligated soybean ferric leghaemoglobin to be a thermal-equilibrium mixture of penta- and hexacoordinate species with low temperature favouring hexaco-ordination and a low-spin electronic structure of heme Fe. With Einstein College of Medicine colleague Jack Peisach, and Bill Blumberg at the Murray Hill Bell Laboratories, Jonathan, Bea and I became involved in a low-temperature (1.6 °K) EPR study of unligated ferric leghaemoglobin and several ligand complexes (Appleby et al. 1976). At low temperature the distal histidine of unligated ferric leghaemoglobin was found to be predominant sixth ligand to heme Fe whereas in the acetate and nicotinate complexes, as mentioned above, it was displaced completely.

Taken together, the results of our kinetics, optical spectral, EPR and X-ray structure studies offered strong support for the concept of a flexible, accessible distal pocket in soybean leghaemoglobin that would permit rapid oxygen access. But my fanciful idea that mobile distal histidine, closing like a trapdoor over bound oxygen, could explain the relatively slow oxygen Off rate, had to be abandoned. Mark Hargrove's later mutant studies showed that replacing distal histidine with smaller apolar residues had little effect on this rate constant.

During the course of our 1971 work Bea, Jonathan and I had many discussions about the design of experiments to explore the role (if any) of leghaemoglobin in facilitating the diffusion of oxygen to nitrogen-fixing *Rhizobium* bacteroids. The classic study of Bergersen and Turner (1967) had been the first to demonstrate an ability of isolated, washed *Rhizobium* bacteroids rather than intact root nodules to reduce nitrogen to ammonia. But these authors found that addition of a nodule extract containing unpurified leghaemoglobin diminished slightly rather than enhanced the rate of washed-bacteroid nitrogen fixation. The Wittenbergs asked if this was a problem. I explained that Fraser Bergersen and his equally painstaking technician Graham Turner, despite special abilities in bacteroid isolation and nitrogenase assay, were probably too proud to admit that I could make better leghaemoglobin and had never asked for any. I surmised that when Fraser knew the results of our oxygenation kinetics work the situation would change.

With anticipation of this outcome it was agreed in principle that Jonathan would try to arrange a Canberra visit, bringing haemoglobins and other oxygen-carrier proteins having the widest possible range of oxygen affinities, oxygenation kinetics and molecular size. It was hoped he and Fraser would, with some help from me, work out experimental protocols that merged our abilities with respect to nitrogenase assay and preparation of bacteroids and stable solutions of oxygen-carrier proteins, toward proof of a function for haemoglobin in the facilitated diffusion of oxygen to nitrogen-fixing bacteroids. I would prepare in my laboratory leghaemoglobin and other oxygen carriers that Jonathan could not bring.

Soon after my return to Australia Fraser Bergersen did ask for pure leghaemoglobin and we set up trial experiments in which the stimulation of nitrogenase activity in bacteroid suspensions by oxygenated leghaemoglobin was shown to be much greater than stimulation of oxygen uptake. Also, increased reduction of acetylene to ethylene (a convenient assay for nitrogenase activity) in response to increased agitation of assay flasks was accompanied by increased oxygenation of leghaemoglobin. These were delightful results, but they did not tell us whether leghaemoglobin was acting in facilitated diffusion with final delivery of molecular oxygen near the bacteroid, or whether there was specific, direct interaction between oxyleghaemoglobin and an "efficient" oxidase located on the bacteroid surface. Consequently, when Jonathan arrived in February 1973 with frozen haemerythrin, *Aplysia* and human myoglobins, also human haemoglobin A both "stripped" and loaded with IHP to substantially lower its oxygen affinity, Fraser and I were very pleased. If any of these were able to substitute for leghaemoglobin in stimulation of bacteroid nitrogenase activity, we could eliminate the dubious hypothesis of "specific direct interaction" with an oxidase on the bacteroid surface.

We needed other oxygen carriers to make Bea and Jonathan's 1971 wish list complete. Earthworm haemoglobin was purified from *Lumbricus terrestris* dug out of the Bergersen compost pit and haemocyanin from live *Jasus nova hollandae* lobsters, kindly donated by a supplier at the Sydney Fish Market and sent to us by air express, nestled in seaweed in a chilled container. We had no trouble disposing of the boiled carcasses after bleeding was complete. The getting of raw material for *Gastrophilus* (horse bot fly) and *Ascaris suum* (a pig small-intestine worm) haemoglobin preparations was less pleasant. I located a knackery close to Brisbane where brumbies (Queensland wild horses) were butchered for pet food and while Jonathan, the "Visiting Professor" was on the catwalk explaining the importance of our work to the manager I was on the killing-room floor scraping bot fly larvae from the walls of slit stomachs after their evil-smelling contents had been rinsed out. Similarly, after the supervisor of an abattoir close to Canberra alerted me that a batch of *Ascaris*-infected pigs was about to be slaughtered, Jonathan joined him on the catwalk while I was at the lower level slitting intestines and removing worms. We were lucky to get them because soon afterwards the slaughtering of infected pigs in Australian abattoirs was forbidden.

The paper describing use of these carrier proteins to establish and investigate the nature of facilitated oxygen diffusion to *Rhizobium* bacteroids (Wittenberg et al. 1974) is complex. I summarise its principal findings but suggest that those with a particular interest read through at least its Summary, Introduction and Conclusions and look at figures I and IV. In the standard assay, which represented a nice compromise between Bergersen/Turner and Wittenberg experimental practices, bacteroid suspensions were shaken at 150 cpm at 25°C in serum-stopped flasks under a nitrogen gas phase usually containing 20 mmHg partial pressure of oxygen and 173 mm partial pressure of acetylene. Oxygen consumption was measured at intervals by mass spectrometry of gas samples withdrawn via the serum stopper, and nitrogenase activity by measuring (via gas chromatography) the extent of acetylene reduction to ethylene. This is a convenient, rapid procedure but from time to time the conversion of gas-phase $^{15}N_2$ to bacteroid $^{15}NH_3$ was measured by mass spectrometry.

At 20 mm partial pressure and standard stirring rate of 150 cpm oxygen consumption in the absence of leghaemoglobin or other carrier was moderate but nitrogenase activity low. With faster stirring, oxygen consumption increased and nitrogenase activity more so. At standard stirring rate in the presence of 0.5 mM leghaemoglobin oxygen consumption increased slightly but nitrogenase activity increased substantially, although with little further increment at faster stirring rate. For comparison of the efficiency of leghaemoglobin and alternative oxygen-carrier proteins, the ratio of increment of bacteroid nitrogenase activity on addition of a carrier protein, to increment of oxygen consumption, was taken as an indication of that carrier's "effective" oxygen supply. As shown in table IV of Wittenberg et al. (1974), nine of the ten tested haemoglobins and the two non-heme proteins (hemerythrin and haemocyanin) augmented both oxygen uptake and nitrogenase activity. Hence, the concept of specific direct interaction between leghaemoglobin and an efficient oxidase on the bacteroid surface could be abandoned. For carrier proteins of moderate molecular size and moderate to fast oxygen dissociation rates the augmentation of effective oxygen uptake was greatest with those of highest oxygen affinity. For carriers of larger molecular size, and hence of lower diffusion coefficients, the facilitation of diffusion was less.

All of the above was consistent with the idea that leghaemoglobin facilitated the diffusion of oxygen to nitrogen-fixing bacteroids but in the Summary of our paper (Wittenberg et al. 1974) we made an error. We supposed that in the absence of leghaemoglobin the ineffective oxygen uptake maintained a very low oxygen pressure at the bacteroid surface and that in the presence of leghaemoglobin the operation of facilitated diffusion increased $pO_2$ at that surface, resulting in "effective" oxygen uptake. We should have taken more notice of the opinion of CSIRO mathematician colleague A.N. Stokes, acknowledged in a footnote to our paper: "...that leghemoglobin may stabilize the oxygen pressure at the bacteroid interface, which in the absence of leghemoglobin might fluctuate widely". We know now that the terminal oxidase of the effective, symbiosis-specific respiratory chain of *Rhizobium* bacteroids is of extremely high oxygen affinity. It is cytochrome $cbb_3$ with $KmO_2$ of 7 nM (Preisig et al. 1996) and so fully functional at the bottom of the oxygen-unloading range of leghaemoglobin during facilitated diffusion. It is likely that the "ineffective" respiration of bacteroids observed in our assay system in the absence of leghaemoglobin occurs at higher, not lower, oxygen concentration. Its character remains unknown although the possibility of uncoupled, protective respiration during oxygen surges, as observed in other nitrogen-fixing bacteria, comes to mind.

The Wittenberg-Bergersen-Appleby partnership never did attempt a similar laborious "proof" of leghaemoglobin-facilitated oxygen diffusion in other legume or non-legume symbiosis, nor indeed has any other lab. What I and others did in later years was to purify symbiotic haemoglobins from nodules of a range of legumes and the non-legumes *Casuarina* and *Parasponia*. Quentin Gibson, the Wittenbergs and I then determined their oxygenation kinetics (Gibson et al. 1989). Despite a wide range of On and Off rate constants, the calculated equilibrium dissociation constants ($K_D$) varied only twofold for haemoglobins from *Rhizobium*–legume symbioses, and little more in comparison with haemoglobins from the *Rhizobium–Parasponia* and *Frankia–Casuarina* symbioses: a pleasing example of evolutionary pressures producing something "comfortable" for oxygen-intolerant nitrogenase from various ancestral non-symbiotic haemoglobins.

During my visits to the Wittenberg lab in later years, bringing these other symbiotic haemoglobins for kinetic analysis, it was a pleasure to see Bea's ambition to isolate undamaged cardiac myocytes realised. This achievement, and the first use of such cells in a wide range of experimental programmes (including the initial demonstration of myocyte myoglobin function in the facilitated diffusion of oxygen) are described in two exemplar papers (Wittenberg and Robinson 1981; Wittenberg 1983). Both are a delight to read; the second especially so. Bea's clarity of thought and expression and meticulous experimental skills are very apparent. They match those in what I consider to be Jonathan's exemplar (Wittenberg 1966), where his concept of facilitated oxygen diffusion is elucidated with elegance.

Bea's preparation of rectangular, calcium-tolerant myocytes, with well preserved fine structure, did not beat as spontaneously and irregularly as had others' earlier preparations of leaky cells with damaged membranes. They could be made to contract repeatedly and synchronously in response to electrical stimulation. This property allowed Bea and later collaborators to emulate working-muscle situations yet with the tighter control of environmental conditions achievable in myocyte suspensions. Bea and Tom Robinson used these suspended myocytes to define the extracellular oxygen concentration required to maintain oxidative phosphorylation in heart cells. Respiratory oxygen uptake was found to be limited only when the partial pressure was less than 1 torr – well below the normal venous $pO_2$ of 20 torr. Of this more later.

A subsequent paper, exemplifying the Wittenbergs' remarkable synergy, is "Oxygen pressure gradients in isolated cardiac myocytes" (Wittenberg and Wittenberg 1985). Jonathan had conceived and assembled an exquisite observation chamber using a short length of 16-mm Pyrex tubing painted, except for optical windows, with a reflective white coating on the outside. The chamber floor was a membrane-covered oxygen sensor with very low background current, and its top plugged by a Delrin stopper with O-ring seal. Gas inlet and outlet tubes and a Delrin stirring shaft and propeller, powered by a DC motor, passed through the stopper. An important practical consideration was that this stirrer caused much less myocyte damage than did others' use of magnetic stirrers operated at sufficient speed to minimise unstirred layers adjacent to the myocytes. The chamber was inserted into a thermostatted brass block placed in the light beam of a Cary 17 recording spectrophotometer equipped with a scattered-transmission accessory. The fluid volume of the chamber was 3.5 ml and gas space 5 ml. Spectral data were acquired digitally then absolute and difference spectra generated by a computer. Fractional oxygenation of myocyte myoglobin and fractional oxidation of mitochondrial cytochrome oxidase and cytochrome c were determined from difference spectra at appropriate wavelength pairs. Oxygen uptakes were calculated following measurement of inlet and outlet gas streams at steady state conditions, with daily calibration.

When suspensions of Bea's myocytes, prepared with similar finesse, were in the chamber, these probes of mitochondrial function showed no significant change until extracellular oxygen pressure was less than 2 torr and intracellular myoglobin largely deoxygenated. Sarcosomal oxygen pressure in resting cells was nearly the same as extracellular oxygen pressure but about 2 torr less in cells whose respiration was increased 3.5-fold by mitochondrial uncoupling. With this demonstration that differences within cardiac myocytes were relatively small, the Wittenbergs' conclusion was that the large, ~20 torr, difference in oxygen pressure between capillary lumen and mitochondria must be extracellular.

Another observation system, involving an inverted microscope with epifluorescence attachment and dual photomultiplier tubes functioning as a microfluorimeter, enabled Bea and Philadelphia colleague Roy White to measure the NADH fluorescence of individual, isolated rat cardiac myocytes or myocyte suspensions (White and Wittenberg 1993). In this way the effects of pacing, myoglobin and oxygen supply on the status of mitochondrial NAD could be determined. For resting myocytes in a medium equilibrated with 95% $O_2$/5% $CO_2$, NAD reduction was ~27%. This fell reversibly to ~11% and oxygen consumption was stimulated five fold in contracting myocytes paced at 5 Hz. However, NAD reduction increased reversibly to ~37% when cells were paced after myoglobin was inactivated with sodium nitrite, and NAD reduction increased to ~48% when cells were more densely layered in medium equilibrated with 20% $O_2$ / 5% $CO_2$. Their conclusions were that the ratio NADH/NAD is decreased in well oxygenated cells during increased work, that steady-state NAD reduction is increased during increased work when oxygen is limited and that functional myoglobin ensures an oxygen supply to the mitochondria of working cells.

It must not be supposed that the Wittenberg concept of a function for myoglobin in facilitated diffusion of oxygen within heart and skeletal muscle was accepted without challenge. I was at a 1998 Asilomar conference on oxygen-binding proteins, and indeed in Bea's company, when she heard that R.S. Williams and colleagues at the University of Texas had found that mice without myoglobin, generated by gene knockout technology, could survive, exercise and meet the demands of pregnancy (Garry et al. 1998). I was appalled on her and Jonathan's behalf, but Bea said: "Don't worry Cyril; things will turn out alright". And they did! Several laboratories, including Schrader's in

Düsseldorf (see e.g. Gödecke et al. 1999) showed that the disruption of myoglobin induces multiple compensatory mechanisms. This new work is summarised within a Wittenberg and Wittenberg (2003) review entitled "Myoglobin function reassessed". A supportive paper, cited by them with evident approval, is: "Myoglobin facilitates oxygen diffusion" by Merx et al. (2001). In that work carbon monoxide was used as a blocker of myoglobin function in perfused hearts having oxygen supply limited such that myoglobin was only partly oxygenated – a requirement for facilitated diffusion to be operative. They observed a significant decrease in efficiency of left-ventricular function as carbon monoxide was added to the perfusion medium of wild-type myoglobin-containing hearts, but no decrease as this inhibitor was added to the perfusion medium of hearts from myoglobin-knockout mice – a satisfying control. The Wittenbergs' and others' earlier studies where nitrite, carbon monoxide or other potential inhibitor of myoglobin had been added to perfused hearts, muscle fibres or myocyte suspensions had to endure the criticism that mitochondrial oxidases might have been an unintended second target.

Bea and Jonathan have not retired into oblivion. In "Myoglobin-enhanced oxygen delivery to isolated cardiac mitochondria" (Wittenberg and Wittenberg 2007) they determined the conditions required at the surface of actively respiring, state III, tightlycoupled mitochondria to enhance oxygen flow to cytochrome oxidase. Their mitochondria were liberated from minced pigeon-heart ventricle using just three strokes of an undersized gently-rotating Teflon pestle in a small Arthur H. Thomas homogeniser. In this way, damage to the mitochondrial outer membrane was minimised; in fact their description of the overall isolation procedure and subsequent experimental conditions reveals the exquisite care applied to their work. These mitochondria were incubated at low pressure of free oxygen where respiration is oxygen-limited. Then, by the separate use of six monomeric haemoglobins with very different oxygen affinities and kinetics, whose fractional oxygenations were being monitored spectrophotometrically, the oxygen partial pressures at half-maximal rate of mitochondrial respiration were measured as oxygen was being depleted in unstirred suspensions. These pressures were similar for all carriers, leading to the conclusion that there was no specific, direct interaction between a myoglobin docking site and another site on the mitochondrial outer membrane.

The oxygen pressure required to support mitochondrial state III respiration was only 0.04 torr. This is very small compared with that measured (see above) in the sarcoplasm of the working heart (~2 torr), leading to their conclusion that: "in the normal steady states of contraction of the myoglobin-containing heart, oxygen utilization by mitochondrial cytochrome oxidase is not limited by oxygen availability". In contrast, Merx et al. (2001) had shown that for the resuspended myocytes from myoglobin-knockout hearts, but not from normal hearts, respiration rate began to decline below 8 torr gaseous oxygen.

While Wittenberg and Wittenberg (2007) was being written, Jonathan and Bea must have seen the accepted or on-line version of a new paper appearing later as Lin, Kreutzer and Jue (2007). Its authors, evidently longtime

Wittenberg sparring partners, had used some recent NMR experiments to again challenge Wittenberg facilitated-diffusion dogma. Jonathan responded in his best polemic style: "Major difficulties in the predictive use of (their) equations are that the model does not correspond to conditions thought to exist in living muscle…and that the parameters employed can be approximated with only poor precision." I am sure that Bea, ever the diplomat, caused the next sentence to be written: "Perhaps, however, the discordance between prediction and actuality serves to warn that there is more to be learned about myoglobin function than we already know."

## *References*

Appleby, C. A. 1962. The oxygen equilibrium of leghemoglobin. Biochim. Biophys. Acta 60:226–235.

Appleby, C. A. 1984. Leghemoglobin and Rhizobium respiration. Ann. Rev. Plant Physiol. 35:443–478.

Appleby, C. A. 1992. The origin and functions of haemoglobin in plants. Sci. Progress Oxford. 76:365–398.

Appleby, C. A., Blumberg, W. E., Peisach, J., Wittenberg, B. A., and Wittenberg, J. B. 1976. Leghemoglobin: An electron paramagnetic resonance and optical spectral study of the free protein and its complexes with nicotinate and acetate. J. Biol. Chem. 251:6090–6096.

Bergersen, F. J., and Turner, G. L. 1967. Nitrogen fixation by the bacteroid fraction of breis of soybean root nodules. Biochim. Biophys. Acta 141:507–515.

Ellis, P. J., Appleby, C. A., Guss, J. M., Hunter, W. N., Ollis, D. L., and Freeman, H. C. 1997. Structure of ferric soybean leghemoglobin a nicotinate at 2.3 Å resolution. Acta Crystallogr. D 53:302–310.

Garry, D. J., Ordway, G. A., Lorenz, J. N., Radford, N. B., Chin, E. R., Grange, R. W., Bassel-Duby, R., and Williams, R. S. 1998. Mice without myoglobin. Nature 395:905–908.

Gibson, Q. H., Wittenberg, J. B., Wittenberg, B. A., Bogusz, D., and Appleby, C. A. 1989. The kinetics of ligand binding to plant hemoglobins: Structural implications. J. Biol. Chem. 264:100–107.

Gödecke, A., Flögel, U., Zanger, K., Ding, Z., Hirchenhain, J., Decking, U. K. M., and Schrader, J. 1999. Disruption of myoglobin in mice induces multiple compensatory mechanisms. Proc. Natl. Acad. Sci. U S A 96:10495–10500.

Lin, P.-C., Kreutzer, U., and Jue, T. 2007. Myoglobin translational diffusion in rat myocardium and its implication on intracellular oxygen transport. J. Physiol. 578:595–603.

Merx, M. W., Flögel, U., Stumpe, T., Gödecke, A., Decking, U. K. M., and Schrader, J. 2001. Myoglobin facilitates oxygen diffusion. FASEB J. 15:1077–1079.

Preisig, O., Zufferey, R., Thöny-Myer, L., Appleby, C. A., and Hennecke, H. 1996. A high-affinity cbb3-type cytochrome oxidase terminates the symbiosis-specific respiratory chain of *Bradyrhizobium japonicum*. J. Bacteriol. 178:1532–1538.

Scholander, P. F. 1960. Oxygen transport through hemoglobin solutions. Science 131:585–590.

White, R. L., and Wittenberg, B. A. 1993. NADH fluorescence of isolated ventricular myocytes: Effects of pacing, myoglobin, and oxygen supply. Biophys. J. 65:196–204.

Wittenberg, B. A. 1983. Isolated mammalian adult heart cells: A model system. Einstein Quarterly J. of Biol. and Med. 1:117–120.

Wittenberg, B. A., and Robinson, T. F. 1981. Oxygen requirements, morphology, cell coat and membrane permeability of calcium-tolerant myocytes from hearts of adult rats. Cell Tissue Res. 216:231–251.

Wittenberg, B. A., and Wittenberg, J. B. 1985. Oxygen pressure gradients in isolated cardiac myocytes. J. Biol. Chem. 260:6548–6554.

Wittenberg, J. B. 1959. Oxygen transport – a new function for myoglobin. Biol. Bull. 117:402–403.

Wittenberg, J. B. 1966. The molecular mechanism of hemoglobin-facilitated oxygen diffusion. J. Biol. Chem. 241:104–114.

Wittenberg, J. B., Appleby, C. A., and Wittenberg, B. A. 1972. The kinetics of the reactions of leghemoglobin with oxygen and carbon monoxide. J. Biol. Chem. 247:527–531.

Wittenberg, J. B., Bergersen, F. J., Appleby, C. A., and Turner, G. L. 1974. Facilitated oxygen diffusion: The role of leghemoglobin in nitrogen fixation by bacteroids isolated from soybean root nodules. J. Biol. Chem. 249:4057–4066.

Wittenberg, J. B., and Wittenberg, B. A. 2003. Myoglobin function reassessed. J. Exp. Biol. 206:2011–2020.

Wittenberg, J. B., and Wittenberg, B. A. 2007. Myoglobin-enhanced oxygen delivery to isolated cardiac mitochondria. J. Exp. Biol. 210:2082–2090.

# 4

# A Crystallographer's Perspective on the 2/2Hb Family

Alessandra Pesce, Mario Milani, Marco Nardini
and Martino Bolognesi

## Abstract

The discovery of 2/2Hbs as short haemoproteins structurally related to haemo-globins, in the 1990s, was complemented by extensive sequence and crystallo-graphic investigations in the early 2000s. Amino acid sequences first provided a clear indication that the 2/2Hb family (formerly known as "truncated Hbs") is composed of three main protein groups (I, II and III) that display low primary structure conservation relative to vertebrate Hbs. Crystal structures showed that a simple protein fold, essentially composed of four α-helices, is common to members of all three groups. Specific structural features can however be recognised in each 2/2Hb group. Among these, a tightly intertwined network of hydrogen bonds stabilising the heme exogenous ligand, based on group-specific distal site residues, is a landmark of all 2/2Hbs. We present here a review of the different structural aspects discovered for the 2/2Hb family, in the light of the currently known three-dimensional structures.

## Introduction

Widespread investigations spanning the last 15 years have substantially changed our views on the roles of haemoglobin (Hb) in nature, on its structure and on the composition of the Hb protein homology family. Hbs, or related haemoproteins (globins), have been found in virtually all kingdoms of life, suggesting an ancient origin for their genes; IIbs, particularly from prokaryotes, have increasingly been shown to cover functions well beyond that of simple oxygen carriers; far homologues of Hb have been identified through genomic searches and at least partly characterised (Wittenberg and Wittenberg 1990; Hardison 1998; Imai 1999; Minning et al. 1999). A first class of Hb-related proteins that was discovered in bacteria and fungi include single-chain flavo-haemoglobins (hosting a flavoprotein reductase domain; Bonamore et al. 2003; Gardner 2005) and the dimeric *Vitreoscilla* sp. Hb (Bolognesi et al. 1997; Tarricone et al. 1997). Both were shown to be built on the classical (non-)vertebrate globin fold, composed of seven/eight α-helices per haeme (Perutz 1979;

Holm and Sander 1993; see also Boffi et al. in this Volume).

A second class of haemoproteins, referred to as "truncated Hbs", was later identified in unicellular prokaryotes and eukaryotes and in some higher plants (Iwaasa, Takagi and Shikama 1989; Potts et al. 1992; Takagi 1993; Couture et al. 1994; Couture et al. 1999; Thorsteinsson, Bevan and Potts 1999; Wittenberg et al. 2002). Members of this branch in the Hb superfamily are small (monomeric or dimeric) haemoproteins, characterised by medium to very high oxygen affinities, with occasional ligand-binding cooperativity (Couture et al. 1999). The term "truncated Hb" was first introduced to refer to the size of these small Hbs, whose amino acid sequences span approximately 115–140 residues, and display lower than 15% residue identity to (non-)vertebrate Hbs (Vuletich and Lecomte 2006; Vinogradov et al. 2007); the term "truncated Hb" was recently replaced by "2/2Hbs" (read 2-on-2 Hbs) in relation to specific features of their fold (see below; Nardini et al. 2007). The functional role of 2/2Hbs is still a matter of debate; enzymatic functions have been proposed in addition to the simple role of oxygen carriers. Remarkably, inspection of genomes indicates that certain eubacteria may contain both flavohaemoglobins and 2/2Hbs, hinting at specific functions for each of the two classes.

Analysis of the current protein sequence data bases indicates that more than 300 2/2Hbs have been identified and sequenced (Vuletich and Lecomte 2006; see also Vuletich and Lecomte in this Volume). Structural alignments of vertebrate Hbs *vs.* 2/2Hbs indicates that the latter are mostly characterised by substantial residue deletions at either N- and C-termini and in the CD-D region of the (non-)vertebrate globin fold (α-helices of the classical globin fold are conventionally labelled A, B, …, H according to their sequential order; topological residue sites are numbered sequentially within each α-helix of sperm whale myoglobin; Perutz 1979). Deletions at the N-terminal region suggest that the A-helix may be fully absent in 2/2Hbs; similarly, the CD-D region, connecting the C- and E-helices, is particularly short in several 2/2Hb sequences, hinting at full deletion of the D-helix. Moreover, the CD1 residue, which is invariably Phe in (non-)vertebrate Hbs, may be substituted by Tyr or Leu. The heme distal residue at the E7 position, almost invariably His or Gln in (non-)vertebrate Hbs where it stabilises the heme-bound ligand (i.e., $O_2$) through hydrogen bonding, is found as Gln or small apolar residues not capable of hydrogen bonding, in 2/2Hbs. A second distal residue, at the B10 site, is invariably Tyr; it may play a role in promoting ligand stabilisation by complementing the hydrogen bonding capability of the E7 residue (Yang et al. 1995; Couture et al. 1999). The amino acid sequences show unequivocally that the heme proximal (and heme-Fe coordinating) residue HisF8 is fully conserved in 2/2Hbs, thus reducing to just one the number of residues that are invariant throughout the Hb homology superfamily. Several 2/2Hbs display a conserved Trp residue in the first half of the expected G-helix, at site G8, in a sequence position likely to promote interactions within the distal side of the heme porphyrin ring. Finally, and most relevant for the following discussion on the 2/2Hb three-dimensional fold, three Gly-based sequence motifs can be recognised as strongly conserved (although with variations) in the aligned sequences. The first motif is generally recognised as a Gly-Gly pair located in the N-terminal region of 2/2Hbs; given the known conformational properties of Gly residues, it may mark the boundary between the short-variable N-terminal segment and the protein's core region, rich in α-helical structure. A second Gly-Gly motif is found approximately in the middle of the 2/2Hb sequences, where it may code for the termination of the E-helix, or for specific backbone conformations. The third Gly-motif (one Gly residue only) is found 5–10 residues after the second motif, in the sequence region belonging to the F-helix of (non-)vertebrate Hbs (Pesce et al. 2000; Vuletich and Lecomte 2006; Nardini et al. 2007). The three motifs are strictly connected to the achievement of the 2/2Hb three-dimensional fold, as will be shown below.

An additional main issue emerging from amino acid sequence analysis concerns internal groupings within the 2/2Hb sub-family. In particular, inspection of specific sequence trends at the B10, CD1 and E7 sites (all contributing to the size and properties of the heme distal cavity), but also the loss of part of the Gly-based motifs, allows three main phylogenetic groups in 2/2Hbs to be distinguished. Group I 2/2Hbs display full conservation of the Gly-based motifs, and host predominantly TyrB10, PheCD1 and GlnE7; group I 2/2Hbs are also identified by the suffix "N" (HbN). In group II the CD1 residue is often Tyr, while the E7 site is often occupied by a small residue (Ala, Ser, Thr); additionally, residue G8 is invariably Trp. The Gly-based motifs are essentially conserved in group II, whose members are also identified as HbOs. Group III is the least represented so far, being characterised by TyrB10, PheCD1, HisE7 and TrpG8; in this case (members of the group are identified as HbPs) the conservation of Gly-based motifs appears to be quite relaxed. Additionally, significant residue insertions are present at both N- and C-termini, and in the CD region (Nardini et al. 2006; Vuletich and Lecomte 2006; Nardini et al. 2007). From the functional and evolutionary viewpoints, it is worth noting that in certain bacteria 2/2Hbs from different groups can coexist.

## 2/2Hb Fold and Fold Variation in Groups I, II, III

The globin fold of 2/2Hb (Fig. 1) has been described as consisting of a simplified version of the "classical" globin fold (a 3-on-3 α-helical sandwich; Perutz 1979) typical of sperm whale myoglobin (Mb) (Pesce et al. 2000; Nardini et al. 2007). The topology of the 2/2Hb fold is characterised by a 2-on-2 α-helical sandwich based on four α-helices, corresponding to the B-, E-, G- and H-helices of the classical globin fold. The helix pairs B/E and G/H are arranged each in antiparallel fashion and assembled in a sort of α-helical bundle which surrounds and protects the heme group from the solvent. Although the G- and H-helices generally match the globin fold topology, they may be much shorter or bent as compared to sperm whale Mb. The most striking differences between the 2/2 and the full-length globin folds are (i) the drastically shortened A-helix

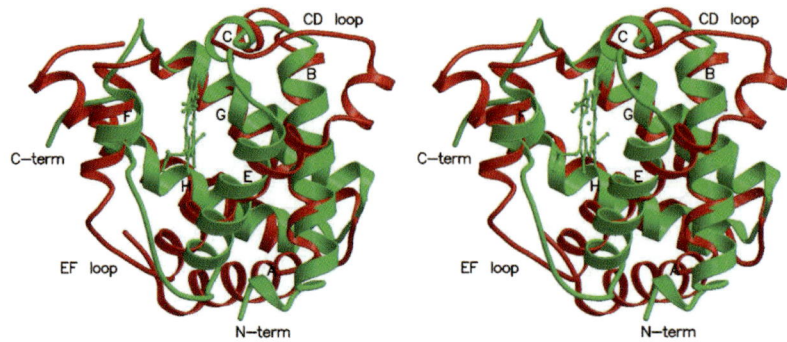

FIG. 1. Stereo view of the 2/2 α-helical sandwich fold achieved in *Chlamydomonas eugametos* 2/2HbN (green), together with the heme group, superimposed to the classical 3/3 α-helical sandwich fold of sperm whale Mb (red). The main topological regions (helices and hinges) are labelled according to the (non-)vertebrate globin fold nomenclature. Structural deviations of the two folds are evident at the N-terminal A-helix, at the CD loop region, in the F-helix and at the C-terminus

(completely deleted in group III 2/2HbP), (ii) the absence of the D-helix, (iii) the presence of a long polypeptide segment (pre-F) in extended conformation and (iv) a variable-length F-helix (reduced to a one-turn-helix in group I and III 2/2Hbs) that properly supports the haeme-coordinated proximal HisF8 residue (Pesce et al. 2000; Milani et al. 2003; Nardini et al. 2006) (Fig. 1). Structural differences are evident also on the heme distal side, where the 2/2 fold CD-D region differs in length relative to (non-)vertebrate globins.

Structural superposition of 2/2Hbs of known structure highlights a general good conservation of the α-helical scaffold among the three groups, with the overall fold of group III 2/2HbP equally diverging in its $C_\alpha$ trace from group I and group II 2/2Hbs. Interesting group-specific structural variability/plasticity can be recognised at well defined sites and correlated to attainment and stabilisation of the compact 2/2Hb fold. At the N- and C-termini, the A-helix can be either very short or fully absent (as in group III), while the H-helix is highly variable in length and linearity, being kinked in group I, short in group II and unusually long in group III. Other important structural variations are localised in the core of the protein. For instance, the polypeptide stretch bridging the C- and E-helices is usually trimmed to about three residues in group I 2/2HbNs and group II 2/2HbOs; on the contrary a 3–7 amino acid insertion is invariably found in group III 2/2HbPs. Such elongation of the CD region has structural implications on the spanning of the C- and E-helices and on the $3_{10}$ helical character of helix C. Indeed, in group III 2/2HbP the C- and E-helices are elongated by one additional turn at their C- and N-termini, respectively, relative to the corresponding helices in group I and group II 2/2Hbs, not affecting, however, the position of the E-helix relative to the heme distal site. Additionally, in group III 2/2HbP the C-helix displays a clear α-helical character, whereas it is a $3_{10}$ helix in group I and group II 2/2Hbs and in (non-)vertebrate globins (Bolognesi et al. 1997). Despite the group-specific structural variations, in all 2/2Hbs the CD region and the E-, F- and G-helices build the protein crevice where the haeme is shielded from the solvent and stabilised by well conserved polar/electrostatic interactions involving the porphyrin propionates.

A stable and properly structured heme crevice in the context of such a short polypeptide chain has been correlated to the presence of three glycine motifs, conserved among sequences of group I 2/2HbNs and group II 2/2HbOs. These Gly motifs are located at the AB and EF inter-helical corners, and just before the short F-helix, and they are thought to provide the protein backbone flexibility needed to stabilise the short A-helix in a conformation locked onto the B- and E-helices, and to support the pre-F segment in building the heme pocket (Pesce et al. 2000; Milani et al. 2001). A similar stabilisation, however, cannot be achieved in group III 2/2HbP, where the AB Gly-Gly motif is missing, due to the full deletion of the A-helix. As a consequence the 2/2HbP amino-terminus cannot face the BE inter-helical region, and the protein residues preceding the B-helix extend towards the GH region; such conformation is opposite to that found in group I and II 2/2Hbs. Similarly, a clear Gly-Gly motif cannot be recognised in the EF region of group III 2/2HbP, although scattered Gly residues are present in this region of the sequence. Despite the absence of a clear EF Gly-Gly motif, B- and E-helices in group III 2/2HbP are oriented as in groups I and II, their stabilisation being achieved through group-specific hydrophobic contacts at the B/E helical interface.

Other group-specific structural variations are localised inside the haeme binding pocket, with group III 2/2HbP having a high degree of similarity in sequence and structure of the distal region to group II 2/2HbOs and, simultaneously, sharing a proximal side extended EF region typically found in group I 2/2HbNs. Structural differences and group-specific residues at the haeme distal and proximal sites are correlated to different ligand-binding properties of 2/2Hbs.

## The Haeme Environment

Despite its small size, the 2/2 helical sandwich provides a scaffold that allows efficient incorporation of the heme group (Fig. 1). Besides the HisF8-Fe coordination bond, in all 2/2Hbs the heme is stabilised by a network of van der Waals contacts with hydrophobic residues at positions C6, C7, CD1, E14, F4, FG3, G8 and H11. Further stabilising interactions are provided by hydrogen bonds between the heme and polar residues, involving Thr/Tyr at sites E2, E5 and EF6, and by salt bridges involving heme propionates and Arg/Lys residues located at position E10 in all 2/2Hbs, at position F2 in group I 2/2HbNs, and at position F7 in group II 2/2HbOs (where F7 is invariantly Arg) and in group III 2/2HbP (Nardini et al. 2006, 2007). The crystallographic studies on group I

2/2HbNs and group II 2/2HbOs have shown that heme isomerism may be present (Milani et al. 2005).

The conformation of the Fe-coordinated proximal HisF8 is typical of an unstrained imidazole ring that facilitates haeme in-plane location of the iron atom, with the imidazole plane lying in a staggered azimuthal orientation relative to the heme pyrrole nitrogen atoms, supporting fast oxygen association (Bolognesi et al. 1997; Wittenberg et al. 2002) and electron donation to the bound distal ligand (Fig. 2). The 2/2Hb heme distal site cavity, hosting the exogenous ligands, is characterised by unusual residues, as compared to (non-) vertebrate globins. It should be noted that in all 2/2Hbs, the E-helix falls close to the heme distal face due to the "pulling action" of the shortened CD region, thus causing side-chain crowding of the distal site residues at topological positions B10, CD1, E7, E11, E14, E15 and G8. Among these, group-specific selections of residues display polar character, allowing the formation of networks of hydrogen bonds functional to the stabilisation of the diatomic heme ligand, and to the rebinding kinetics of dissociated ligands (Samuni et al. 2003). In group I the hydrogen bond network involves residues at B10, E7 and E11 topological positions, with some variations (Milani et al. 2005). In 2/2HbN from *Mycobacterium tuberculosis* a direct TyrB10-$O_2$ hydrogen bond occurs, stabilised by GlnE11 that interacts with TyrB10 (Fig. 2) (Couture et al. 1999; Yeh et al. 2000; Milani et al. 2001, 2004). Comparable ligand-stabilising interactions are observed in the cyanide derivative of the same 2/2HbN in the ferric state (Milani et al. 2004). In *Paramecium caudatum* 2/2HbN and in *Chlamydomonas eugametos* 2/2HbN, residue TyrB10, buried in the inner part of the haeme pocket, is properly oriented through hydrogen bonds toward residues GlnE7 and Thr/GlnE11, to provide stabilisation of the heme-bound

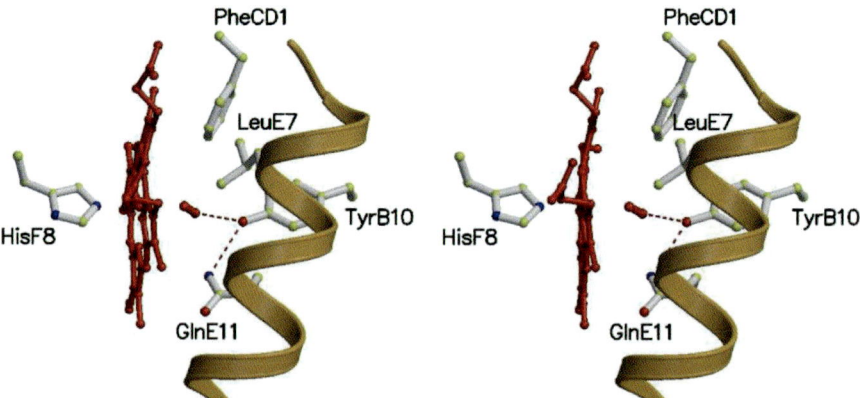

FIG. 2. Stereo view of the distal site of oxy *M. tuberculosis* 2/2HbN showing the oxygen ligand and the network of hydrogen bonds (dashed) that stabilise the ligand in the protein distal site. Part of the E-helix defining the heme distal site is shown as a orange ribbon (right); the location of the proximal HisF8 residue is also shown

distal ligand (Pesce et al. 2000). In all cases, the strongly conserved TyrB10 plays a pivotal role in ligand stabilisation through a direct hydrogen bond to the heme ligand. In general, when a hydrogen bonding residue is present at B10 a Gln is located at E7 or E11, or at both these sites, likely completing the distal site hydrogen-bonded network. On the contrary, when a side chain devoid of hydrogen-bonding capabilities is (rarely) hosted at B10, in general large hydrophobic residues are coupled at the E7 and E11 sites (Vuletich and Lecomte 2006; Nardini et al. 2007).

Although 2/2Hbs do not show a prominent trend for bis-histidine hexacoordination (Couture et al. 2000; Falzone et al. 2002; Scott et al. 2002; Hoy et al. 2004), few examples have been reported. The ferric derivative of *Synechocystis* sp. group I 2/2HbN displays a bis-histidine coordination of the haeme-Fe-atom (involving the proximal/distal residues HisF8 and HisE10, respectively), where binding of an exogenous ligand to the heme requires the dissociation of the Fe-coordinated HisE10 from the haeme and a large conformational change of the B- and E-helices (Trent et al. 2004; Vu et al. 2004). Endogenous heme hexaco-ordination has also been observed, under specific conditions, in other 2/2Hbs (Couture and Guertin 1996; Couture et al. 1999; Visca et al. 2002).

Specific residue substitutions characterise the group II 2/2HbO distal site environment relative to group I 2/2HbNs. In *Bacillus subtilis* 2/2HbO cyano-met crystal structure TyrB10 is directly hydrogen bonded to the cyanide ligand. Residues GlnE11, TrpG8 and ThrE7 complete the distal polar residues frame, with GlnE11 side chain and the TrpG8 indole NE1 atom at hydrogen-bonding distance from the bound ligand (Giangiacomo et al. 2005). In *M. tuberculosis* 2/2HbO the presence of a TyrCD1 residue modifies the hydrogen-bonding distal network; unexpectedly, TyrCD1 is the residue responsible for hydrogen bonding to the diatomic ligand (Milani et al. 2003). Further ligand-stabilising interactions are provided by TrpG8, whose indole NE1 atom is hydrogen bonded to the haeme-bound ligand and to TyrCD1 OH. Thus, the group II conserved TrpG8 residue may stabilise the ligand and modulate its escape rate out of the distal pocket (Milani et al. 2003; Giangiacomo et al. 2005). However, when a Phe residue is present at the CD1 position TyrB10 assumes again the role of hydrogen bond donor (Ouellet et al. 2003). It should also be recalled that the crystal structure of *M. tuberculosis* 2/2HbO has shown that the simultaneous presence of Tyr residues at the B10 and CD1 sites may trigger the formation of a covalent (iso-dityrosine like) bond between the two side chains (Milani et al. 2003); however, no hints on the functional role played by an iso-dityrosine in this 2/2HbO have so far been provided.

Structure and sequence analyses suggest that the properties of residues at the CD1 and E11 sites are correlated in group II 2/2HbOs. All 2/2HbOs that display a Tyr residue at position CD1 bear a nonpolar residue at the E11 site (Milani et al. 2003; Giangiacomo et al. 2005; Vuletich and Lecomte 2006). Conversely, when a Phe residue is found at CD1, a hydrogen bond donor is present at the E11 site (Gln or Ser). Alternatively, a His residue at the topological position CD1 is matched by a hydrophobic E11 residue (Leu or Phe). Thus, one of the necessary hydrogen-bonding elements involved in ligand stabilisation is alternatively located either at the CD1 position or at the E11 position, but never simultaneously.

TyrB10 and TrpG8 are not only invariant in group II, but also in group III 2/2Hbs. In group III *Campylobacter jejuni* 2/2HbP the heme distal pocket residues TyrB10, PheCD1, HisE7, IleE11, PheE14 and TrpG8 surround the heme-bound ligand (Nardini et al. 2006). Among these, only HisE7 is specific and fully conserved in group III (Vuletich and Lecomte 2006). Contrary to group II, group III 2/2HbPs display Phe (or hydrophobic) residue at position CD1 and a hydrophobic residue at site E11. In group III *C. jejuni* 2/2HbP the only hydrogen-bonding residues involved in ligand stabilisation are TyrB10 and TrpG8, while no ligand-stabilising interactions may be provided by residues at CD1 and E11 positions, or by HisE7. The latter residue is observed in two alternate conformations, 'open' and 'closed', corresponding to the side chain pointing towards the solvent or to the haeme distal site, respectively. Unexpectedly, no polar contacts (either to the heme ligand or to TyrB10) characterise HisE7 (Nardini et al. 2006), at difference from sperm whale Mb, where the gating-residue HisE7 is actively involved in locking the heme-Fe(III)-bound cyanide through hydrogen bonding (Bolognesi et al. 1999).

# Cavities in 2/2Hb Protein Matrix

Group I 2/2Hb fold is characterised by the presence of a protein matrix tunnel system offering a potential path for ligand diffusion to the haeme distal site. Such a peculiar feature may be related to the orientation of the CD-D region that forces positioning of the E-helix close to the haeme distal face thus preventing ligand access to the distal site cavity through the classical E7. The protein matrix tunnel, linking the protein surface to the haeme distal site, appears to be conserved in group I 2/2Hbs, suggesting a functional role in small ligand diffusion or storage (Milani et al. 2004). In particular, in *C. eugametos* 2/2HbN and *M. tuberculosis* 2/2HbN, the tunnel is composed of two roughly orthogonal branches converging at the haeme distal site from two distinct protein surface access sites (Fig. 3). On one hand, a 20 Å long tunnel branch connects the protein region nestled between the AB and GH hinges to the haeme distal site. On the other, a path of about 8 Å connects an opening in the protein structure between G- and H-helices to the haeme. The tunnel branches display inner diameters of about 5–7 Å, corresponding to a ligand accessible volume of about 330–360 Å$^3$. In *P. caudatum* 2/2HbN the heme site is connected to the solvent region by a three-cavity system (overall volume of 180 Å$^3$), topologically distributed along the tunnel long branch described above (Milani et al. 2004). Residues lining the tunnel branches are hydrophobic and are substantially conserved throughout group I (Vuletich and Lecomte 2006). PheE15, a well conserved residue, adopts two conformations in *M. tuberculosis* 2/2HbN; one of these may gate the tunnel, limiting ligand access to the heme group

(Milani et al. 2001, 2004) (Fig. 3). As an exception, the hexacoordinated *Synechocystis* sp. 2/2HbN does not display a cavity/tunnel system typical of group I 2/2Hbs, likely because of the conformational transitions supporting heme bis-His hexacoordination. An alternative haeme distal site access through the exposed 8-methyl edge of the haeme group and near the propionates has been proposed (Falzone et al. 2002). In agreement with the availability of cavities in the protein matrix, it has been shown that at least three group I 2/2Hbs can bind Xe atoms in the crystalline state. The Xe atoms map experimentally, at multiple sites and with comparable topology, the tunnel/cavity path in these 2/2Hbs (Milani et al. 2004).

Contrary to group I, group II 2/2Hbs do not show an evident tunnel/cavity system connecting the protein surface to the haeme distal pocket. The protein matrix tunnel observed in 2/2HbNs is dramatically restricted in 2/2HbOs, where different relative orientations of the G- and H-helices, and increased side chain volumes at topological sites B1, B5, G8, G9, G12 and H12, mostly fill the protein matrix tunnel space (Milani et al. 2003; Giangiacomo et al. 2005). In particular, most of the 2/2HbN short tunnel branch and the deeper part of the distal site pocket are filled by TrpG8 in 2/2HbOs. The long tunnel branch retains only two cavities in 2/2HbOs, both fully shielded from solvent contact. The highly restricted protein matrix tunnel seems to be mirrored by the general presence of a small distal site E7 residue in group II 2/2HbOs (Vuletich and

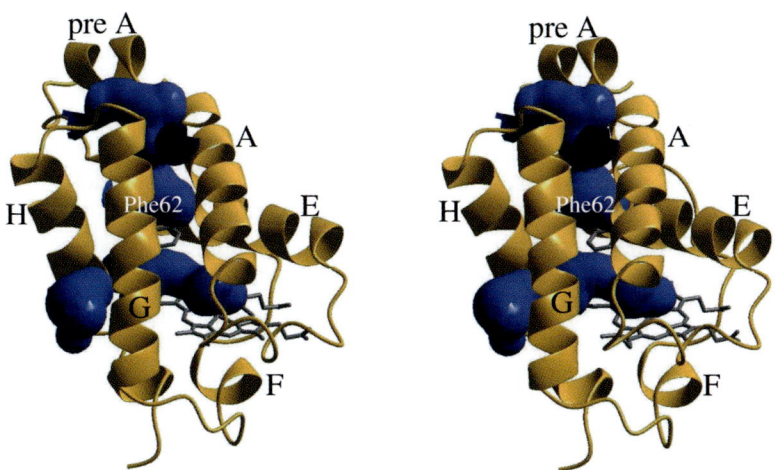

FIG. 3. Stereo view of *M. tuberculosis* 2/2HbN (shown as ribbons) displayed in an orientation allowing a comprehensive view of the protein matrix tunnel system highlighted by the crystal structures. The heme is shown as a stick model; the surface of the tunnel/cavity system is highlighted by a blu envelope traversing the protein matrix; the location of Phe62 (E15) residue that may act as a gate within the tunnel system is shown. Helices are identified as described above

Lecomte 2006), contrary to what was observed in group I where a Gln/Leu residue is highly conserved at position E7. Therefore, accessibility of diatomic ligands (such as $O_2$, CO and NO) to the 2/2HbO haeme distal site may be favoured by the small apolar E7 residue, which does not hinder entrance to the heme distal cavity through an E7 route.

Inspection of group III *C. jejuni* 2/2HbP structure shows no evident protein matrix tunnel/cavity system. Indeed, the obstruction of the (group I specific) protein matrix tunnel is related to the peculiar backbone conformation of the pre-B-helix residues in *C. jejuni* 2/2HbP, and to bulky side chain substitutions (conserved among members of group III) at residues lining the tunnel/cavities walls in group I and II 2/2Hbs. Since HisE7 (conserved in group III) adopts two alternate conformations ('open' and 'closed') in *C. jejuni* 2/2HbP, E7 haeme-distal-site gating may play a functional role for ligand diffusion to the heme, in the absence of a protein matrix tunnel/cavity system (Nardini et al. 2006).

## Conclusions

The functional roles of 2/2Hbs are currently understood only in very few cases, and may vary substantially through the three groups or in different organisms. All 2/2Hbs investigated so far bind $O_2$, with affinities covering the nanomolar to micromolar range. Because 2/2Hbs occur with a high frequency in bacteria, they are held to be of very ancient origin, probably older than the vertebrate globins. 2/2Hbs from more than one group have been shown to coexist in some bacteria (for example, *Mycobacterium avium* encodes 2/2HbN, 2/2HbO and 2/2HbP; *Mycobacterium bovis*, *Mycobacterium smegmatis* and *Mycobacterium tuberculosis* carry both 2/2HbN and 2/2HbO, displaying 18% amino acid sequence identity), whereas *Mycobacterium leprae* displays only 2/2HbO, suggesting a wide diversification of functions. Consistent with the known biochemical/biophysical properties, 2/2Hb functions may include long-term ligand storage, NO detoxification, $O_2$/NO sensing, redox reactions and $O_2$ delivery under hypoxic conditions (Wittenberg et al. 2002). For example, group I 2/2HbN of the unicellular green alga *Chlamydomonas eugametos* is induced in response to active photosynthesis and is localised partly along the chloroplast thylakoid membranes (Couture and Guertin 1996). Group I 2/2HbN from the ciliated protozoa *Paramecium caudatum* may supply $O_2$ to the mitochondria (Wittenberg et al. 2002). Moreover, group I 2/2HbN from the *Nostoc* sp. cyanobacterium is thought to protect the nitrogen-fixation apparatus from oxidative damage by scavenging $O_2$ (Hill et al. 1996). In *Mycobacterium bovis BCG*, group I 2/2HbN promotes an efficient dioxygenase reaction, whereby NO is converted to nitrate by the oxygenated haeme (Ouellet et al. 2002). A similar function has been proposed for *M. tuberculosis* where 2/2HbN may act as a defence system against NO toxicity *in vivo*, based on the observation that the protein catalyses *in vitro* the rapid oxidation of NO to nitrate (2/2HbN-$Fe^{2+}O_2$+NO→2/2HbN-$Fe^{3+}$+$NO_3^-$) (Ouellet et al. 2002; Pathania et al. 2002). In contrast, the physiological function of 2/2HbO from *M. tuberculosis* is still unknown. In *C. jejuni*, a predominant pathogenic agent in bacterial gastrointestinal disease, a group III 2/2HbP defective bacterium was found to be disadvantaged with respect to wild-type cells when grown under high aeration, achieving lower growth yields and consuming $O_2$ at approximately half the rate displayed by wild-type cells. Interestingly, the mutated cells did not show increased sensitivity to NO or oxidative stress, suggesting that this 2/2HbP may play a role in cell respiration (Wainwright et al. 2005, 2006). Given the widespread distribution of 2/2Hbs among bacteria, and the recent discovery of chimaeric proteins bearing a recognisable 2/2Hb domain coupled to a structurally unrelated protein domain, the issue of functionality within this relatively young haemoprotein family becomes more and more complex. In consideration of the varying $O_2$ affinity displayed, likely it will be only through the analysis of phenotypes in deletion mutants that individual roles for 2/2Hbs will be unravelled.

The foundations for the structural work on 2/2Hbs were laid down in a lunch-break discussion that Jonathan and Beatrice Wittenberg "catalysed" in the Summer of 1996, at the Invertebrate Dioxygen Binding Proteins meeting, held in Padova (Italy). That meeting was indeed seminal to the field of 2/2Hbs, since many of the actors who later developed the field were attending, and their interest on these very uncommon/small globins was just beginning to emerge. It is obvious now that Jonathan and Beatrice foresaw that something radically new was emerging, and operated with friendly but firm style to arrange a meeting where our group (working at the time on *Vitreoscilla* Hb) got in touch with the Laval group of Michel Guertin (who at the time was shedding the first light on *C. eugametos* 2/2HbN). That discussion was a true spark: our groups started a joint collaboration that is today still quite active, and spread soon to include the Antwerp group (Luc Moens and collaborators), which was independently moving into this new field. It took more than three years for the first structural results to become available, as discussed in this Volume Chapter. Many distinguished scientists joined the 2/2Hb club at about the time the first two crystal structures (*C. eugametos* and *P. caudatum* 2/2HbNs) were solved (Pesce et al. 2000), bringing stimulating new contributions, many of which are reported in this Volume. A memorable writing session was held at the Wittenbergs' home in Woods Hole in the summer of 2001. The work of the team turned into a review on 2/2Hbs that is still a highly cited reference paper (Wittenberg et al. 2002). We all owe Jonathan and Beatrice much of what today is the scientific significance of the 2/2Hb field.

## *References*

Bolognesi, M., Bordo, D., Rizzi, M., Tarricone, C., and Ascenzi, P. 1997. Nonvertebrate hemoglobins: structural bases for reactivity. Prog. Biophys. Mol. Biol. 68:29–68.
Bolognesi, M., Rosano, C., Losso, R., Borassi, A., Rizzi, M., Wittenberg, J. B., Boffi, A.,

and Ascenzi, P. 1999. Cyanide binding to *Lucina pectinata* hemoglobin I and to sperm whale myoglobin: an x-ray crystallographic study. Biophys. J. 77:1093–1099.

Bonamore, A., Gentili, P., Ilari, A., Schininà, M. E., and Boffi, A. 2003. *Escherichia coli* Flavohemoglobin Is an Efficient Alkylhydroperoxide Reductase. J. Biol. Chem. 278:22272–22277.

Couture, M., Chamberland, H., St-Pierre, B., Lafontaine, J., and Guertin, M. 1994. Nuclear genes encoding chloroplast hemoglobins in the unicellular green alga *Chlamydomonas eugametos*. Mol. Gen. Genet. 243:185–197.

Couture, M., and Guertin, M. 1996. Purification and spectroscopic characterization of a recombinant chloroplastic hemoglobin from the green unicellular alga *Chlamydomonas eugametos*. Eur. J. Biochem. 242:779–787.

Couture, M., Yeh, S. R., Wittenberg, B. A., Wittenberg, J. B., Ouellet, Y., Rousseau, D. L., and Guertin, M. 1999. A cooperative oxygen-binding hemoglobin from *Mycobacterium tuberculosis*. Proc. Natl. Acad. Sci. U.S.A. 96:11223–11228.

Couture, M., Das, T. K., Savard, P. Y., Ouellet, Y., Wittenberg, J. B., Wittenberg, B. A., Rousseau, D. L., and Guertin, M. 2000. Structural investigations of the hemoglobin of the cyanobacterium *Synechocystis* PCC6803 reveal a unique distal heme pocket. Eur. J. Biochem. 267:4770–4780.

Falzone, C. J., Vu, B. C., Scott, N. L., and Lecomte, J. T. 2002. The solution structure of the recombinant hemoglobin from the cyanobacterium *Synechocystis* sp. PCC 6803 in its hemichrome state. J. Mol. Biol. 324:1015–1029.

Gardner, P. R. 2005. Nitric oxide dioxygenase function and mechanism of flavohemoglobin, hemoglobin, myoglobin and their associated reductases. J. Inorg. Biochem. 99:247–266.

Giangiacomo, L., Ilari, A., Boffi, A., Morea, V., and Chiancone, E. 2005. The truncated oxygen-avid hemoglobin from *Bacillus subtilis*: X-ray structure and ligand binding properties. J. Biol. Chem. 280:9192–9202.

Hardison, R. 1998. Hemoglobins from bacteria to man: evolution of different patterns of gene expression. J. Exp. Biol. 201:1099–1117.

Hill, D. R., Belbin, T. J., Thorsteinsson, M. V., Bassam, D., Brass, S., Ernst, A., Boger, P., Paerl, H., Mulligan, M. E., and Potts, M. 1996. GlbN (cyanoglobin) is a peripheral membrane protein that is restricted to certain *Nostoc* spp. J. Bacteriol. 178:6587–6598.

Holm, L., and Sander, C. 1993. Structural alignment of globins, phycocyanins and colicin A. FEBS Lett. 315:301–306.

Hoy, J. A., Kundu, S., Trent, J. T. 3rd, Ramaswamy, S., and Hargrove, M. S. 2004. The crystal structure of *Synechocystis* hemoglobin with a covalent heme linkage. J. Biol. Chem. 279:16535–16542.

Imai, K. 1999. Physiology: The haemoglobin enzyme. Nature 401:437–439.

Iwaasa, H., Takagi, T., and Shikama, K. 1989. Protozoan myoglobin from *Paramecium caudatum*. Its unusual amino acid sequence. J. Mol. Biol. 208:355–358.

Milani, M., Pesce, A., Ouellet, Y., Ascenzi, P., Guertin, M., and Bolognesi, M. 2001. *Mycobacterium tuberculosis* hemoglobin N displays a protein tunnel suited for $O_2$ diffusion to the heme. EMBO J. 20:3902–3909.

Milani, M., Savard, P. Y., Ouellet, H., Ascenzi, P., Guertin, M., and Bolognesi, M. 2003. A TyrCD1/TrpG8 hydrogen bond network and a TyrB10TyrCD1 covalent link shape the heme distal site of *Mycobacterium tuberculosis* hemoglobin O. Proc. Natl. Acad. Sci. U.S.A. 100:5766–5771.

Milani, M., Ouellet, Y., Ouellet, H., Guertin, M., Boffi, A., Antonini, G., Bocedi, A., Mattu, M., Bolognesi, M., and Ascenzi, P. 2004. Cyanide binding to truncated hemoglobins: a crystallographic and kinetic study. Biochemistry 43:5213–5221.

Milani, M., Pesce, A., Nardini, M., Ouellet, H., Ouellet, Y., Dewilde, S., Bocedi, A., Ascenzi, P., Guertin, M., Moens, L., Friedman, J. M., Wittenberg, J. B., and Bolognesi, M. 2005. Structural bases for heme binding and diatomic ligand recognition in truncated hemoglobins. J. Inorg. Biochem. 99:97–109.

Minning, D. M., Gow, A. J., Bonaventura, J., Braun, R., Dewhirst, M., Goldberg, D. E., and Stamler, J. S. 1999. *Ascaris* haemoglobin is a nitric oxide-activated 'deoxygenase' Nature 401:497–502.

Nardini, M., Pesce, A., Labarre, M., Richard, C., Bolli, A., Ascenzi, P., Guertin, M., and Bolognesi, M. 2006. Structural determinants in the group III truncated hemoglobin from *Campylobacter jejuni*. J. Biol. Chem. 281:37803–37812.

Nardini, M., Pesce, A., Milani, M., and Bolognesi, M. 2007. Protein fold and structure in the truncated (2/2) globin family. Gene 398:2–11.

Ouellet, H., Ouellet, Y., Richard, C., Labarre, M., Wittenberg, B., Wittenberg, J., Guertin, M., 2002. Truncated hemoglobin HbN protects *Mycobacterium bovis* from nitric oxide. Proc. Natl. Acad. Sci. U.S.A. 99:5902–5907.

Ouellet, H., Juszczak, L., Dantsker, D., Samuni, U., Ouellet, Y. H., Savard, P. Y., Wittenberg, J. B., Wittenberg, B. A., Friedman, J. M., and Guertin, M. 2003. Reactions of *Mycobacterium tuberculosis* truncated hemoglobin O with ligands reveal a novel ligand-inclusive hydrogen bond network. Biochemistry 42:5764–5774.

Pathania, R., Navani, N. K., Gardner, A. M., Gardner, P. R., and Dikshit, K. L., 2002. Nitric oxide scavenging and detoxification by the *Mycobacterium tuberculosis* haemoglobin, HbN in *Escherichia coli*. Mol. Microbiol. 45:1303–1314.

Perutz, M. F. 1979. Regulation of oxygen affinity of hemoglobin: influence of structure of the globin on the heme iron. Annu. Rev. Biochem. 48:327–386.

Pesce, A., Couture, M., Dewilde, S., Guertin, M., Yamauchi, K., Ascenzi, P., Moens, L., and Bolognesi, M. 2000. A novel two-over-two alpha-helical sandwich fold is characteristic of the truncated hemoglobin family. EMBO J. 19:2424–2434.

Potts, M., Angeloni, S. V., Ebel, R. E., and Bassam, D. 1992. Myoglobin in a cyanobacterium. Science 256:1690–1691.

Samuni, U., Dankster, D., Ray, A., Wittenberg, J. B., Wittenberg, B. A., Dewilde, S., Moens, L., Ouellet, Y., Guertin, M., and Friedman, J. 2003. Kinetic modulation in carbonmonoxy derivatives of truncated hemoglobins: the role of distal heme pocket residues and extended apolar tunnel. J. Biol. Chem. 278:27241–27250.

Scott, N. L., Falzone, C. J., Vuletich, D. A., Zhao, J., Bryant, D. A., and Lecomte, J. T. 2002. Truncated hemoglobin from the cyanobacterium *Synechococcus* sp. PCC 7002: evidence for hexacoordination and covalent adduct formation in the ferric recombinant protein. Biochemistry 41:6902–6910.

Takagi, T. 1993. Hemoglobin from single-celled organisms. Curr. Opin. Struct. Biol. 3:413–418.

Tarricone, C., Galizzi, A., Coda, A., Ascenzi, P., and Bolognesi, M. 1997. Unusual structure of the oxygen-binding site in the dimeric bacterial hemoglobin from *Vitreoscilla* sp. Structure 5:497–507.

Thorsteinsson, M. V., Bevan, D. R., and Potts, M. 1999. A cyanobacterial hemoglobin with unusual ligand binding kinetics and stability properties. Biochemistry 38:2117–2126.

Trent, J. T. 3rd., Kundu, S., Hoy, J. A., and Hargrove, M. S. 2004. Crystallographic analysis of synechocystis cyanoglobin reveals the structural changes accompanying ligand binding in a hexacoordinate hemoglobin. J. Mol. Biol. 341:1097–1108.

Vinogradov, S. N., Hoogewijs, D., Bailly, X., Mizuguchi, K., Dewilde, S., Moens, L., and Vanfleteren, J. R. 2007. A model of globin evolution. Gene 398:132–142.

Visca, P., Fabozzi, G., Petrucca, A., Ciaccio, C., Coletta, M., De Sanctis, G., Bolognesi, M., Milani, M., and Ascenzi, P. 2002. The truncated hemoglobin from *Mycobacterium leprae*. Biochem. Biophys. Res. Commun. 294:1064–1070.

Vu, B. C., Nothnagel, H. J., Vuletich, D. A., Falzone, C. J., and Lecomte, J. T. J. 2004. Cyanide binding to hexacoordinate cyanobacterial hemoglobins: hydrogen-bonding network and heme pocket rearrangement in ferric H117A *Synechocystis* hemoglobin. Biochemistry. 43:12622–12633.

Vuletich, D. A., and Lecomte, J. T. 2006. A phylogenetic and structural analysis of truncated hemoglobins. J. Mol. Evol. 62:196–210.

Wainwright, L. M., Elvers, K. T., Park, S. F., and Poole, R. K. 2005. A truncated haemoglobin implicated in oxygen metabolism by the microaerophilic food-borne pathogen *Campylobacter jejuni*. Microbiology 151:4079–4091.

Wainwright, L. M., Wang, Y., Park, S. F., Yeh, S. R., and Poole, R. K. 2006. Purification and spectroscopic characterization of Ctb, a group III truncated hemoglobin implicated in oxygen metabolism in the food-borne pathogen *Campylobacter jejuni*. Biochemistry 45:6003–6011.

Wittenberg, J. B., and Wittenberg, B. A. 1990. Mechanisms of cytoplasmic hemoglobin and myoglobin function. Annu. Rev. Biophys. Chem. 19:217–241.

Wittenberg, J. B., Bolognesi, M., Wittenberg, B. A., and Guertin, M. 2002. Truncated hemoglobins: a new family of hemoglobins widely distributed in bacteria, unicellular eukaryotes, and plants. J. Biol. Chem. 277: 871–874.

Yang, J., Kloek, A. P., Goldberg, D. E., and Matthews, F. S. 1995. The structure of *Ascaris* hemoglobin domain I at 2.2 A resolution: molecular features of oxygen avidity. Proc. Natl. Acad. Sci. U.S.A. 92:4224–4228.

Yeh, S. R., Couture, M., Ouellet, Y., Guertin, M., and Rousseau, D. L. 2000. A cooperative oxygen binding hemoglobin from *Mycobacterium tuberculosis*. Stabilization of heme ligands by a distal tyrosine residue. J. Biol. Chem. 275:1679–1684.

# 5

# The Phylogeny and Structural Properties of 2/2 Haemoglobins

David A. Vuletich and Juliette T.J. Lecomte

## Abstract

In 2002, Jonathan and Beatrice Wittenberg inspected the 42 protein sequences then known to belong to the lineage of truncated (two-on-two) globins. They pointed out that these proteins parted into three distinct phylogenetic groups. The classification allowed for the identification of essential residues surrounding the heme group and guided subsequent structural studies. It also provided clear targets for further experimental work and set the stage for analyses of gene history. In this chapter, the status of the two-on-two globin lineage is updated. It is shown that the 2002 observations withstand the test of additional sequences and structures. The two-on-two family, which has grown practically ten-fold since the original study, conforms to the group-specific mechanisms of ligand stabilisation and fold features anticipated in the seminal Wittenberg contribution.

## Introduction

The three-dimensional structures of vertebrate myoglobin (Kendrew et al. 1958, 1960) and haemoglobin (Perutz 1960) were determined half a century ago. Since then, biochemistry textbooks have represented the globin fold with seven or eight $\alpha$-helices, six of which form a 3/3 orthogonal bundle shown in Fig. 1a. These extraordinary structures revealed basic principles of protein architecture; specifically, they showed how the ~150-residue globin chain wrapped around the single $b$ heme group and formed a tidy hydrophobic binding site. Clearly, the canonical 3/3 fold had the advantage of excellent shape complementarity with the cofactor. It also allowed an axial ligation bond to form with good geometry between the iron centre and a histidine, referred to as proximal (HisF8).

In a typical haemoglobin, the iron performs its function in the ferrous state and binds dioxygen as a second axial ligand, on the distal side of the heme. It is well known that modern vertebrate globins have evolved to control multiple properties simultaneously, including affinity for the heme group, iron redox potential, exogenous ligand discrimination, holoprotein oligomerisation state and dependence of oxygen affinity – importantly the dissociation rate constant – on pH, temperature and effector concentration. Even the monomeric myoglobin satisfies complex functional requirements (Wittenberg and Wittenberg 2003; Wittenberg 2007). To rationalise the fine adjustment of globin properties using amino acid sequences and three-dimensional structures, however, remains a daunting task.

Although the 3/3 globin fold is well suited for sequestering the heme and modulating its chemical properties, it is not the only α-helical fold capable of such achievement. In 1937, Sato and Tamiya noted the presence of haemoglobin in *Paramecium caudatum* (Sato and Tamiya 1937). This discovery was extended to *Tetrahymena pyriformis* by Keilin and Ryley (1953). The protozoan haemoglobins were sequenced by Iwaasa nearly four decades later (Iwaasa et al. 1989; Iwaasa, Takagi and Shikama, 1990) and were confirmed to be approximately 40 residues shorter than the 3/3 globins. The sequence identity with the vertebrate globins was quite low, but the proteins did bind dioxygen, an ability that, combined with the reduced size of the heme domain, earned these and related proteins the name of "truncated haemoglobins" or trHbs (Moens et al. 1996).

The unusual primary structure of the protozoan haemoglobins attracted limited attention until a similar globin was identified in the cyanobacterium *Nostoc* (Potts et al. 1992) and shortly thereafter in the unicellular alga *Chlamydomonas eugametos* (Couture et al. 1994). In the *Mycobacterium tuberculosis* genome, two trHb genes were detected, *glbN* and *glbO* (Cole et al. 1998), which hinted at great sequence diversity within the new family. Since these early reports, many trHb genes have been found, and, because most occur in bacterial genomes, they are thought to be of ancient origin (Vinogradov et al. 2005; Vuletich and Lecomte 2006).

The functions of trHbs are mostly uncharacterised. The properties of the protein from the three source organisms suggested different roles. *Nostoc* trHb was proposed to serve as an oxygen scavenger (Hill et al. 1996), whereas *P. caudatum* trHb seemed to perform traditional oxygen-binding tasks connected to oxidative metabolism. As for the light-inducible algal trHb, its oxygen affinity appeared too high for efficient oxygen delivery (Couture et al. 1999).

In 2000, Bolognesi and coworkers determined the structure of the trHbs from *P. caudatum* and *C. eugametos* (Pesce et al. 2000). The topology of the abbreviated chain was best described as a 2/2 orthogonal bundle[1] (Fig. 1b). Meanwhile, the ligand-binding properties of *P. caudatum* trHb were being inspected in detail by Wittenberg and colleagues (Das et al. 2000). Dioxygen association and dissociation was indeed consistent with a respiratory role despite the presence of distal residues associated with slow oxygen release in

---

[1] The name 2/2 Hb is now favoured over trHb. In this review, both 2/2 Hb and trHb are used for convenience.

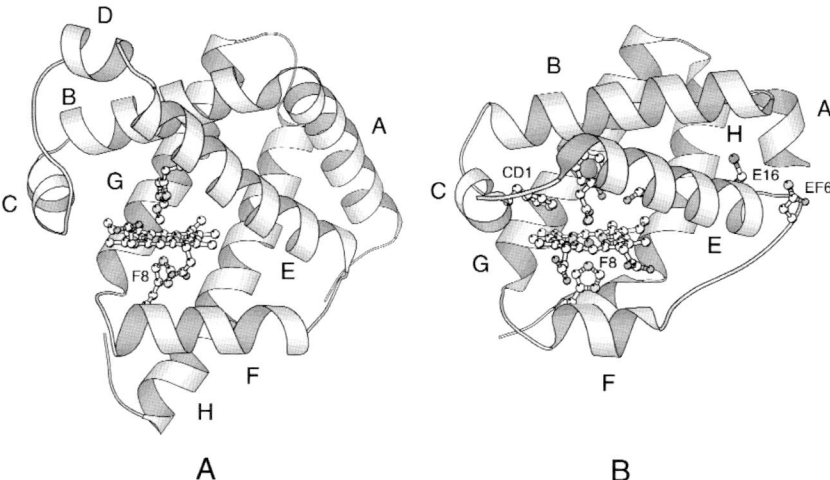

FIG. 1. (a) The X-ray structure of sperm whale myoglobin in the cyanomet form (1ebc, Bolognesi et al. 1999). (b) The X-ray structure of *Paramecium caudatum* trHb in the aquomet form (1dlw, Pesce et al. 2000). The Perutz nomenclature is applied to the helices. The proximal histidine is identified with F8 in both structures. Other represented residues are in (a) HisE7 and in (b) TyrB10, PheCD1, GlnE7, ThrE11, CysE16 and AsnEF6

invertebrate proteins (Bolognesi et al. 1997; Pesce et al. 2001; Weber and Vinogradov 2001). To interpret the kinetic data, two forms of the protein, open and closed, were necessary. Establishing robust relationships between structure and function for this family was obviously hindered by the limited number of characterised 2/2 Hbs and the lack of systematic study of their amino acid conservation patterns.

Two years later, Jonathan and Beatrice Wittenberg and their coworkers compiled the sequences of available putative trHbs, which then totalled 42, and performed an analysis (Wittenberg et al. 2002) that immediately redirected how these proteins were studied. They noted that trHbs split into three phylogenetically distinct groups (I, II and III, or N, O, P, respectively) and that the patterns of amino acid conservation in the heme pocket of each of these groups differed from that of 3/3 globins, be they of vertebrate or invertebrate origin. The recognition of three distinct groups of trHbs was pivotal: it provided a much needed framework within which to organise structural and kinetic data for this new lineage of globins, and it guided the selection of representative experimental targets. It also implied that trHbs had evolved to generate diverse schemes of interaction with exogenous ligands. In this family, one could envision a wide range of chemistry, including enzymatic activity. When examined separately, the three groups of trHbs were bound to offer a rich source of material for genetic history.

We reinspected the phylogeny in 2004, using 105 trHbs (Vuletich and Lecomte 2006) to test the generality of sequence-based conclusions. With this expanded set, the main three groups identified by Wittenberg et al. (2002) were now defined by bootstrap values high enough to leave no doubt about their significance. Within each group, the sequences were orthologous to one another; across groups, they were paralogous. This further supported the proposition that the groups originated as the result of a duplication event early in prokaryotic history.

As a testament to the rapid increase in the number of documented trHb genes, a new family was recently added to Pfam (Bac_globin, PF01152), which to-date contains 365 entries. Given this larger than 3-fold increase in the number of sequences since 2004, it is instructive to reassess the patterns of amino acid conservation and the trHb gene history. This Chapter, which is written as four Group I, four Group II and one Group III wild-type structures are available in the protein data bank, provides a brief update on the state of the 2/2 family.

## Overall Phylogeny

Figure 2 displays a phylogenetic tree consisting of 30 representative trHb sequences (ten from each group, alignments shown in Fig. 3). The figure also includes the identity of the residues found at key positions in the 2/2 fold. E11, E15 and G8 are added to the essential set presented in the 42-sequence analysis (figure 1 in Wittenberg et al. 2002), and the strictly conserved HisF8 is omitted. Thirty sequences were selected to cover a wide species distribution and to encompass all major branches of the family. Despite these restrictive criteria, the three groups hold and can be discussed separately. One important addition is that of two previously unknown archaeal trHbs, from *Haloarcula marismortui* and *Natronomonas pharaonis*.

## *Group I trHb (trHbN)*

When the proteins of Group I are examined alone, a number of interesting conservation patterns arise. In the analysis based on 24 trHbN sequences, the B10 position appeared to harbour a Tyr in most instances and, if not a Tyr, a hydrophobic side-chain. Residue B10 influences exogenous ligand binding in invertebrate globins (Moens et al. 1996; Bolognesi et al. 1997; Pesce et al. 2001; Weber and Vinogradov 2001). In Group I trHbs, TyrB10 establishes a hydrogen bond with exogenous ligand, stabilising its bound state; in turn, TyrB10 is stabilised by H-bonding interactions with the side chains at either E7 or E11, or both (Pesce et al. 2000; Egawa and Yeh 2005; Milani et al. 2005). The sequences lacking a Tyr at B10, however, also lacked H-bonding residues at E7 and E11, and the residue at position B9, typically a Phe, seemed to vary

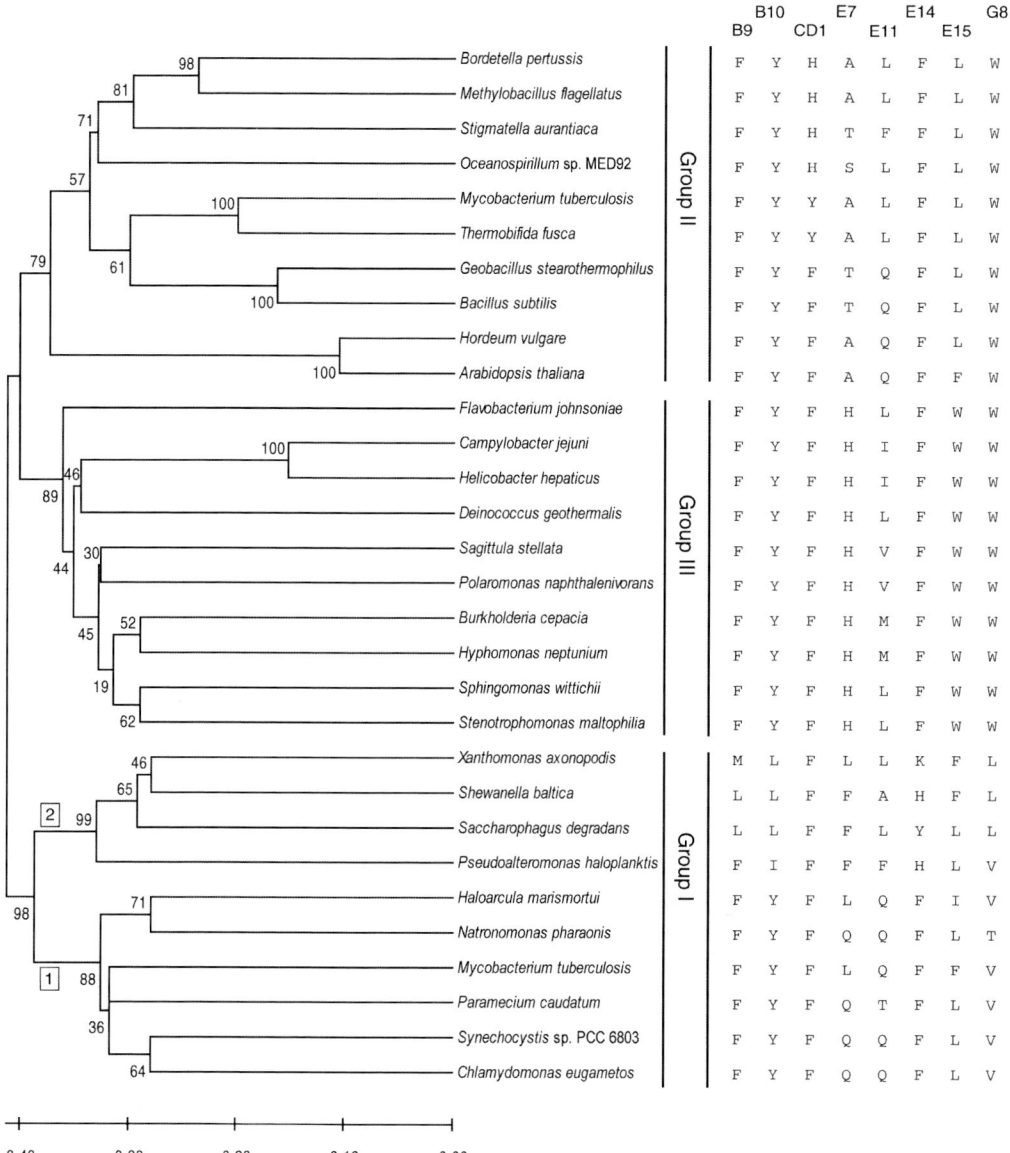

FIG. 2. Minimum evolution tree using alignment of 30 trHb sequences based on p-distance. The tree was constructed with MEGA2 software (Kumar et al. 2001). Phylogeny was tested with 1000 bootstrap replications. The three main trHb groups are indicated. The conservation at key positions derived from the alignment is shown on the right. The boxed numbers indicate the two classes of Group I proteins

36                                    D. A. Vuletich and J. T. J. Lecomte

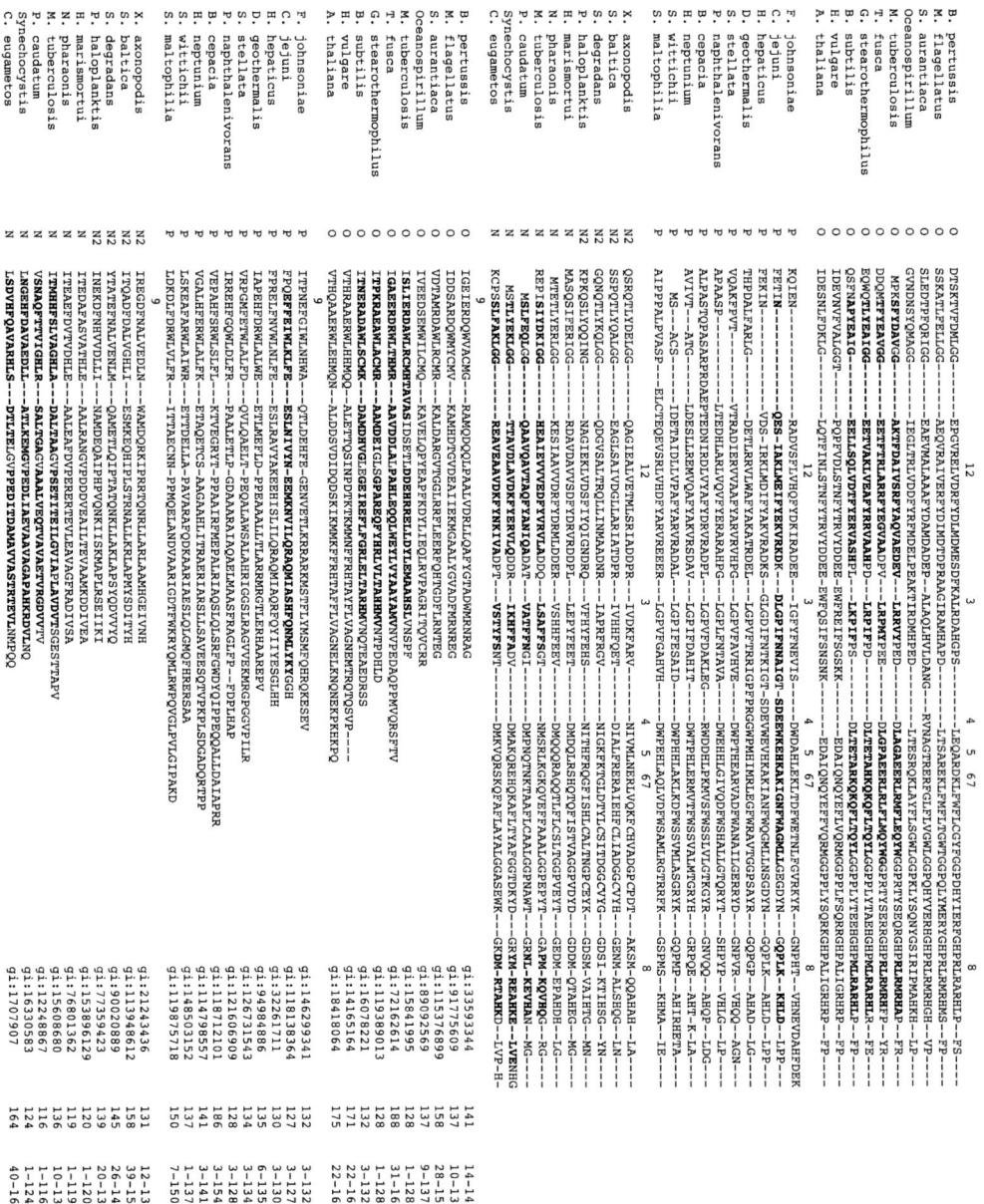

FIG. 3. ClustalX (Thompson et al. 1997) alignment using 30 trHb sequences. Alignment was completed using a Gonnet250 matrix. Alignments were done for five randomly selected sequences at a time
out of the 30 used. After initial alignment, all sequences were realigned over the globin domain. The
gi number, protein length and represented portion are given at the end of each sequence. Source organisms are as named and in the same order as in Fig. 2. The numbers above the sequences refer to 1, B9;
2, B10; 3, CD1; 4, E7; 5, E11; 6, E14; 7, E15; 8, F8; 9, G8. Helical regions are in bold face. The letters O, P and N refer to the three groups; N2 indicates Group I class 2 proteins

as well. Thus, within Group I trHbs, two classes (designated 1 and 2) were distinguishable by the composition of the distal heme pocket (Vuletich and Lecomte 2006). The two archaeal trHbs belong to Group I and exhibit unambiguous Class 1 conservation patterns.

The number of Class 2 proteins has grown significantly, and one can now discern a well conserved pair of cysteines, at positions E16 and EF6 (Fig. 3). Spatial proximity and the feasibility of disulphide bond formation are confirmed by inspection of a Class 1 structure (that of *P. caudatum* trHbN, Fig. 1b) as a model for Class 2 proteins. A disulphide bond may simply increase the thermodynamic stability of the fold. A more interesting scenario, however, is suggested by the coexistence of Class 2 and other trHb genes in the same genome and by Cys pairs in certain neuroglobins and cytoglobins (Hamdane et al. 2003; Nicolis et al. 2007). As in these 3/3 globins, oxidation and reduction of a disulphide bond in response to the cell potential may be coupled to structural changes and modulate exogenous ligand affinity. If such a mechanism is at work, Class 2 trHbNs may give a glimpse of an adaptative response to drastic changes in the early environment of bacteria and cyanobacteria.

## *Group II trHb (trHbO)*

Group II remains the most populated group and the only one found in multicellular organisms, namely plants. Despite the large number of sequences and wide species distribution, trHbOs show a higher level of conservation than trHbNs. In this group, TyrB10 is strictly conserved, and so is TrpG8. The four available structures (Milani et al. 2003; Bonamore et al. 2005; Giangiacomo et al. 2005; Ilari et al. 2007) show this residue to be in direct contact with the heme on the distal side. It has been speculated that TrpG8 is involved in ligand stabilisation, ligand discrimination or even redox properties (Milani et al. 2003; Giangiacomo et al. 2005).

Prior to the discovery of trHbOs, the CD1 position was thought to harbour a strictly conserved Phe (Kapp et al. 1995) whose role was to fasten the heme in its binding site (Ptitsyn and Ting 1999). Instead of a Phe, several Group II trHbs displayed a His or Tyr suggesting additional roles for the residue at CD1 (Wittenberg et al. 2002). From the original alignments, it was also apparent that the branching pattern within Group II trHbs was correlated with the nature of the CD1 residue. These initial observations, combined with ligand-bound structures, mutagenesis studies and an increased number of sequences, provided the means to classify Group II trHbs further (Vuletich and Lecomte 2006).

In *M. tuberculosis* trHbO, TyrCD1 forms a hydrogen bond to the exogenous ligand and controls oxygen association and dissociation rates (Milani et al. 2003; Ouellet et al. 2003). However, the Group II trHb from *Bacillus subtilis* lacks a hydrogen-bonding residue at CD1, and the structure (Giangiacomo et al. 2005) shows that TyrB10 is responsible for the interactions with exogenous ligand, with aid from GlnE11 and ThrE7. These features belong to a general pattern of heme pocket conservation, which presents the CD1 and E11 positions as related. Most trHbO sequences partition into distinct types according to CD1–E11 pairing: TyrCD1 or HisCD1 with a nonpolar residue at E11, PheCD1 with GlnE11 as in several Group I trHbs, and a few other combinations of bulky residues. TrHbOs also tend to contain a small residue at E7, typically Ala, Ser or Thr (Fig. 2).

The expanded sequence data confirm that a Thr at E7 is often present with the PheCD1/GlnE11 pair. ThrE7 interacts with the bound ligand in *Bacillus subtilis*, perhaps imparting increased $O_2$ affinity (Giangiacomo et al. 2005). An alternative combination, PheCD1/AlaE7/GlnE11, is well represented in plants. Also occurring with high frequency are the pairs HisCD1/LeuE11 and HisCD1/PheE11. The structures solved so far are of the TyrCD1/LeuE11 (Milani et al. 2003; Bonamore et al. 2005) and PheCD1/GlnE11 (Giangiacomo et al. 2005; Ilari et al. 2007) varieties. Since at first sight, one expects the divide in the phylogeny and conservation patterns to be reflected in different ligand-binding properties, it will be interesting to examine representatives of the other types, in particular those with a histidine at CD1.

## Group III trHb (trHbP)

The number of Group III trHbs increased from 10 in 2002 to over 30 at present. Group III sequences display the largest extent of conservation and the narrowest distribution among species, as originally noted (Wittenberg et al. 2002). All sequences are bacterial; all contain PheB9, TyrB10, PheCD1, HisE7, PheE14, TrpE15, and TrpG8 (Fig. 2). Such a large number of strictly conserved residues near the heme group strongly suggests a narrow range of chemical properties. Overall, the high sequence identity shows Group III trHbs to form a single homogeneous class.

The phylogenetic analysis of the three groups led to structural expectations (Vuletich and Lecomte 2006): involvement of TyrB10 and TrpG8 but not PheCD1 with the ligand, as derived from Group II trHbs. The conspicuous HisE7 could serve either as an axial ligand to the iron, as observed in certain Group I trHbs (Couture et al. 2000; Scott et al. 2002), or as a H-bonding residue stabilising bound exogenous ligand, as in many 3/3 globins. It was not until 2006 that the structure of a Group III trHb was solved (Nardini et al. 2006). The source organism, *Campylobacter jejuni*, appears to use the globin as a respiratory protein (Wainwright et al. 2006). The new data put to a test the predictions outlined above. TyrB10 and TrpG8 are indeed the only residues that interact with the bound ligand. The role of HisE7, however, is more complex than anticipated. Although not involved in the hydrogen-bonding network, HisE7 is found in the heme pocket and occurs in two distinct conformations. Alternative position for HisE7 may illustrate a ligand-gating mechanism similar to that described in *P. caudatum* trHbN (Das et al. 2000).

# Gene History

The haemoglobin superfamily splits into three main lineages: two 3/3 lineages and one made up of the three groups of 2/2 proteins (Vinogradov et al. 2005). Detailed phylogenetic analysis indicates that the ancestral haemoglobin arose in Bacteria and was horizontally transferred to Archaea and Eukaryotes (Vinogradov et al. 2007). With globins from more than 500 sequenced genomes, it now appears that the ancestral haemoglobin was a 3/3 protein and that 2/2 Hbs originated later (Vinogradov et al. 2007). Within the 2/2 lineage, the three orthologous groups provide evidence that there was at one time a single type of trHb gene and that this ancestral 2/2 gene underwent duplication to form the other groups. The identity of the ancestral sequence is unclear and two different histories of the trHb family have been proposed (Vuletich and Lecomte 2006; Vinogradov et al. 2007).

Table 1 shows the current distribution of the three groups among various types of organisms. The approximate numbers of sequences belonging to each group are included. At first glance, the table shows that only Group I trHbs are present in all three superkingdoms; Group III proteins are restricted to Bacteria; and Group II proteins are so far absent from Archaea but present in Plantae. Both gene histories agree that Group III proteins are likely the result of the most recent duplication event, as their number is low, they are narrowly distributed, and they show the highest levels of conservation.

Although Group I and II proteins exist in organisms other than bacteria, their presence in these superkingdoms is limited. It is therefore probable that horizontal gene transfer events were responsible for dissemination from bacterial sources. One interpretation of the data holds that the ancestral trHb was most like a Group II protein (Vuletich and Lecomte 2006). Two observations support this view: Group II is the largest group, and trHbOs are almost always present in organisms that have more than one trHb Group in their genome

TABLE 1. Distribution of trHb sequences across kingdoms of life, adapted from Vinogradov et al. 2007. The approximate number of sequences of each group in Pfam is shown in parentheses

| Kingdom | trHb | | |
| | Group I (~130) | Group II (~200) | Group III (~30) |
| --- | --- | --- | --- |
| *Archaea* | X | | |
| *Bacteria* | X | X | X |
| *Protista* | | | |
|   Chlorophytes | X | | |
|   Ciliates | X | | |
|   Stramenophiles | | X | |
| *Plantae* | | X | |

(Vuletich and Lecomte 2006; Vinogradov et al. 2007). The other interpretation suggests relatedness of the ancestral 2/2 protein to a Group I trHb (Vinogradov et al. 2007). The results of PSI-Blast searches favour this interpretation: when Group II and III proteins are used as query sequences, Group I proteins are the first to be recognised on subsequent iterations in both cases (Vinogradov et al. 2007). The updated analysis does not resolve the issue; however, as the number of Group I Class 2 proteins increases, these proteins are seen to add greater variability to the heme pocket composition than observed elsewhere in the family. This hints at a longer divergence time for Group I than Group II.

# Concluding Remarks

When confronted with a family of proteins that have widely dissimilar sequences and large distribution through the kingdoms of life, no single course of study is expected to be effective for the family as a whole. Novel insights into adaptation and the role of specific residues in the 3/3 fold are coming to light (Weber and Fago 2004; Berenbrink 2006; di Prisco et al. 2007) with the steady accumulation of biophysical, physiological and genomic data. Without a doubt, the same approach will be necessary for the 2/2 proteins and their full integration into the haemoglobin superfamily.

Initially, trHb structural work was restricted to Group I proteins, for no other reason than that they were the first to be identified. Since the recognition of three distinct groups, experimental trHb studies have been broadened to encompass representative members of each group and have accelerated our understanding of the versatile 2/2 alternative to the 3/3 fold. The data also illustrate that the information derived solely from sequence and structural alignment lacks the level of accuracy required for the reliable prediction of chemical properties. The inspection of the properties conveyed by each type of trHb heme environment highlighted above will be critical for improving the set of rules relating structure and function. The characterisation of Group I Class 2 proteins and additional types of Group II proteins, for example, will help in refining the determinants of heme reactivity and revising the evolutionary history of 2/2 globins.

If one measures contributions to a particular field by the significance of the results and how these shape subsequent rounds of investigation, those interested in globins in the context of physiology, evolution and structural biology will find the body of work published by the Wittenbergs outstanding. The 2002 study (Wittenberg et al. 2002), which captured many of the essential features of the trHb family, will figure as a stepping stone in that area. But there are countless unpublished contributions that advance a scientific field. Jonathan and Beatrice have facilitated and continue to foster interactions among haemoglobin researchers while encouraging newcomers to join in. Our own studies have been marked by their scholarly guidance and enthusiasm for the deepest of inquiry into a fascinating realm of life sciences.

*Acknowledgement*
This work was supported by National Science Foundation grant MCB-0349409. Figure 1 was prepared with Molscript (Kraulis 1991).

# References

Berenbrink, M. 2006. Evolution of vertebrate haemoglobins: Histidine side chains, specific buffer value and Bohr effect. Respir. Physiol. Neurobiol. 154:165–184.

Bolognesi, M., Bordo, D., Rizzi, M., Tarricone, C., and Ascenzi, P. 1997. Nonvertebrate hemoglobins: structural bases for reactivity. Prog. Biophys. Mol. Biol. 68:29–68.

Bolognesi, M., Rosano, C., Losso, R., Borassi, A., Rizzi, M., Wittenberg, J. B., Boffi, A., and Ascenzi, P. 1999. Cyanide binding to *Lucina pectinata* hemoglobin I and to sperm whale myoglobin: an x-ray crystallographic study. Biophys. J. 77:1093–1099.

Bonamore, A., Ilari, A., Giangiacomo, L., Bellelli, A., Morea, V., and Boffi, A. 2005. A novel thermostable hemoglobin from the actinobacterium *Thermobifida fusca*. FEBS J. 272:4189–4201.

Cole, S. T., Brosch, R., Parkhill, J., Garnier, T., Churcher, C., Harris, D., Gordon, S. V., Eiglmeier, K., Gas, S., Barry, C. E., 3rd, et al. 1998. Deciphering the biology of *Mycobacterium tuberculosis* from the complete genome sequence. Nature 393:537–544.

Couture, M., Chamberland, H., St-Pierre, B., Lafontaine, J., and Guertin, M. 1994. Nuclear genes encoding chloroplast hemoglobins in the unicellular green alga *Chlamydomonas eugametos*. Mol. Gen. Genet. 243:185–197.

Couture, M., Das, T. K., Lee, H. C., Peisach, J., Rousseau, D. L., Wittenberg, B. A., Wittenberg, J. B., and Guertin, M. 1999. *Chlamydomonas* chloroplast ferrous hemoglobin. Heme pocket structure and reactions with ligands. J. Biol. Chem. 274:6898–6910.

Couture, M., Das, T. K., Savard, P. Y., Ouellet, Y., Wittenberg, J. B., Wittenberg, B. A., Rousseau, D. L., and Guertin, M. 2000. Structural investigations of the hemoglobin of the cyanobacterium *Synechocystis* PCC6803 reveal a unique distal heme pocket. Eur. J. Biochem. 267:4770–4780.

Das, T. K., Weber, R. E., Dewilde, S., Wittenberg, J. B., Wittenberg, B. A., Yamauchi, K., Van Hauwaert, M. L., Moens, L., and Rousseau, D. L. 2000. Ligand binding in the ferric and ferrous states of *Paramecium* hemoglobin. Biochemistry 39:14330–14340.

di Prisco, G., Eastman, J. T., Giordano, D., Parisi, E., and Verde, C. 2007. Biogeography and adaptation of *Notothenioid* fish: hemoglobin function and globin-gene evolution. Gene 398:143–155.

Egawa, T., and Yeh, S.R. 2005. Structural and functional properties of hemoglobins from unicellular organisms as revealed by resonance Raman spectroscopy. J. Inorg. Biochem. 99:72–96.

Giangiacomo, L., Ilari, A., Boffi, A., Morea, V., and Chiancone, E. 2005. The truncated oxygen-avid hemoglobin from *Bacillus subtilis*. X–ray structure and ligand binding properties. J. Biol. Chem. 280:9192–9202.

Hamdane, D., Kiger, L., Dewilde, S., Green, B. N., Pesce, A., Uzan, J., Burmester, T., Hankeln, T., Bolognesi, M., Moens, L., et al. 2003. The redox state of the cell regulates the ligand binding affinity of human neuroglobin and cytoglobin. J. Biol. Chem. 278:51713–51721.

Hill, D. R., Belbin, T. J., Thorsteinsson, M. V., Bassam, D., Brass, S., Ernst, A., Boger, P., Paerl, H., Mulligan, M. E., and Potts, M. 1996. GlbN (cyanoglobin) is a peripheral membrane protein that is restricted to certain *Nostoc* spp. J. Bacteriol. 178:6587–6598.

Ilari, A., Kjelgaard, P., von Wachenfeldt, C., Catacchio, B., Chiancone, E., and Boffi, A. 2007. Crystal structure and ligand binding properties of the truncated hemoglobin from *Geobacillus stearothermophilus*. Arch. Biochem. Biophys. 457:85–94.

Iwaasa, H., Takagi, T., and Shikama, K. 1989. Protozoan myoglobin from *Paramecium caudatum*. Its unusual amino acid sequence. J. Mol. Biol. 208:355–358.

Iwaasa, H., Takagi, T., and Shikama, K. 1990. Protozoan hemoglobin from *Tetrahymena pyriformis*. Isolation, characterization, and amino acid sequence. J. Biol. Chem. 265:8603–8609.

Kapp, O. H., Moens, L., Vanfleteren, J., Trotman, C. N., Suzuki, T., and Vinogradov, S. N. 1995. Alignment of 700 globin sequences: extent of amino acid substitution and its correlation with variation in volume. Protein Sci. 4:2179–2190.

Keilin, D., and Ryley, J. F. 1953. Haemoglobin in Protozoa. Nature 172:451.

Kendrew, J. C., Bodo, G., Dintzis, H. M., Parrish, R. G., Wyckoff, H., and Phillips, D. C. 1958. A three-dimensional model of the myoglobin molecule obtained by x-ray analysis. Nature 181:662–666.

Kendrew, J. C., Dickerson, R. E., Stranberg, B. E., Hart, R. G., Davies, D. R., Phillips, D. C., and Shore, V. C. 1960. Structure of Myoglobin. Three-dimensional Fourier synthesis at 2Å. resolution. Nature 185:422–427.

Kraulis, P. 1991. MOLSCRIPT: A program to produce both detailed and schematic plots of protein structures. J. Appl. Crystallogr. 24:946–950.

Kumar, S., Tamura, K., Jakobsen, I. B., and Nei, M. 2001. MEGA2: molecular evolutionary genetics analysis software. Bioinformatics 17:1244–1245.

Milani, M., Pesce, A., Nardini, M., Ouellet, H., Ouellet, Y., Dewilde, S., Bocedi, A., Ascenzi, P., Guertin, M., Moens, L., Friedman, J. M., Wittenberg, J.B., and Bolognesi, M. 2005. Structural bases for heme binding and diatomic ligand recognition in truncated hemoglobins. J. Inorg. Biochem. 99:97–109.

Milani, M., Savard, P. Y., Ouellet, H., Ascenzi, P., Guertin, M., and Bolognesi, M. 2003. A TyrCD1/TrpG8 hydrogen bond network and a TyrB10TyrCD1 covalent link shape the heme distal site of *Mycobacterium tuberculosis* hemoglobin O. Proc. Natl. Acad. Sci. U.S.A. 100:5766–5771.

Moens, L., Vanfleteren, J., Van de Peer, Y., Peeters, K., Kapp, O., Czeluzniak, J., Goodman, M., Blaxter, M., and Vinogradov, S. 1996. Globins in nonvertebrate species: dispersal by horizontal gene transfer and evolution of the structure-function relationships. Mol. Biol. Evol. 13:324–333.

Nardini, M., Pesce, A., Labarre, M., Richard, C., Bolli, A., Ascenzi, P., Guertin, M., and Bolognesi, M. 2006. Structural determinants in the group III truncated hemoglobin from *Campylobacter jejuni*. J. Biol. Chem. 281:37803–37812.

Nicolis, S., Monzani, E., Ciaccio, C., Ascenzi, P., Moens, L., and Casella, L. 2007. Reactivity and endogenous modification by nitrite and hydrogen peroxide: does human neuroglobin act only as a scavenger? Biochem. J. 407:89–99.

Ouellet, H., Juszczak, L., Dantsker, D., Samuni, U., Ouellet, Y. H., Savard, P. Y., Wittenberg, J. B., Wittenberg, B. A., Friedman, J. M., and Guertin, M. 2003. Reactions of *Mycobacterium tuberculosis* truncated hemoglobin O with ligands reveal a novel ligand-inclusive hydrogen bond network. Biochemistry 42:5764–5774.

Perutz, M. F. 1960. Structure of hemoglobin. Brookhaven Symp. Biol. 13:165–183.

Pesce, A., Couture, M., Dewilde, S., Guertin, M., Yamauchi, K., Ascenzi, P., Moens, L., and Bolognesi, M. 2000. A novel two-over-two α-helical sandwich fold is characteristic of the truncated hemoglobin family. EMBO J. 19:2424–2434.

Pesce, A., Dewilde, S., Kiger, L., Milani, M., Ascenzi, P., Marden, M. C., Van Hauwaert, M. L., Vanfleteren, J., Moens, L., and Bolognesi, M. 2001. Very high resolution structure of a trematode hemoglobin displaying a TyrB10-TyrE7 heme distal residue pair and high oxygen affinity. J. Mol. Biol. 309:1153–1164.

Potts, M., Angeloni, S. V., Ebel, R. E., and Bassam, D. 1992. Myoglobin in a *Cyanobacterium*. Science 256:1690–1691.

Ptitsyn, O. B., and Ting, K. L. 1999. Non-functional conserved residues in globins and their possible role as a folding nucleus. J. Mol. Biol. 291:671–682.

Sato, T., and Tamiya, H. 1937. Ueber die Atmungsfarbstoffe von *Paramecium*. Cytologia (Tokyo) Fujii-Jubilaei Volume:1133–1138.

Scott, N. L., Falzone, C. J., Vuletich, D. A., Zhao, J., Bryant, D. A., and Lecomte, J. T. J. 2002. Truncated hemoglobin from the cyanobacterium *Synechococcus* sp. PCC 7002: evidence for hexacoordination and covalent adduct formation in the ferric recombinant protein. Biochemistry 41:6902–6910.

Thompson, J. D., Gibson, T. J., Plewniak, F., Jeanmougin, F., and Higgins, D. G. 1997. The CLUSTAL_X windows interface: flexible strategies for multiple sequence alignment aided by quality analysis tools. Nucleic Acids Res. 25:4876–4882.

Vinogradov, S. N., Hoogewijs, D., Bailly, X., Arredondo-Peter, R., Guertin, M., Gough, J., Dewilde, S., Moens, L., and Vanfleteren, J. R. 2005. Three globin lineages belonging to two structural classes in genomes from the three kingdoms of life. Proc. Natl. Acad. Sci. U.S.A. 102:11385–11389.

Vinogradov, S. N., Hoogewijs, D., Bailly, X., Mizuguchi, K., Dewilde, S., Moens, L., and Vanfleteren, J.R. 2007. A model of globin evolution. Gene 398:132–142.

Vuletich, D. A., and Lecomte, J. T. J. 2006. A phylogenetic and structural analysis of truncated hemoglobins. J. Mol. Evol. 62:196–210.

Wainwright, L. M., Wang, Y., Park, S. F., Yeh, S. R., and Poole, R. K. 2006. Purification and spectroscopic characterization of Ctb, a group III truncated hemoglobin implicated in oxygen metabolism in the food-borne pathogen *Campylobacter jejuni*. Biochemistry 45:6003–6011.

Weber, R. E., and Fago, A. 2004. Functional adaptation and its molecular basis in vertebrate hemoglobins, neuroglobins and cytoglobins. Respir. Physiol. Neurobiol. 144:141–159.

Weber, R. E., and Vinogradov, S. N. 2001. Nonvertebrate hemoglobins: functions and molecular adaptations. Physiol. Rev. 81:569–628.

Wittenberg, J. B. 2007. On optima: the case of myoglobin-facilitated oxygen diffusion. Gene 398:156–161.

Wittenberg, J. B., Bolognesi, M., Wittenberg, B. A., and Guertin, M. 2002. Truncated hemoglobins: a new family of hemoglobins widely distributed in bacteria, unicellular eukaryotes, and plants. J. Biol. Chem. 277:871–874.

Wittenberg, J. B., and Wittenberg, B. A. 2003. Myoglobin function reassessed. J. Exp. Biol. 206:2011–2020.

# 6
# Reexamining Data from the Past Related to the Evolution of the Functional Properties of the Haemoglobins of the Teleost Fish

Robert W. Noble

## Abstract

The beautiful work of Berenbrink and his coworkers (2005) has renewed interest in the relationships among the evolution of the Root effect and of the choroid and swimbladder retia. Their work has confirmed, and gone well beyond earlier indications, that the Root effect appeared in evolution in conjunction with the choroid rete, and that the swimbladder rete was a later development that look advantage of the previously developed Root effect. However, it is important to note that evolution of the Root effect appears to have continued after its application to filling swimbladders using a rete. The ligand affinities of the minimum affinity states of the haemoglobins of abyssal fish, which have swimbladders, correlate with the depth of habitat. The minimum ligand affinity decreases as habitat depth increases, consistent with an adaptation to facilitate the unloading of oxygen at ever higher partial pressures.

At the urging of Jonathan Wittenberg, I accepted an invitation to be one of the theme leaders for a round table discussion of the Root effect at the XIV International Conference on Dioxygen Binding and Sensing Proteins. It was apparent that there was particular interest in the evolution of the Root effect and its relationship to the oxygen pumping organelles, the choroid and swimbladder rete. This renewed interest was the result of the beautiful work of Michael Berenbrink and his coworkers (2005), which had been reported recently in *Science*. My own research related to this topic had taken place at least two decades before, but a brief review of the current knowledge made it clear that there was something that I could add to the discussion. The studies of Berenbrink et al. confirmed the far less complete study in which I had collaborated with Marty Farmer, and Hans and Uni Fynn (Farmer et al. 1979) while on the Alpha Helix Expedition to the Amazon for the Study of Fish Bloods and Hemoglobins (Riggs 1970). Marty and I designed this study, but the vast majority of the work was done by Marty. The Fynns furnished electrophoretic data on the haemolysates so that we knew the number of components present in each sample. The haemolysates of representatives of 56 genera of Amazonian fish were examined. It was found that within this sample the presence of the Root effect correlated primarily with the presence of a choroid rete, and only secondarily with the swimbladder rete. It was proposed that the former is the more primitive structure which is associated with the origin of the Root effect. Neither of these retia were found to be present in the absence of a Root effect haemoglobin, but a swimbladder rete was never present in the absence of the choroid rete. These patterns have been beautifully confirmed in the far more precise and extensive work of Berenbrink et al.

However, a second of the early studies in which I participated showed clearly that the evolution of the Root effect did not end with its involvement with the choroid rete and its subsequent use in filling swimbladders with gaseous oxygen. Two decades ago I coauthored an article on the Root effects of haemoglobins from a number of bottom-dwelling, deep-sea fish (Noble et al. 1986). Among these fish were species that had swimbladders and swimbladder retia, *Coryphaenoides armatus*, *C. brevibarbis*, *Antimora rostrata*, *C. rupestris* and *Macrourus berglax*, and species lacking both, *Bathysaurus mollis* and *Alepocephalus sp.* Because of the very low minimum oxygen affinities displayed by many of these haemoglobins, ligand affinities were compared by measuring equilibria of carbon monoxide binding. The haemoglobins of the two deep-dwelling fish lacking swimbladders displayed ordinary Root effects with little subunit heterogeneity and at their minimum affinities they had $p(50)_{CO}$ values of approximately 3 Torr. The haemoglobins of the fish possessing swimbladders exhibited very different properties. At their minimum affinities they exhibited great subunit heterogeneity, with two apparently equal heme populations differing in ligand affinity by as much as 500-fold. The haemoglobins could be grouped into two functionally distinguishable sets. For both sets the subunits with the higher ligand affinities had $p(50)_{CO}$ values of approximately 0.2 Torr. For the higher affinity set, the $p(50)_{CO}$ value for the low affinity heme groups was 30 Torr while that for the lower affinity set was 100 Torr. The latter was the lowest CO affinity ever reported for a haemoglobin molecule. The ocean depths normally inhabited by these fish differ. The deepest dwelling of the species possessing swimbladder retia are *C. armatus* and *C. brevibarbis*, whose depths of greatest abundance are 2900 and 2600 m respectively. *A. rostrata* occurs at lesser depth, with its greatest abundance found at 1700 m. The depth of greatest abundance of *C. rupestris* is 800 m, while that for *M. berglax* is 400 m. It is notable that the lowest CO affinity, $p(50)_{CO}=100$ Torr, is displayed by the haemoglobins of the three species living at the greater depths. It is the two species whose greatest abundance occurs at less that 1000 m that display the $p(50)_{CO}$ value of 30 Torr. There is little doubt that the ability to reduce the ligand affinity of haemoglobin to such an extent has the potential to facilitate the release of oxygen from haemoglobin into a gas phase at very high pressure. It appears likely that these very low affinities represent evolutionary adaptations that served to increase the depths at which species that rely on swimbladders for the maintenance of neutral buoyancy can survive.

It was particularly appropriate that it was Jonathan Wittenberg who urged me to attend this meeting in order to revisit these studies. Jonathan was instrumental in making it possible for these studies to be completed. He invited me on my first expedition aboard the *Oceanus*, an expedition aimed at obtaining some of the haemoglobins examined in this study. Richard Haedrich, an authority on the habitats and distributions of these fish on the ocean floor, was also a participant in this trip. He had developed techniques that permitted him to obtain these fish with great success. This was the first of a series of trips, some on the *Oceanus* with Jonathan and some on other ships with other investigators. However, I doubt that this study would have been started without the help and encouragement of Jonathan. This was not the only time that Jonathan facilitated my work or scientific development. The first occasion was over 40 years ago in Rome where I was working as a postdoctoral fellow with Jeffries Wyman. Jonathan arrived with Beatrice and their two sons for a sabbatical. He kindly offered me the chance to participate in a study of the reactions of horseradish peroxidase with ligands. In the course of this work we discovered the formation of oxyperoxidase in the reaction of reduced, ferrous peroxidase with molecular oxygen. During this work Jonathan taught me many techniques for carrying out anaerobic experiments, techniques which I found to be of great utility throughout my career. I know that I am only one of many scientists who have benefited from Jonathan's and Bea's mentoring, enthusiasm and great knowledge.

*Acknowledgement*
This work was supported by the resources and facilities of the VA Western New York Health Care System.

# References

Berenbrink, M., Koldkjaer, P., Kepp, O., and Cossins, A. R. 2005. Evolution of oxygen secretion in fishes and the emergence of a complex physiological system. Science 307:1752–1757.

Farmer, M., Fynn, H. J., Fynn, U. E. H., and Noble, R. W. 1979. Occurrence of Root effect hemoglobins in Amazonian fishes. Comp. Biochem. Physiol. 62A:115–124.

Riggs, A., ed. 1970. The Alpha Helix Expedition to the Amazon for the study of Fish Bloods and Hemoglobins. Comp. Biochem. Physiol. 62A.

Noble, R. W., Kwiatkowski, L. D., De Young, A., Davis, B. J., Haedrich, R. L., Tam, L.-T., and Riggs A. F. (1986) Functional properties of hemoglobins from deep-sea fish: correlation with depth distribution and presence of a swimbladder. Biochim. Biophys. Acta 870:552–563.

# 7

# Evolutionary Physiology of Oxygen Secretion in the Eye of Fishes of the Suborder *Notothenioidei*

Cinzia Verde, Michael Berenbrink and Guido di Prisco

## Abstract

We wish to tackle a survey on the overall understanding of the molecular properties, biological occurrence, physiological role and evolutionary origin of Root-effect Hbs.

Because high-Antarctic notothenioids still have Hbs endowed with Root effect also when the choroid *rete* is absent, this function may undergo neutral selection. Moreover, the deleterious effects of acidosis can be prevented by increase in the buffering capacity of Hb. Alternatively, high Hb buffer values may be related to the lower Hb content in the blood of notothenioids. As Hb is the main non-bicarbonate buffer in many vertebrates, a decrease in its concentration may entail detrimental consequences for blood acid-base regulation, which could be overcome by an increase in the number of buffering amino-acid residues per molecule. Whether these residues are the cause of the reduced Root effect, or the consequence of altered selection pressure on Hb buffer properties once the Root effect was diminished, remains an open question.

## Introduction

Bea and Jonathan Wittenberg have been performing research for decades at the Albert Einstein College of Medicine, in parallel with their work at Woods Hole.

Guido di Prisco, the senior author of this chapter, moved his first steps into the research scenario at the Department of Biochemistry of AECOM in 1962; for two years he was a post-doc under the guidance of Prof Harold J. Strecker, a famous scientist and a wonderful person. At that time, the only science building was the Forchheimer (also associated with the Jacobi Hospital across the road). GdP was again at AECOM for most of 1968, working in the same department, now hosted in the new adjacent Ullman Building. Harold's expertise was mostly focussed on oxidases and dehydrogenases; at that time, GdP was directing his attention to NADH oxidase and then L-glutamate dehydrogenase. In his green years, he was not familiar with (or even aware of) the exciting research of Bea and Jonathan.

In the early 1960s, GdP was very young and his experience was in the embryonic stage. The almost three years spent at AECOM left a deep mark on his subsequent development as a researcher, but they lapsed without enjoying the proximity of Bea and Jonathan. Perhaps it would have been possible to draw large benefits from interacting with them and becoming familiar with their field of research. Or perhaps not – there is no way one can tell; in fact the respective fields of interest were quite different.

Certainly, GdP did not suspect that their pioneering work on the Root effect in haemoglobin (Hb) would become a constant reference for his team's work. The inception of his interest in Hb began in 1983 with the first isolation and characterisation of Hbs from Antarctic fish, with subsequent development during expeditions and field and bench work in Antarctica occurring annually. From that moment on, although there was no further reason for visiting AECOM, also in view of the sad, untimely departure of Harold from this world, the suite of fundamental papers published by Bea and Jonathan (Wittenberg and Wittenberg 1961; Wittenberg, Schwend and Wittenberg 1964; Wittenberg, Briehl and Wittenberg 1965; Wittenberg et al. 1965) on the secretion of $O_2$ into the swimbladder and the choroid *rete* of fish literally became the bible for GdP's team. In fact, one of our main research lines is currently centred on the molecular basis and evolution of the Root effect.

Time went by, and life is indeed strange. In 2003, we were invited to give a lecture on the structure, function and evolution of the Hbs of Antarctic fish at the XIII International Conference on Invertebrate Dioxygen Binding Proteins (iO$_2$BiP), which took place in Mainz, Germany. The Wittenbergs were amongst the participants. Their work had become quite familiar to us by then, and that was the occasion for the first direct contact with Bea and Jonathan. We promptly realised how deep their scientific and human qualities were. Following the Mainz Conference, we undertook to organise the following one of the series in Naples, at the "Anton Dohrn" Zoological Station in 2006. Again, we had the opportunity to interact with Bea and Jonathan, to take advantage of their immense knowledge on the $O_2$ carriers, and to deepen our personal relationship.

It is remarkable that about 40 years later, a crowded, exciting research domain now shelters a research pathway which merges in a balanced and multidisciplinary way into grounds in common with Bea and Jonathan. We consider this development a great, great stroke of luck.

After this foreword, let us go back to the scope of this volume assembled in honour of this outstanding couple: Bea and Jonathan Wittenberg.

We wish to tackle a survey on the overall understanding of the molecular properties, biological occurrence, physiological role and evolutionary origin of Root-effect Hbs.

# Oxygen Secretion: The Root Effect

## *The Physiological Function*

Hb carries $O_2$ from the environment to tissues; in vertebrates, it is located in specialised cells, the erythrocytes. Hbs are endowed with specific ligand-binding affinity and generally exhibit cooperativity in ligand binding. These properties (affinity and cooperativity) may be adaptively modified in different Hbs to provide adequate $O_2$ supply under a wide variety of environmental and physiological conditions. Many of the functional differences observed in vertebrate Hbs may be interpreted in terms of substitutions of amino acid residues, although many others are due to changes in the composition of the medium in which the protein works. Lastly, change in the concentration of Hbs in the erythrocyte provides another strategy for environment adaptations.

Fish Hbs, similar to other vertebrate Hbs, are tetrameric proteins consisting of two identical pairs of $\alpha$ and $\beta$ subunits; each subunit contains one $O_2$-binding heme group. The subunits are paired in two dimers, $\alpha_1\beta_1$ and $\alpha_2\beta_2$. Within different species, the transport of $O_2$ can be modulated by changes in the Hb structure and allosteric-ligand concentration (ATP for most teleost fish), and by changes in the expression of multiple Hbs likely to display different functional features.

The decreased $O_2$ affinity of Hb at lower pH values in the physiological range is known as the alkaline Bohr effect, reviewed by Riggs (1988). In many Hbs from teleost fishes, when the pH is lowered, the $O_2$ affinity decreases to such an extent that Hbs cannot be fully saturated even at very high $O_2$ pressure. In addition, cooperativity is totally lost and the $O_2$ capacity of blood undergoes reduction of 50% or more of its value at alkaline pH. This feature is known as the Root effect, reviewed by Brittain (2005). Root-effect Hbs are so strongly pH dependent that they are able to unload a large amount of bound $O_2$ at low pH and against a pressure gradient.

The Root effect dictates to what extent the $O_2$ tension can be raised in acid-producing tissues. The physiological role of the Root effect is to secrete $O_2$ against high $O_2$ pressures into the swimbladder (when present) and the poorly vascularised retina, following local acidification of the blood in a counter-current capillary system (Wittenberg and Wittenberg 1961; Wittenberg, Schwend and Wittenberg 1964). Lactic acid as well as $CO_2$ are released into the blood stream and acidify the blood. Both the swimbladder and the retina possess a small $O_2$ concentration *rete mirabile* (known as choroid *rete*, in the eye). It is likely that the eye choroid *rete* represents the most ancient anatomical structure associated with the presence of Root-effect Hbs.

Among non-teleosts, the choroid *rete* is present only in *Amia* (Wittenberg and Haedrich 1974) and evolved just once in fishes, in the clade *Amia + Teleostei* (Berenbrink et al. 2005). In spite of its physiological importance, the choroid *rete* is evolutionarily labile and has been frequently lost during fish evolution (Wittenberg and Haedrich 1974).

The perciform suborder Notothenioidei, mostly confined within Antarctic and sub-Antarctic waters, is the dominant component of the southern ocean fauna. Notothenioids appeared in the early Tertiary and began to diversify on the Antarctic continental shelf in the middle Tertiary, gradually adapting to progressive cooling. These fishes are distributed along very wide latitudes, as they include high-Antarctic, sub-Antarctic and temperate species. Thus they offer a remarkable opportunity to study the physiological and biochemical characters gained and, conversely, lost during thermal evolutionary history. Investigating the evolutionary processes leading to adaptations in the oxygen transport requires a well resolved phylogenetic hypothesis. Luckily, current molecular and morphological hypotheses of notothenioid phylogeny are strongly congruent (Near, Pesavento and Cheng 2004; di Prisco et al. 2007).

The choroid *rete* is well developed in the basal notothenioid families (Eastman 2006). In the phyletically derived Antarctic families, many species have lost the *rete*, and the discovery of reduced and microscopic *retia* suggests that this loss might occur gradually. In Notothenioidei, the randomness of the loss remains open to investigation. However, among notothenioids, the few species possessing Hbs without a Root effect, as well as Hb-less Channichthyidae (Ruud 1954), also lack the choroid *rete*. The molecular phylogenetic studies identifying the Clade X, which assembles, among other fish, notothenioids, zoarcoids and percoids, also infer at least three distinct origins for the Antarctic components of the clade. The whole Antarctic lineages lack the swimbladder, in contrast with their respective sister groups (yet unidentified), in which this structure is certainly maintained (Dettaï and Lecointre 2004, 2005).

In many fish, the swimbladder functions as a hydrostatic organ filled with gas from the blood to achieve neutral buoyancy. The gas partial-pressure gradients necessary for the diffusional gas transport from the blood to the swimbladder are achieved by a decreased gas-carrying capacity caused by blood acidification (Pelster 1997). The latter is produced by specialised epithelial cells, known as gas-gland cells. These cells produce lactic acid *via* the glycolytic pathway, and $CO_2$ *via* the pentose phosphate pathway. Under these conditions, the $O_2$ affinity of Root-effect Hb is strongly decreased and $O_2$ is delivered into the swimbladder. Therefore, the physiological role of Root-effect Hbs is to inflate swimbladders at considerable depths or maintain high levels of oxygenation in the poorly vascularised retinal tissues.

## Interspecific Distribution

The Bohr and Root effects are almost exclusively present in teleost Hbs (Noble et al. 1986); there is currently no explanation for the lack of these physiological characters in some advanced teleost groups.

When addressing ancestral organisms, studies on temperate-elasmobranch Hbs have shown that their functional properties do not always include well developed cooperativity or proton and organophosphate regulation. Elasmobranchs may have diverged from the mammalian line prior to stabilisation of the homotropic and heterotropic interactions typical of mammalian Hbs. In *Torpedo marmorata* Hb, the replacement Asp β97→Lys (Huber and Braunitzer 1989) causes the loss of the salt bridge that is crucial for the Bohr effect in human HbA (Perutz and Brunori 1982). Hence, this mutation may be responsible for the lack of Bohr and Root effects also in *Torpedo nobiliana* Hb, which binds $O_2$ with low cooperativity and shows no appreciable pH dependence or response to organophosphates (Bonaventura, Gillen and Riggs 1974). The polar skates *Bathyraja eatonii* and *Amblyraja hyperborea* (both of the class Elasmobranchii, order Rajiformes, family Rajidae) have a single major Hb not displaying Bohr and Root effects (Verde et al. 2005). The replacement Asp 94β→Leu in *B. eatonii* Hb and Asp 94β?Thr in *A. hyperborea* Hb causes the loss of salt bridges that are essential for the Bohr effect in human HbA. In general, in sharks, lungfish and tetrapods, the Root effect is strongly reduced, as indicated by over 90% $O_2$ saturation also at low pH values (Berenbrink et al. 2005).

A recent study on temperate fish (Berenbrink et al. 2005) has shown that the Root effect apparently evolved 100 million years before the appearance of the choroid *rete*, whereas the swimbladder evolved independently at least four times during evolution. According to this report, the swimbladder, the choroid *rete* and the Root effect have been lost several times in some fish groups. Most important, the weakening of the Root effect is noticed in lineages where the choroid *rete* has been lost, whereas the loss of swimbladder $O_2$ secretion does not affect the magnitude of the Root effect when the choroid *rete* is still present. However, some Antarctic notothenioid species of the family Bathydraconidae lack the choroid *rete*, but their Hbs are endowed with significant Root effect (see below).

## *The Choroid Rete and the Hb Buffer Properties in* Notothenioidei

The comparison of two phylogenetically related groups of notothenioids, one including the non-Antarctic and the other the high-Antarctic species, showed that the choroid *rete* is present in the first group, whilst the high-Antarctic group displays a more variable scenario. In the family Nototheniidae, in fact, *Aethotaxis mitopteryx*, *Pagothenia borchgrevinki* and *Trematomus newnesi* have lost the choroid *rete*. In contrast, *Trematomus bernacchii* has the *rete*; in *Gobionotothen gibberifrons* it is markedly reduced. The two Artedidraconidae *Artedidraco orianae* and *Pogonophryne scotti*, as well as the Bathydraconidae *Gymnodraco acuticeps*, *Cygnodraco mawsoni* and *Bathydraco marri*, also lack the *rete*, whilst the bathydraconid *Racovitzia glacialis* still retains a vestigial form. Similar to other teleosts, a pseudobranch is present in both groups of fish.

The Hbs of the sub-Antarctic notothenioids *Cottoperca gobio* and *Bovichthus diacanthus* are characterised by a very high Root effect (Fig. 1). It must be noted that the choroid *rete* is very well developed in this lineage, which is the most basal of the suborder.

A general reduction in the Root effect is noticed during the evolution of the Antarctic notothenioids (di Prisco et al. 2007).

The clade of Artedidraconidae+Bathydraconidae reveals some intriguing features. While the Root effect drops to low values (as expected) in the Artedidraconidae lineage, as well as in one Bathydraconidae (*G. acuticeps*), it is found at unexpectedly high levels in three species of the latter family (*C. mawsoni*, *B. marri* and *R. glacialis*), all displaying a similar sluggish lifestyle (di Prisco et al. 2007). The question whether the maintenance of the Root effect in Hbs of high-Antarctic notothenioids is related or not to environmental conditions remains to be answered.

Because high-Antarctic notothenioids still have Hbs endowed with Root effect also when the choroid *rete* is absent, this function may undergo neutral selection, not representing a disadvantage for the species. Moreover, a viable hypothesis is that, as in the early ray-finned fish, the deleterious effects of acidosis can be prevented by high Hb buffer values (Berenbrink et al. 2005; Berenbrink 2006, 2007). On the other hand, the weakening of the Root effect in Hbs of many high-Antarctic notothenioids indicates that the Root effect is not an all-or-nothing phenomenon, suggesting that it may be generated by a combination of several factors.

Teleosts that display the Root effect typically have Hbs with a reduced number of physiologically relevant buffer groups when compared to elasmobranch sharks or skates, lobe-finned coelacanths or lungfishes, and tetrapods (Berenbrink et al. 2005; Berenbrink 2006). This reduction has been suggested as an important step in the evolution of the $O_2$-transport system (Berenbrink et al. 2005), as less acid needs to be released into the blood to achieve the low pH values necessary for eliciting the Root effect.

The N-terminal amino groups of the α chains of the major Hbs of teleosts are chemically blocked by acetylation and the number of His in both α and β chains is reduced. The conserved, heme-associated distal His E7 and proximal His F8 are generally considered non-titratable under physiological conditions (Ito, Komiyama and Fermi 1995; Mazzarella et al. 2006b). Of the remaining three His that typically occur in α chains of the major Hbs of teleosts, His G10 and H5 are buried in the $α_1β_1$ contact site, leaving only His CD3 as a predominant potential hydrogen-ion buffer site. In the β chains, surface-accessible His are typically restricted to FG4 and HC3, although some higher teleosts also possess His E13.

Analysis of the amino-acid sequences of the major Hbs of notothenioid fishes indicates that all residues mentioned above are present in most species. However, the selective pressure maintaining a low number of conserved His residues appears to have been relaxed, if not reversed, in these fishes.

The evolutionary changes in physiologically titratable His of the major Hbs are mapped onto a phylogenetic tree of notothenioids (Fig. 1).

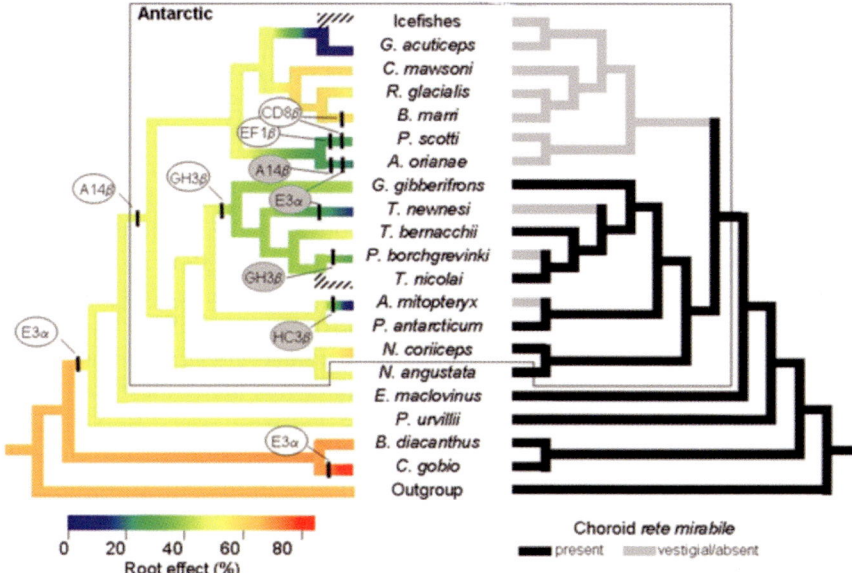

FIG. 1. Reconstruction of the evolutionary history of the Root effect, changes in surface His residues of major Hbs (left panel), and choroid *rete* mirabile (right panel). Character evolution was traced using MacClade 4 (Maddison and Maddison 2003). The Root effect in extant species was obtained from literature sources tabulated in di Prisco et al. (2007). Its magnitude was traced as a continuous-valued character using linear parsimony with the MAXSTATE option. Information on the presence and absence of His at different positions in the major α and β globin chains of notothenioids is based on the amino-acid alignments in Stam et al. (1997) and C. Verde and G. di Prisco (unpublished). Physiologically buffering His residues were predicted as in Berenbrink (2006) and as described in the text. Residue positions are given in helical annotation; gains (unfilled labels, continuous lines) and losses (grey labels, dashed lines) of predicted physiologically buffering His only are traced by maximum parsimony. Choroid *rete* information was obtained from Table 2 in di Prisco et al. (2007). The relationships among major clades of Notothenioidei have been derived from Fig. 4 in Near, Pesavento and Cheng (2004) and for the subfamily Trematominae from Fig. 3 in Sanchez et al. (2007). Outgroups are *Perca fluviatilis* for the Root effect and choroid *rete*, and *Thunnus thynnus* and *Chelidonichthys kumu* for Hb His. The absence of His E3α in *B. diacanthus* can be explained equally parsimoniously either by its independent gain in *C. gobio* and in an immediate common ancestor of all non-bovichthyid notothenioids, as illustrated, or by its gain in the immediate ancestor of all notothenioids and subsequent loss in *B. diacanthus* (not shown). Cross-hatched branches indicate not applicable/missing information. Antarctic species are boxed in

With the exception of *B. diacanthus*, all basal notothenioids, Antarctic or non-Antarctic, have His E3α, which is titratable, based on studies on *T. bernacchii* (Mazzarella et al. 2006a), and not found in other non-notothenioid teleosts. In addition, most Antarctic notothenioids have His A14β, which again is titratable in *T. bernacchii* and extremely rare in other teleosts. Furthermore, all

*Trematomus* species and *G. gibberifrons* have His GH3β, which is fully solvent exposed in *T. bernacchii* and absent in other teleost Hbs.

Fig. 1 also shows the evolutionary reconstruction of the Root effect in the haemolysate of notothenioids on each of the branches of the tree. This indicates that the gains of His E3α, His A14β and His GH3β all occurred in the presence of a Root-effect level of 50–65% and suggests that, other factors being equal, more acid needs to be released into the blood of notothenioids for a given pH decrease compared to other teleosts. In other words, it appears more difficult in notothenioids to achieve the low blood pH values causing $O_2$ to be released from Root-effect Hbs. This may be important for maintaining full $O_2$ loading by Hb at the gills during exercise-induced acidosis, as some of the few investigated notothenioids appear to lack the powerful adrenergic stress response that helps to maintain high intracellular pH in erythrocytes under these conditions in many modern teleosts (Egginton 1997; Lowe and Wells 1997; Berenbrink et al. 2005). It would be interesting to quantify the adrenergic stress response in erythrocytes of additional notothenioids at their low physiological temperature and to relate this to the buffer properties of Hb.

Alternatively, the increase in the buffering capacity of Hb may be related to the trend of lower total Hb concentrations in the blood of notothenioids. As Hb is the main non-bicarbonate buffer in many vertebrates, a decrease in its concentration may entail detrimental consequences for blood acid–base regulation. This could be overcome by an increase in the number of buffering amino-acid residues per Hb molecule. Interestingly, the blood of some Hb-less icefishes shows non-bicarbonate buffer capacities similar to those of red-blooded teleosts (Acierno et al. 1997), but in this case a plasma immunoglobulin-like protein, which contains twice as many His residues as related proteins in other teleosts, has taken over as principal non-bicarbonate buffer (Feller et al. 1994).

As well as a strong Root effect, a choroid *rete* was also present when a higher number of surface His occurred in notothenioids (Fig. 1). This suggests that retinal metabolism was still dependent on the mechanism of ocular $O_2$ secretion. The repeated independent loss of the choroid *rete*, and thereby the ability to increase $O_2$ tensions at the retina above environmental levels, provides an ideal test case to explore the selection pressures shaping the pattern of $O_2$ supply to the retina.

Thus, although the Root effect decreased considerably (by more than 20%) on four independent occasions, all these reductions occurred after, or concomitantly with, the loss of the choroid *rete*. The Root effect does not always disappear or even decrease when the choroid *rete* is lost, as seen in several bathydraconids, and it is not clear whether this is due to some evolutionary time lag, to the Root effect becoming evolutionarily neutral, or to some other function served by the Root effect. At least a role of the Root effect for swimbladder $O_2$ secretion can be excluded, as all notothenioids lack the swimbladder. As there currently appear to be no supporting data, alternative roles of the Root effect remain speculative.

Estimation of potential evolutionary time lags between loss of the choroid *rete* and reduction/disappearance of the Root effect requires a more precise and absolute timing of the choroid *rete* loss in the four cases identified in Fig. 1, which is currently not available. It should become obtainable by including choroid *rete* information on many more notothenioid species in evolutionary reconstructions and by using a molecular-clock approach to estimate divergence times. However, this may not be so straightforward, because Fig. 1 suggests that the loss of the choroid *rete* in the ancestors of Artedidraconidae, Bathydraconidae and icefishes has resulted in very different Root-effect magnitudes, ranging between 5% and 60% in the extant descendants. Thus, different rates of evolution need to be postulated in different lineages to uphold the hypothesis of evolutionary time lags.

Regarding neutrality, it has been argued that the possession of the Root effect may undergo neutral selection pressure in the simultaneous absence of *retia mirabilia* and presence of high Hb buffer values, as in some basal ray-finned fishes and in the ancestors of teleosts (Berenbrink et al. 2005). This may generally also be the case in notothenioids Hbs with increased surface His contents, especially in *B. marri*, which has a Root effect of 60% but an additional His substitution in the external position CD8β. This applies even more to *P. scotti*, with a Root effect of only 25% and two additional His substitutions in the external positions CD8β and EF1β.

Interestingly, in the Hb of a close relative of *P. scotti*, namely *A. orianae* (which has a similar-sized Root effect of 20% and has also lost the choroid *rete*), two of the additional surface His residues of Antarctic notothenioids have been secondarily lost again. This pattern is even more striking in Nototheniidae; in the three species of this family that have independently lost the choroid *rete*, a different solvent-accessible His has been lost as well. In all cases the Root effect has also been reduced, although only to 30% in the blood of *P. borchgrevinki*. A role for some His residues as modulators of the Root effect has recently been postulated (Mazzarella et al. 2006a). It remains to be shown to what extent each of these substitutions is the mechanistic cause of the reduced Root effect or the consequences of an altered selection pressure on Hb buffer properties once the Root effect was diminished.

Whatever the answer to the above questions may turn out to be, it seems that the multiple losses of the ocular $O_2$-secretion mechanism in notothenioids are not necessarily associated with degenerate eyes or less visually oriented lifestyles, as an alternative $O_2$ supply route to the retina by a system of hyaloid capillaries is especially well developed in several notothenioid species that have lost the choroid *rete* (e.g. Eastman and Lannoo 2004; Wujcik et al. 2007).

## The Search for Links with the Molecular Structure

The structural basis of the Root effect has been addressed in many studies. Nevertheless, several aspects of it remain a puzzle. Originally, it was proposed that intersubunit salt bridges stabilise the T quaternary structure, lower the $O_2$ affinity of T relative to R, and are responsible for the Bohr and Root effects (Perutz and Brunori 1982). The ongoing debate on the structural interpretation of the Root effect envisages the possibility that in fish Hbs the classical model (MWC), proposed by Monod, Wyman and Changeux (1965) is an oversimplified explanation of the pH modulation.

C-terminal His in 146β appears to be involved in the Root effect in some Hbs but not in all. Moreover, a fundamental difference between Root-effect Hbs and Hbs with normal Bohr effect (and no Root effect) is that in the former Hbs the $\alpha_1\beta_2$ interface remains stable in the T state upon oxygenation, whereas in the latter the switch to the R state occurs.

Recent studies on the Root-effect Hb of the spot *Leiostomus xanthurus* have suggested that destabilisation of the R state (rather than stabilisation of the T state) at low pH, inducing an R→T transition, may provide a molecular basis for the Root effect (Mylvaganam et al. 1996). In the R state of spot Hb, the central cavity is more narrow than in human HbA, due to steric effects caused by some residues in the β chains. The same feature was also found in the Root-effect Hb of the Antarctic notothenioid *T. bernacchii* (Ito, Komiyama and Fermi 1995). However, these structural properties by themselves are not sufficient to explain the absence of the Root effect in Hb 1 of another Antarctic notothenioid, *T. newnesi* (D'Avino et al. 1994), which in fact has all the residues believed important to elicit the Root effect. Interestingly, the sequence identity between these two major cold-adapted Hbs is very high.

Recently, $O_2$-binding experiments under a wide range of medium conditions have brought about a new model of allosteric regulation of Hb, the *Global Allostery Model*, proposed by Yonetani et al. (2002). This model highlights the role of heterotropic effectors in altering the tertiary structures of both T and R states and consequently in modulating the $O_2$ affinity and the Bohr/Root effect. These authors used bezafibrate and inositol hexaphosphate (IHP) on human HbA over a wide range of pH, obtaining clear indications that allosteric effectors interact with Hb in both the T and R states whereas, in the absence of the effectors, the conformational transition does not seem strictly related to the release of protons. Current evidence now supports the view that in Root-effect Hbs the steric influences on the tertiary structure act in synergy with the quaternary transition between R and T conformations. Allosteric effectors (organophosphates and protons) may modulate these steric effects within both conformations.

Thus it is feasible that the Root effect not only depends on the primary structure, but also on how and to what extent the Hb molecule binds effectors in the T and R states (Yonetani et al. 2002). As pointed out by Brittain (2005), a detailed definition of determinants for the Root effect requires structural information on R and T states at different pH values.

In summary, two main hypotheses, essentially based on the stereochemical model [structural translation of the concerted model (Perutz et al. 1987)], have been proposed to interpret the loss of the $O_2$-driven cooperative T→R transition related to the Root effect, the first one invoking destabilisation of the R state at acidic pH, due to a cluster of positive charges at the $\alpha_1\beta_2$ interface (Mylvaganam et al. 1996), and the second one suggesting overstabilisation of the low-affinity T state (Ito, Komiyama and Fermi 1995; Mazzarella et al. 1999).

Initially, the crystal structure at moderate resolution of deoxy Hb from *T. bernacchii* was resolved, revealing an inter-Asp hydrogen bond at the $\alpha_1\beta_2$ interface, which could stabilise the T state (Ito, Komiyama and Fermi 1995). The hypothesis of pH modulation of the R state was supported by the crystal structure of carbomonoxy Hb of *L. xanthurus*; Mylvaganam et al. (1996) proposed that some protons released in the T?R transition appear from two positively charged clusters formed across the allosteric $\beta_1\beta_2$ interface of the R structure. These clusters would justify two of the Root protons as well as the destabilisation of the R state at acidic pH. Nevertheless, other Hb structures in the R state do not support this hypothesis. In fact, two Root-effect Hbs from Antarctic *T. bernacchii* (Camardella et al. 1992) and tuna (Yokoyama et al. 2004), as well as the non-Root-effect major Hb 1 of *T. newnesi* (Mazzarella et al. 1999), have the same cluster in the carbomonoxy form. Relevantly, as the two Antarctic fish Hbs in the R state display close similarity of the quaternary association, their differences in pH dependence do not seem related to modulation of the R state (Mazzarella et al. 1999). Therefore, the pH modulation of the functional properties of Antarctic fish Hbs was attributed to changes in the tertiary structure within the T state, supporting the *Global Allostery Model* (Yonetani et al. 2002). Crystal structures of Root-effect Hbs in the deoxy form from *T. bernacchii* (Mazzarella et al. 2006b), tuna (Yokoyama et al. 2004) and *T. newnesi* Hb C (Mazzarella et al. 2006a) are available in the literature.

The crystal structure of the deoxy form of Hb C from *T. newnesi* (endowed with the Root effect) has been particularly illuminating to explain the proton heterotropic effect. The amino-acid sequence is characterised by extremely low histidyl content (D'Avino et al. 1994). The deoxy structure exhibits a hydrogen bond between Asp 95$\alpha$ and Asp 101$\beta$, stabilised by Asp 99$\beta$ (Mazzarella et al. 2006a). Therefore, this structural motif *per se* is sufficient to generate the Root effect, and can be considered as the minimal structural requirement needed for designing Root-effect Hbs. Consistent with the findings in *T. newnesi* Hb C, also in tuna and *T. bernacchii* Hbs two Root protons released upon oxygenation at physiological pH are due to the breakage of an Asp-Asp hydrogen bond at the $\alpha_1\beta_2$ interface (Fig. 2).

The role played by the histidyl residues in modulating the strength of the Root effect has been highlighted by the crystal structures of tuna (Yokoyama et al. 2004) and *T. bernacchii* (Mazzarella et al. 2006b) Hbs, both analysed at two different pH values, 5.0 and 7.5, and 6.2 and 8.4, respectively. Berenbrink (2006), using a different approach and perspective, also highlights the role of His in the molecular mechanisms of the Bohr and Root effects.

In conclusion, a general explanation of the Root effect in Hbs of all fish (both temperate and Antarctic) is still unavailable. From the structural analyses

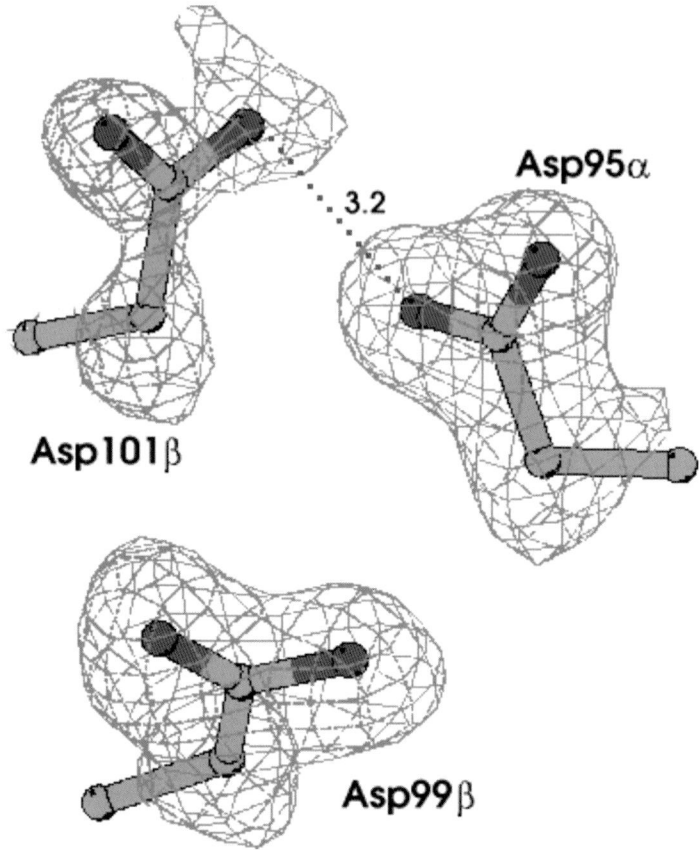

FIG. 2. Electron density map of the $\alpha_2\beta_1$ interface of deoxy *T. bernacchii* Hb at pH 6.2 (a) and 8.4 (b). Distances are in Å. Reproduced from Verde et al. 2007 with permission from Antarctic Science Ltd

it appears that the Root effect is a diffuse structural phenomenon, linked to the interplay of a number of factors in the protein architecture. Nevertheless, in Antarctic fish Hbs, the most convincing hypothesis is based on overstabilisation of the T state, mainly induced by the inter-Asp hydrogen bond at the $\alpha_1\beta_2$ interface (Mazzarella et al. 2006a), possibly modulated by salt bridges (Fig. 3) involving histidyl residues (Mazzarella et al. 2006b).

## Concluding Remarks

After more than three decades of studies, it is virtually impossible as yet to ascribe the presence or absence of the Root effect to a single explanation, for

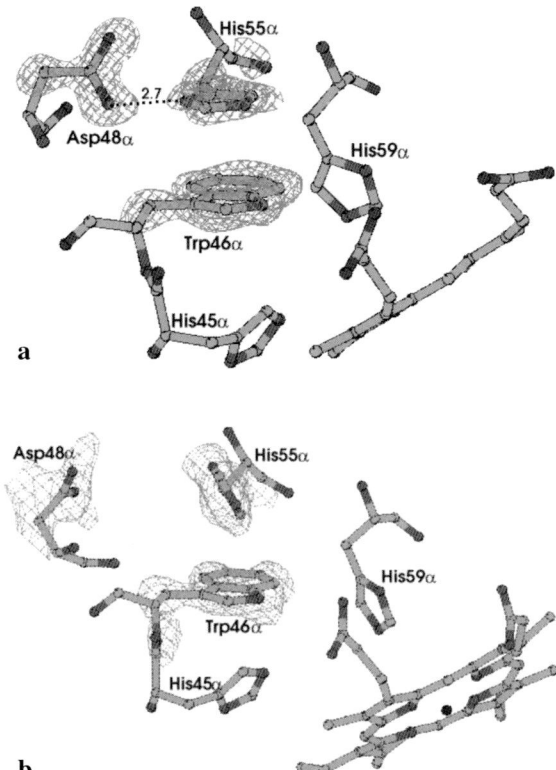

FIG. 3. Electron density map of residues Asp 48α, His 55α and Trp 46α in deoxy *T. bernacchii* Hb at pH 6.2 (a) and at pH 8.4 (b). The α heme and proximal α His are also indicated. Distances are in Å. Reproduced from Verde et al. 2007 with permission of Antarctic Science Ltd

instance to substitutions of a few amino-acid residues. Indeed, the situation seems highly complex, and is probably linked to the combination and interplay of a number of factors in the architecture of the globin tetramer. A multidisciplinary approach has become mandatory. The new structural and functional information (Yokoyama et al. 2004; Mazzarella et al. 2006a, Mazzarella et al. 2006b) will help us, in general, to understand the evolution of the Root effect in fish Hbs, and to question whether the maintenance of the effect in Antarctic notothenioid Hbs is also related to environmental conditions. It ought to be considered that a large number of notothenioids have lost (or are perhaps losing) the choroid *rete*, which in Antarctic fish (lacking the swimbladder) is the morphological structure thought to be linked to the Root effect (Eastman 2006). We suggest that the Hb phenotype of notothenioids is undergoing dynamic changes in response to temperature adaptation over evolutionary time, and that the fish morphological changes and the Hb functional changes are indeed not synchronised.

Because high-Antarctic notothenioids still have Hbs endowed with Root effect also when the choroid *rete* is absent, this function may undergo neutral selection. Moreover, the deleterious effects of acidosis can be prevented by increase in the buffering capacity of Hb (Berenbrink et al. 2005; Berenbrink 2006, 2007). Alternatively, high Hb buffer values may be related to the lower Hb content in the blood of notothenioids. As Hb is the main non-bicarbonate buffer in many vertebrates, a decrease in its concentration may entail detrimental consequences for blood acid–base regulation, which could be overcome by an increase in the number of buffering amino-acid residues per molecule. Whether these residues are the cause of the reduced Root effect, or the consequence of altered selection pressure on Hb buffer properties once the Root effect was diminished, remains an open question.

Once identified, the sister group of the suborder Notothenioidei will probably turn out to be a most important additional tool to study the evolutionary dynamics of the gain and loss of the Root effect.

*Acknowledgements*

This study is financially supported by the Italian National Programme for Antarctic Research (PNRA). It is in the framework of the former SCAR programme Evolution of Antarctic Organisms (EVOLANTA), which merged into the ongoing SCAR programme Evolution and Biodiversity in the Antarctic (EBA). We thank the captain, crew and personnel of Raytheon Polar Services aboard the RVIB *Nathaniel B. Palmer* for their excellent assistance during the ICEFISH 2004 cruise. The cruise was supported by National Science Foundation grant OPP 01-32032 to H. William Detrich (Northeastern University, Boston, USA).

# References

Acierno, R., Maffia, M., Rollo, M., and Storelli, C. 1997. Buffer capacity in the blood of the hemoglobinless Antarctic fish *Chionodraco hamatus*. Comp. Biochem. Physiol. A 118:989–992.

Berenbrink, M. 2006. Evolution of vertebrate haemoglobins: Histidine side chains, specific buffer value and Bohr effect. Respir. Physiol. Neurobiol. 154:165–184.

Berenbrink, M. 2007. Historical reconstructions of evolving physiological complexity: O2 secretion in the eye and swimbladder of fishes. J. Exp. Biol. 209:1641–1652.

Berenbrink, M., Koldkjær, P., Kepp, O., and Cossins, A. R. 2005. Evolution of oxygen secretion in fishes and the emergence of a complex physiological system. Science 307:1752–1757.

Bonaventura, J., Gillen, R. G., and Riggs, A. 1974. The hemoglobin of the Crosspterygian fish, *Latimeria chalumnae* (Smith). Arch. Biochem. Biophys. 163:728–734.

Brittain, T. 2005. Root effect hemoglobins. J. Inorg. Biochem. 99:120–129.

Camardella, L., Caruso, C., D'Avino, R., di Prisco, G., Rutigliano, B., Tamburrini, M., Fermi, G., and Perutz, M. F. 1992. Haemoglobin of the Antarctic fish *Pagothenia bernacchii*. Amino acid sequence, oxygen equilibria and crystal structure of its carbonmonoxy derivative. J. Mol. Biol. 224:449–460.

D'Avino, R., Caruso, C., Tamburrini, M., Romano, M., Rutigliano, B., Polverino de

Laureto, P., Camardella, L., Carratore, V., and di Prisco, G. 1994. Molecular characterization of the functionally distinct hemoglobins of the Antarctic fish *Trematomus newnesi*. J. Biol. Chem. 269:9675–9681.

Dettaï, A., and Lecointre, G. 2004. In search for *Notothenioid* (Teleostei) relatives. Antarctic Sci. 16:71–85.

Dettaï, A., and Lecointre, G. 2005. Further support for the clades obtained by multiple molecular phylogenies in the acanthomorph bush. Comp. Rend.-Biol. 328:674–689.

di Prisco, G., Eastman, J. T., Giordano, D., Parisi, E., and Verde, C. 2007. Biogeography and adaptation of Notothenioid fish: hemoglobin function and globin-gene evolution. Gene 398:143–155.

Eastman, J. T. 2006. Aspects of the morphology of phyletically basal bovichtid fishes of the Antarctic suborder *Notothenioidei* (Perciformes). Polar Biol. 29:754–763.

Eastman, J. T., and Lannoo, M. J. 2004. Brain and sense organ anatomy and histology in hemoglobinless Antarctic icefishes (Perciformes: Notothenioidei: Channichthyidae). J. Morphol. 260:117–140.

Egginton, S. 1997. A comparison of the response to induced exercise in red- and white-blooded Antarctic fishes. J. Comp. Physiol. 167:129–134.

Feller, G., Poncin, A., Aittaleb, M., Schyns, R., and Gerday, C. 1994. The blood proteins of the Antarctic icefish *Channichthys rhinoceratus*: biological significance and purification of the two main components. J. Comp. Physiol. B 109:89–97.

Huber, F., and Braunitzer, G. 1989. The primary structure of electric ray haemoglobin (*Torpedo marmorata*). Bohr effect and phosphate interaction. Biol. Chem. Hoppe-Seyler 370:831–838.

Ito, N., Komiyama, N. H., and Fermi, G. 1995. Structure of deoxyhemoglobin of the Antarctic fish *Pagothenia bernacchii* with an analysis of the structural basis of the Root effect by comparison of the liganded and unliganded hemoglobin structures. J. Mol. Biol. 250:648–658.

Lowe, T. E., and Wells, R. M. G. 1997. Exercise challenge in Antarctic fishes: do haematology and muscle metabolite levels limit swimming performance? Polar Biol. 17:211–218.

Maddison, D. R., Maddison, W. P. 2003. MacClade 4: Analysis of phylogeny and character evolution. Version 4.06. Sinauer Associates, Sunderland, MA.

Mazzarella, L., D'Avino, R., di Prisco, G., Savino, C., Vitagliano, L., Moody, P. C. E., and Zagari, A. 1999. Crystal structure of *Trematomus newnesi* hemoglobin re-opens the Root effect question. J. Mol. Biol. 287:897–906.

Mazzarella, L., Bonomi, G., Lubrano, M., Merlino, A., Riccio, A., Vergara, A., Vitagliano, L., Verde, C., and di Prisco, G. 2006a. Minimal structural requirements for Root effect: crystal structure of the cathodic hemoglobin isolated from the Antarctic fish *Trematomus newnesi*. Proteins 62:316–321.

Mazzarella, L., Vergara, A., Vitagliano, L., Merlino, A., Bonomi, G., Scala, S., Verde, C., and di Prisco, G. 2006b. High-resolution crystal structure of deoxy haemoglobin from *Trematomus bernacchii* at different pH values: the role of histidine residues in modulating the strength of the Root effect. Proteins 65:490–498.

Monod, J., Wyman, J., and Changeux, J. P. 1965. On the nature of allosteric transitions: a plausible model. J. Mol. Biol. 12:88–118.

Mylvaganam, S. E., Bonaventura, C., Bonaventura, J., and Getzoff, E. D. 1996. Structural basis for the Root effect in haemoglobin. Nature Struct. Biol. 3:275–283.

Near, T. J., Pesavento, J. J., and Cheng, C.-H. C. 2004. Phylogenetic investigations of Antarctic notothenioid fishes (Perciformes: Notothenioidei) using complete gene sequences of the mitochondrial encoded 16S rRNA. Mol. Phylogenet. Evol. 32:881–891.

Noble, R. W., Kwiatkowski, L. D., De Young, A., Davis, B. J., Haedrich, R. L., Tam, L.T., and Riggs, A. F. 1986. Functional properties of hemoglobins from deep-sea fish: correlations with depth distribution and presence of a swimbladder. Biochim. Biophys. Acta 870:552–563.

Pelster, B. 1997. Buoyancy at depth. In Deep-Sea Fish, eds. D. Randall and A. P. Farrell, pp. 195–237. San Diego: Academic Press.

Perutz, M. F., and Brunori, M. 1982. Stereochemistry of cooperative effects in fish and amphibian hemoglobins. Nature 229:421–442.

Perutz, M. F., Fermi, G., Luisi, B., Shanan, B., and Liddington, R. C. 1987. Stereochemistry of cooperative mechanisms in hemoglobin. Acc. Chem. Res. 20:309–321.

Riggs, A. 1988. The Bohr effect. Annu. Rev. Physiol. 50:181–204.

Ruud, J. T. 1954. Vertebrates without erythrocytes and blood pigment. Nature 173:848–850.

Sanchez, S., Dettai, A., Bonillo, C., Ozouf-Costaz, C., Detrich, H. W. III., and Lecointre, G. 2007. Molecular and morphological phylogenies of Antarctic teleostean family Nototheniidae, with emphasis on the Trematominae. Polar Biol. 30:155–166.

Stam, W. T, Beintema, J. J., D'Avino, R., Tamburrini, M., and di Prisco, G. 1997. Molecular evolution of hemoglobins of Antarctic fishes (Notothenioidei). J. Mol. Evol. 45:437–445.

Verde, C., De Rosa, M. C., Giordano, D., Mosca, D., de Pascale, D., Raiola, L., Cocca, E., Carratore, V., Giardina, B., and di Prisco, G. 2005. Structure, function and molecular adaptations of haemoglobins of the polar cartilaginous fish Bathyraja eatonii and Raja hyperborea. Biochem. J. 389:297–306.

Verde, C., Vergara, A., Giordano, D., Mazzarella, L., and di Prisco, G. 2007. The Root effect – a structural and evolutionary perspective. Antarctic Sci 19:271–278.

Wittenberg, J. B., Schwend, M. J., and Wittenberg B. A. 1964. The secretion of oxygen into the swim-bladder of fish III. The role of carbon dioxide. J. Gen. Physiol. 48:337–355.

Wittenberg, B. A., Briehl, R. W., and Wittenberg, J. B. 1965. Haemoglobins of invertebrate tissues. Nerve haemoglobins of Aphrodite, Aplysia and Halosydna. Biochem. J. 96:363–371.

Wittenberg, B. A., Brunori, M., Antonini, E., Wittenberg, J. B., and Wyman, J. 1965. Kinetics of the reactions of Aplysia myoglobin with oxygen and carbon monoxide. Arch. Biochem. Biophys. 111:576–579.

Wittenberg, J. B., and Haedrich, R. L. 1974. The choroid rete mirabile of the fish eye. II. Distribution and relation to the pseudobranch and to the swim-bladder rete mirabile. Biol. Bull. 146:137–156.

Wittenberg, J. B., and Wittenberg B. A. 1961. The secretion of oxygen into the swim-bladder of fish. II. The transport of molecular oxygen. J. Gen. Physiol. 44:527–542.

Wujcik, J. M., Wang, G., Eastman, J. T., and Sidell, B. D. 2007. Morphometry of retinal vasculature in Antarctic fishes is dependent upon the level of hemoglobin in circulation. J. Exp. Biol. 210:815–824.

Yokoyama, T., Chong, K. T., Miyazaki, G., Morimoto, H., Shih, D. T. B., Unzai, S., Tame, J. R. H., and Park S.-Y. 2004. Novel mechanisms of pH sensitivity in tuna hemoglobin: a structural explanation of the Root effect. J. Biol. Chem. 279:28632–28640.

Yonetani, T., Park, S., Tsuneshige, A., Imai, K., and Kanaori, K. 2002. Global allostery model of hemoglobin. J. Biol. Chem. 277:34508–34520.

# 8

# Mutagenic Studies on the Origins of the Root Effect

Satoru Unzai, Kiyohiro Imai, Sam-Yong Park, Kiyoshi Nagai, Tom Brittain and Jeremy R.H. Tame

## Abstract

Unlike the majority of mammals, which produce only a single major haemoglobin (Hb) component (>90% of the Hb content of the red blood cell), many fish species have multiple Hb components which show considerable differences in sequence and functional properties. Functional heterogeneity of several Hb types within the red cell can extend the range of conditions under which oxygen can be transported effectively around the blood stream, and permits a division of labour between the various components which can each fulfil a specific role. A number of fish Hbs are known to show very low oxygen affinity as pH drops, but the structural basis for this effect is only recently becoming clear. Site-directed mutagenesis and functional studies are described, comparing the Hbs of *Trematomus bernachii* and *Trematomus newnesi*.

## Introduction

Air-breathing vertebrates mostly produce only a single major haemoglobin (Hb), since they normally experience an essentially constant oxygen supply and so a single form of the protein, optimised for the particular habitat of the animal, is sufficient. Many fish species, however, have multiple Hb components which show considerable differences in sequence and functional properties. This functional heterogeneity of several Hbs within the red cell extends the range of conditions under which oxygen can be effectively transported, and permits each of the various components to fulfil a specific role. Unlike land animals, fish require oxygen transport for two unique purposes, namely to fill swim bladders for buoyancy and to provide oxygen to poorly vascularised retina, against extreme hydrostatic pressures. It was Biot (1807) who first described active oxygen secretion into the fish swim bladder and Woodland (1911) who first described the gas gland and *rete mirabile* structures which promote oxygen secretion. In 1922 Haldane suggested that oxygenated Hb was the source of the gaseous oxygen found in the swim bladder (Haldane, 1922). The particular nature of the Hb which performed this task was investigated in a series of papers by Root in the 1930s and 1940s (Root 1931; Root and Irving 1941, 1943). It was in the late 1950s and early 1960s that Jonathan and Beatrice Wittenberg began publishing their investigations into the active secretion of oxygen into the swim bladder of fishes (Wittenberg 1958, 1961; Wittenberg and Wittenberg 1961; Wittenberg, Schwend and Wittenberg 1964), and in 1962 they began their studies on active secretion of oxygen into the fish eye. This led to their detailed description of the anatomical structure of the choroid *rete mirabile* (Wittenberg and Wittenberg 1962; Wittenberg and Haedrich 1974; Wittenberg and Wittenberg 1974), which established that the same Hb function was responsible for not only filling the fish swim bladder but also maintaining oxygenation of the fish retina. During this period, based on the seminal structural work of Perutz and his colleagues, major advances were being made in our understanding of the structural origins of Hb function. A number of fish Hbs have now been shown to exhibit a very low oxygen affinity and loss of cooperativity at low pH, a phenomenon called the Root effect after its discoverer (Root 1931). This is the driver of oxygen transport against high hydrostatic pressures, but the structural origin of this effect has only recently started to become clear. Even though the crystal structures of many vertebrate Hbs are now known, and the sequences of many fish Hbs with and without the Root effect have been determined, the key amino acid residues remain largely unidentified. This is partly due to the fact, revealed by detailed analyses of fish Hb sequences, that the Root effect has evolved (and been lost) repeatedly over hundreds of millions of years (Berenbrink et al. 2005). A number of residues have been suggested to play a role in the Root effect, but almost all of these prove to be of no importance when modified by site-directed mutagenesis (Perutz and Brunori 1982). Only within the last few years have any Root effect residues been positively identified by X-ray crystallography, and few of these have been tested by artificial mutants. As more sequence data have become available, it has become clear that the Root effect is a physiological phenomenon, and need not (as has often been assumed) always arise from the same molecular mechanism. Nor is the Root effect simply a "strong Bohr effect". The Root effect is characterised by a complete loss of homotropic cooperativity of oxygen binding. Several Root effect Hbs show Hill coefficients of less than unity at low pH, a feature never seen in human Hb. There is therefore no *a priori* reason for believing the Root effect and alkaline Bohr effect of human HbA share the same mechanism. Given the enormous number of man-years required to achieve our present understanding of the Bohr effect, it should not be surprising that, even with modern methods, the Root effect has also proved difficult to unravel. In order to discover the molecular origins of the Root effect, it is necessary (but not necessarily sufficient) to obtain high-resolution crystallographic structures of the Hbs at both high and low pH.

Hb isolated from vertebrates consists of an $\alpha_2\beta_2$ tetrameric structure. Both for historical reasons and its simplicity the model first proposed by Monod, Wyman and Changeux (1965) has been the one most used to describe the homotropic effect of oxygen binding to Hb. This model assumes that the protein can exist in equilibrium between two quaternary states, which Perutz identified with the crystal structures of oxygenated and deoxygenated Hb. This two-state model assumes that the subunits within any particular tetramer have identical, fixed oxygen affinities determined solely by the quaternary state. The R state has a high oxygen affinity ($K_R$) and the T state a low oxygen affinity ($K_T$). This model accounts for the observed sigmoidal oxygen-binding characteristics of Hb, if, in the absence of oxygen, the T state is overwhelmingly favoured and in the presence of saturating amounts of oxygen the R state is favoured. The model predicts that in human Hb a concerted switch from predominantly T state to R state occurs at around half saturation. The switch point can occur much earlier or later depending on the relative magnitudes of the oxygen affinities of the two states and the equilibrium constant (L). Although the homotropic effect describes the fundamental operation of vertebrate Hbs, the capacity of the oxygen transport system to respond to changing physiological demands requires a further level of modulation of the oxygen-binding properties of Hb beyond that possible simply by the operation of the homotropic effect. In a heterotropic effect, the binding of oxygen to the Hb tetramer is often affected by the presence of a third substance. Both carbon dioxide and protons, produced by aerobic metabolism, lead to a lowering of oxygen affinity in human Hb. Animal Hbs in general also bind to intracellular organic phosphates such as 2,3 diphosphoglycerate (DPG), and these too lower the oxygen-binding affinity. Fish Hbs respond to ATP or GTP rather than DPG however, and the reason for this altered specificity remains unclear (Wells, 1999).

Several crystal structures of Root effect Hbs have now been solved. The first of these was the structure of *Trematomus bernachii* Hb (HbTb) in the carbonmonoxy form at pH 8.4. This structure, and even the subsequent structure of the deoxy form at pH 6.2, offered few clues to the cause of the Root effect however (Camardella et al. 1992; Ito, Komiyama and Fermi 1995). The only obvious proton binding site was found between a pair of aspartate side-chains, Asp $\alpha$95 and Asp $\beta$101, which approach each other closely in the T state, sharing a proton between side-chain carboxyl groups. The group in Yokohama has solved the structure of tuna Hb in the deoxy form at low and moderate pH, and also the carbonmonoxy form at high pH (Yokoyama et al. 2004). Comparison of these models showed significant changes at several sites in the molecule, both with haem ligation and pH. As well as the Asp-Asp pair, tuna Hb shows a marked tertiary structure change in the $\beta$ chains which is linked to a salt-bridge (His 69-Asp 72) in the E helix. The structures of tuna Hb are significant for two main reasons. One is that, unlike HbTb, tuna Hb shows negative cooperativity at low pH. This effect cannot be reconciled with the MWC model, so it is impossible to explain the Root effect merely in terms of T-state stabilisation or R-state destabilisation. Secondly, the tuna Hb structures show that tertiary effects can be as important, if not more so, than quaternary effects. In essence it appears that at low pH the $\beta$ subunits switch to a low oxygen affinity form no matter what the $\alpha$ chains are doing. Such twin–two-state models have been described for Hb, in which the tertiary and quaternary changes no longer necessarily occur together, as they do in the MWC model (Herzfeld and Stanley 1974; Henry et al. 2002). More recently the crystal structure of the cathodic Hb component from *T. newnesi* has appeared (Mazzarella et al. 1999). The structures of HbTb give clear evidence of a novel salt bridge between Asp 48 and His 55 of the α subunits which appears to cause the Root effect in this protein. The crystal structures therefore provide a picture in which very different groups are responsible for the pH sensitivity in different fish Hbs. However, structures of these proteins, while invaluable in identifying the key groups, give little clue to the actual contribution of particular residues in terms of energy or protons. This can only be achieved by functional studies of site-directed mutants.

In order to investigate the molecular mechanism of the Root effect, we have produced the major Hb component of the Antarctic fish *Trematomus bernachii* (HbTb) in the yeast *Saccharomyces cerevisiae*. This expression system acetylates the N terminal serine residue of the α chain, as found in the native protein, and the recombinant protein shows a Root effect identical to the protein isolated from the animal. The major component of *Trematomus newnesii* (HbTn) differs from HbTb at only 14 residues, but entirely lacks the Root effect. To find out which of these 14 changes are involved in the Root effect, we have produced a number of chimaeras, intermediate between HbTn and HbTb.

# Materials and Methods

## Construction of HbTb Expression Vector

The GAL–GAP hybrid promoter was PCR amplified from pGS4088 using a forward primer containing BamH1 sites and a reverse primer with a NcoI site (at the initiation codon) and a HindIII site. The PCR product was digested with bamH1 and HindIII and cloned into pUC-NOT to form pUC-GALGAP. The 3' UTR sequence was also PCR amplified from pGS4088 with a forward primer containing a HindIII site and a backward primer containing a BglII site. The PCR product was first purified using Qiagen PCR purification kit, digested with HindIII and BglII and cloned into pUC-GALGAP linearised with the same enzymes to form pUC-GALGAP-3'UTR. We designed genes encoding the α and β chains of HbTb (SwissProt P80043 and P80044) using the preferred codons of *S. cerevisiae*. The coding and non-coding strands were synthesised from 8 overlapping oligonucleotides; 100 pmol of each was phosphorylated with 1 μM ATP and 100 units of polynucleotide kinase in 20 μl total volume at 37°C for 30 min. The oligonucleotides were nixed and annealed and ligated overnight at 15°C. After ligase inactivation at 65°C the 5' end of both strands was phosphorylated with T4 polynucleotide kinase. The assembled globin genes were then ligated into the NcoI and HindIII site of pUC-GALGAP-3'UTR. The ATG initiation codon overlaps the NcoI site, and the stop codon is immediately followed by the HindIII site. The α and β globin expression cassettes were cloned together into the NotI site of yeast vector pC1N which contains the LEU2d and URA3 markers.

To prepare mutant HbTb, the α and β chains were altered by PCR-based site-directed mutagenesis, or by cassette mutagenesis. The mutant proteins were expressed and purified by the same method as for wild-type.

## Expression of HbTb in Yeast

*S. cerevisiae* strain GSY112 was transformed with pC1N(HbTb) using the Li-acetate method and plated on a Ura-plate. After 2–3 days, a colony was picked and then plated onto Ura-,leu- plate to select clones harbouring multiple copies of pC1N(HbTb). After 2–3 days, a colony was picked and then inoculated into 5 ml Ura-,Leu- liquid selection SD medium containing 2% (w/v) glucose and shaken at 30°C overnight. The 5 ml yeast saturated culture was inoculated into 100 ml YP liquid medium in a sterile flask containing 2% (v/v) ethanol as the carbon source. Streptomycin (20 ng/ml) was added to prevent bacterial contamination. The culture was incubated at 30°C until saturation, when it was used to inoculate 10 l YP medium with 2% ethanol. Additional ethanol (1%, v/v) was added for every increase in $OD_{600}$ of 3. When the cells reached an $OD_{600}$ of 10, the culture was induced by the addition of galactose to 2%. After 24 h growth, the culture was bubbled with carbon monoxide and harvested by centrifugation at 7000g.

## Haemoglobin Purification

All procedures were carried out at 4°C using CO-saturated buffers. Column elutes were monitored at 280 nm. Purification at each step was assessed spectrophotometrically and by SDS-PAGE. Yeast cells (500 g) were resuspended in 20 mM TrisHCl pH 8.2, 0.5 mM EDTA, 0.5 mM EGTA. Resuspended cells were disrupted on a vortex machine using glass beads (~0.4 mm diameter) with cooling on ice, and the resulting extract was clarified by centrifugation at 3000g for 10 min. The yeast extract was diluted 5× with chilled deionised water and its pH was adjusted to 8.7 with 0.1 M ammonia solution. The extract was applied to a DEAE-sepharose column (about 300 ml bed volume) equilibrated with 5 mM TrisHCl pH 8.7. Hb tightly bound to the top of the column. After the column was washed extensively with the same buffer, the Hb was eluted with 50 mM TrisHCl pH 7.4. Fractions containing Hb were pooled, concentrated and gel-filtered on a Sephadex G25 column equilibrated with 5 mM potassium phosphate pH 7.6, The Hb solution was applied to a Ni-NTA column (Qiagen) equilibrated with the same buffer, and eluted with a gradient to 10 mM potassium phosphate and 10 mM imidazole. The main peak was desalted again and applied to a Q-sepharose-fast-flow column equilibrated with 5 mM potassium phosphate pH 8.0, and eluted with a pH gradient to 20 mM potassium phosphate pH 7.6. The main peak corresponding to the recombinant HbTb was concentrated and passed through a Sephadex G25 column in 20 mM TrisCl pH 8.2 before storage in liquid nitrogen. The identity of each HbTb mutant was confirmed by amino acid sequencing. The α chain is blocked, but in each case the correct β chain sequence (VEWTD) was obtained. Fast atom bombardment mass spectrometry was also used to confirm the molecular weight, which was in each case correct within the expected error.

## Oxygen Equilibrium Curve Measurement

Oxygen equilibrium curves were measured at 15°C in 50 mM bis-Tris or Tris buffer with 100 mM chloride using an automatic oxygenation apparatus as described by Imai et al. (Imau 1981; Imai 1994). The Hb concentration was 60 μM (by haem). IHP (2 mM) was used as organic phosphate effector. In order to reduce met-Hb formation, an enzymatic reduction system (Hayashi et al. 1973) was used.

# Results and Discussion

## Oxygen Equilibrium Properties of the Recombinant Wild-type HbTb

The oxygen equilibrium curves of the recombinant HbTb were measured at 15°C in the presence of 100 mM sodium chloride. Figure 1 shows Hill plots for the data between pH 6.5 and 8.4. Both the oxygen affinity and cooperativity are strongly pH dependent, with p50 ranging from 4.8 torr at pH 8.4 to 104 torr at pH 6.5. As the pH was lowered, the slope of the Hill plot dropped below 1. The data are similar to those reported for wild-type HbTb. Figure 2 shows the effect of pH on $n_{max}$ and p50 of HbTb and human HbA. The p50 of HbA increases moderately as pH falls (the alkaline Bohr effect), whereas $n_{max}$ is almost unchanged. In contrast, $n_{max}$ of HbTb falls with pH, dropping below unity at pH 6.5. The p50 value of HbTb is also strongly pH-dependent. The number of protons released on oxygen binding ($\Delta H^+$) can be determined from the slope of the Bohr plot (log p50 vs pH) and the Wyman linkage equations (Wyman 1964). Figure 3 indicates that HbTb releases 0.78 protons per heme at pH 7.0, whereas human HbA releases only 0.52.

Table 1 shows amino acid residue differences between HbTb and TbTn. These 14 changes completely abolish pH sensitivity in HbTn. Most of them are external, and presumably have little effect. Seven sets of mutations were chosen to investigate which residues are important for the Root effect (Table 2). Making the most of these replacements in HbTb left the Root effect intact,

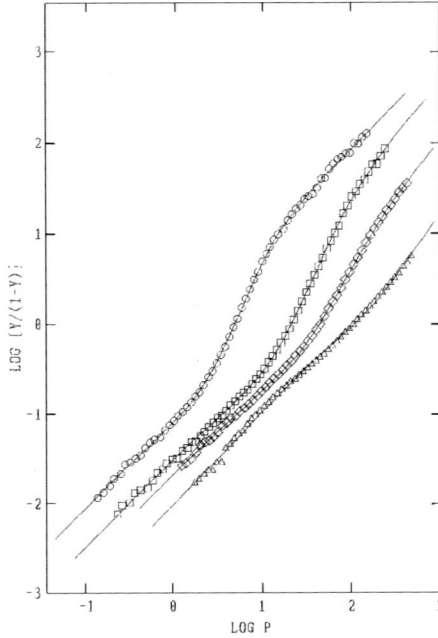

FIG. 1. Hill plots of oxygen equilibrium curves for the recombinant wild-type *T. bernacchii* Hb at pH 8.4 (○), pH 7.4 (□), pH 7.0 (◊), and pH 6.5 (△). Lines were calculated from best-fit values of the four Adair constants. Hb concentration was 60 μM (on a haem basis). The oxygen equilibrium curve measurements were carried out at 15°C in 50 mM bis-Tris or Tris buffer with 100 mM chloride. In order to reduce met-Hb formation, an enzymatic reduction system was added to the samples

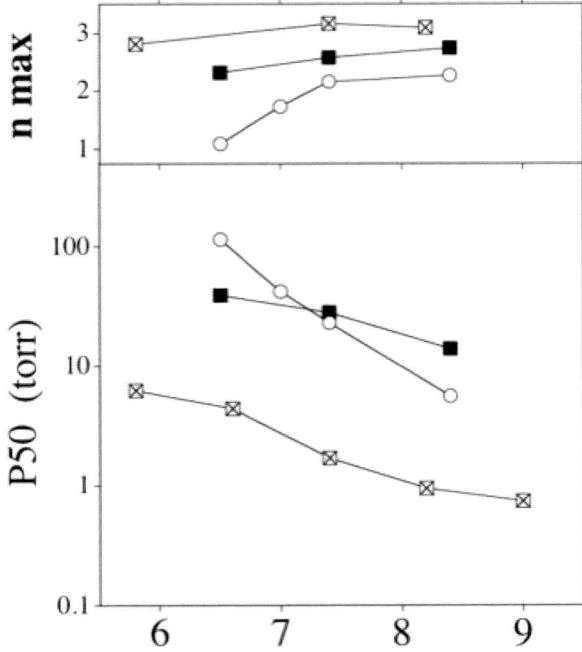

FIG. 2. pH dependence of oxygen equilibrium parameters of wild-type HbTb, mutant HbTb (ii) and human HbA. Experimental conditions were as for Fig. 1. Details of the mutants are shown in Table 2. $n_{max}$ is the maximum slope of the Hill plot, $P_{50}$ is the partial pressure (torr) of oxygen at half saturation. ○, wild-type HbTb; ■, mutant HbTb-(ii); ⊠, human HbA

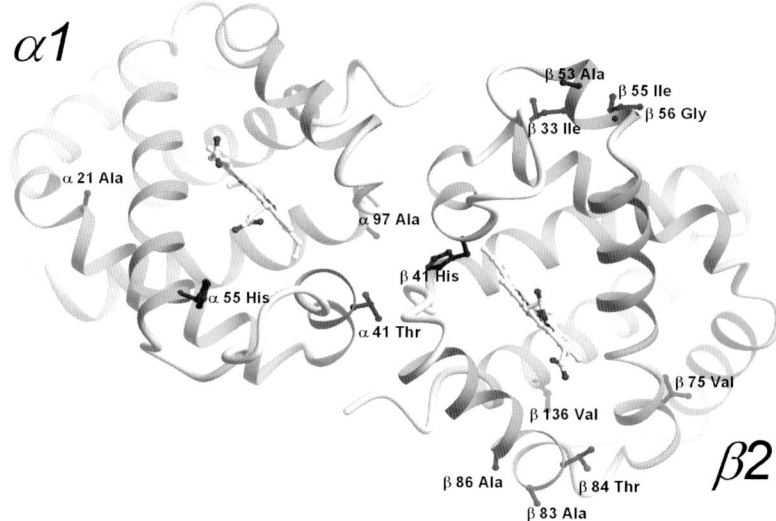

F IG. 3. Schematic representation of the $\alpha_1\beta_2$ subunits of the deoxygenated HbTb structure showing the residues which are substituted in HbTn. HbTb has PDB code 1HBH. The labels show the chain and three-letter amino acid codes

T ABLE 1. Amino acid residues differences between HbTb, HbTn and HbA

|             | HbTb | HbTn | HbA | Location |
|-------------|------|------|-----|----------|
| α21(B2)     | Ala  | Ser  | Ala | External |
| α41(C6)     | Thr  | Ile  | Thr | $\alpha_1\beta_2$ interface |
| α55(E3)     | His  | Asn  | Gln | External |
| α97(G3)     | Ala  | Ser  | Val | $\alpha_1\beta_2$ interface |
| β33(B15)    | Ile  | Val  | Val | $\alpha_1\beta_1$ interface |
| β41(C7)     | His  | Tyr  | Phe | $\alpha_1\beta_2$ interface |
| β53(D4)     | Ala  | Gly  | Ala | External |
| β55(D6)     | Ile  | Met  | Met | External |
| β56(D7)     | Gly  | Ser  | Gly | External |
| β75(E19)    | Val  | Met  | Leu | Internal |
| β83(EF7)    | Ala  | Asp  | Gly | External |
| β84(EF8)    | Thr  | Ala  | Thr | External |
| β86(F3)     | Ala  | Thr  | Ala | External |
| β136(H14)   | Val  | Ala  | Gly | Central cavity |

TABLE 2. ΔH+ values of wild-type HbTb and mutant HbsTb (ii-viii)

|        | α subunit | β subunit | ΔH+ | Root effect |
|--------|-----------|-----------|-----|-------------|
| (i)    | Wild-type | Wild-type | 0.78 | Yes |
| (ii)   | Tα41I, Aα97S | Iβ33V, Hβ41Y, Aβ53G Iβ55M, Gβ56S, Vβ75M Aβ83D, Tβ84A, Aβ86T | 0.15 | No |
| (iii)  | Wild-type | Aβ53G, Iβ55M, Gβ56S Aβ83D, Tβ84A, Aβ86T | 0.62 | Yes |
| (iv)   | Tα41I | Wild-type | 0.79 | Yes |
| (v)    | Wild-type | Hβ41Y | 0.75 | Yes |
| (vi)   | Tα41I | Hβ41Y | 0.73 | Yes |
| (vii)  | Tα41I,Aα97S | Wild-type | 0.67 | Yes |
| (viii) | Tα41I,Aα97S | Hβ41Y | 0.62 | Yes |

showing that these amino acids are not involved. In particular it is notable that replacing α97Ala→Ser had no significant effect. This change is very close to the Asp-Asp pair found in Root effect Hbs, but both aspartate residues are preserved in non-Root effect Hbs such as HbTn. If these residues do indeed play a key role in proton uptake and release, then it remains unclear why proteins such as HbTn have such low pH sensitivity. HbCTn is a Root effect Hb with very few histidines, and its crystal structure shows that Ser α97 does not block formation of the Asp-Asp pair as had been suggested previously.

One HbTb mutant, (ii) in Table 2 with 11 changes, lost the Root effect, and shows an $n_{max}$ of 2.3 at pH 6.5. The number of protons released on oxygenation also dropped significantly, $\Delta H^+$ being only 0.15 at pH 7.0. This mutant is very similar to HbTn, retaining only 3 residues from HbTb, α21(B2)Ala, α55(E3)His and β136(H14)Val. Neither the alanine nor valine has an obvious structural role in the Root effect; the alanine is external and the valine points into the central cavity. Many human Hb variants are known from clinical studies established that surface mutations generally have very little effect on function. None of the mutations between HbTb and HbTb appears significant from the sequence except His α55 →Asn, and recent crystal structures of HbTb refined by the Mazzarella group suggest this mutation is responsible for the loss of the Root effect in HbTb (Mazzarella et al. 2006b). These structures clearly show a salt bridge between Asp α48 and His α55 at pH 6.2 which is absent at pH 8.4. Since the T-R transition causes significant changes at the α1β2 interface, it is possible that amino acid replacements here might also be significant. However, changing α41(C6) Thr →Ile, α97(G3)Ala →Ser and β41(C7)His →Tyr seems from the data presented here to have no appreciable effect on the oxygen affinity.

Based on the crystal structure of spot Hb, it has been suggested that the Root effect arises from two positive charge clusters on the β subunits which strongly destabilise the R state at low pH (Mylvaganam et al. 1996). To explain the absence of the Root effect in HbTn (which has no mutations compared to HbTb in this region), they suggested the β75(E19)Val →Met substitution in HbTn could displace the A helix and the N terminus of the β subunits, preventing this effect occurring. Other structural studies have failed to detect this displacement, and as mentioned above, a destabilisation of the R state alone is unable to explain the Root effect entirely. It cannot explain loss of cooperativity or a severe drop in oxygen affinity for the first oxygen ligand. Nevertheless, comparing mutants (ii) and (iii) in Table 2, it can be seen that mutations β33Ile →Val and β75Val →Met are implicated in the loss of the Root effect. Adding these internal β subunit mutations to mutant (ii) abolishes the Root effect, even though these residues lie far from either the Asp-Asp pair.

## Conclusion

Since the 1930s and 1940s when Root and his colleagues first showed a number of fish Hbs possess an extreme response to acid pH, this phenomenon has been widely studied, yet it remains poorly understood. At low pH these Hbs not only lose cooperative oxygen binding, but may reach less than half saturation with oxygen under air (Root 1931; Brittain 1987). Even though at first sight the Root effect might simply appear to be an exaggerated form of the Bohr effect, in Root effect Hbs the decrease in oxygen binding affinity seen at low pH is accompanied by a loss of cooperativity. This is indeed the defining characteristic of "the Root effect". Many Root effect Hbs exhibit Hill coefficients less than 1.0 at low pH, unlike human HbA, in which cooperativity is maintained well below physiological pH. The loss of cooperativity in Root effect Hb arises from extreme chain heterogeneity, a big difference of oxygen affinity between the α and β chains, and such Hbs lose one-half oxygen capacity at low pH. The extent of the drop in oxygen affinity, and the pH at which it begins, varies hugely among fish Hbs. This has led to enormous confusion when seeking "the Root effect" among sequence alignments of these proteins. It is now realised that the Root effect is a physiological phenomenon which has arisen many times in evolution, and does not invariably arise from the same interactions or sites. Our ongoing mutagenesis experiments with HbTb have highlighted the important point that residues distant from the proton-binding sites may also have a profound effect, and this too makes the identification of key residues from primary structures alone all but impossible. Even where Root effect groups, such as Asp α95 and Asp β101 have been identified, their functional behaviour remains to be explained. The crystallographic results of the

Naples group strongly suggest that these residues may give rise to the Root effect alone (Mazzarella et al. 2006a), but further mutagenesis work is required to understand this surprising result.

## References

Berenbrink, M., Koldkjaer, P., Kepp, O., and Cossins, A.R. 2005. Evolution of oxygen secretion in fishes and the emergence of a complex physiological system. Science 307:1752–1757.

Biot, M. 1807. Sur la nature de l'air continue dans la vessie natatoire des poisons. Mem. Phys. Chim. Soc. D'Arcuiel 1:252.

Brittain, T. 1987. The Root Effect. Comp. Biochem. Physiol. 86:473–481.

Camardella, L., Caruso, C., D'Avino, R., di Prisco, G., Rutigliano, B., Tamburrini, M., Fermi, G., and Perutz, M. F. 1992. Haemoglobin of the antarctic fish *Pagothenia bernacchii*. Amino acid sequence, oxygen equilibria and crystal structure of its carbonmonoxy derivative. J. Mol. Biol. 224:449–460.

Haldane, J. S. 1922. Respiration. New Haven: Yale University Press.

Hayashi, A., Suzuki, T. and Shim, M. 1973. An enzymic reduction system for metmyoglobin and methemoglobin, and its application to functional studies of oxygen carries. Biochim. Biophys. Acta 310:309–316.

Henry, E. R., Bettati, S., Hofrichter, J., and Eaton, W. A. 2002. A tertiary two-state allosteric model for hemoglobin. Biophys. Chem. 98:149–164.

Herzfeld, J., and Stanley, H. E. 1974. A general approach to cooperativity and its application to the oxygen equilibrium of hemoglobin and its effectors. J. Mol. Biol. 82:231–265.

Imai, K. 1981. Measurement of accurate oxygen equilibrium curves by an automatic oxygenation epporatus. Methods Enzymol. 76:438–449

Imai, K. 1994. Adair fitting of oxygen equilibrium curves of hemoglobin. Methods Enzymol. 232:559–576.

Ito, N., Komiyama, N. H., and Fermi, G. 1995. Structure of deoxyhaemoglobin of the antarctic fish *Pagothenia bernacchii* with an analysis of the structural basis of the root effect by comparison of the liganded and unliganded haemoglobin structures. J. Mol. Biol. 250:648–658.

Mazzarella, L., Bonomi, G., Lubrano, M. C., Merlino, A., Riccio, A., Vergara, A., Vitagliano, L., Verde, C., and di Prisco, G. 2006a. Minimal structural requirements for root effect: crystal structure of the cathodic hemoglobin isolated from the antarctic fish *Trematomus newnesi*. Proteins 62:316–321.

Mazzarella, L., D'Avino, R., di Prisco, G., Savino, C., Vitagliano, L., Moody, P.C., and Zagari, A.: Crystal structure of *Trematomus newnesi* haemoglobin re-opens the root effect question. J. Mol. Biol. 287 (1999) 897–906.

Mazzarella, L., Vergara, A., Vitagliano, L., Merlino, A., Bonomi, G., Scala, S., Verde, C., and di Prisco, G. 2006b. High resolution crystal structure of deoxy hemoglobin from *Trematomus bernacchii* at different pH values: the role of histidine residues in modulating the strength of the root effect. Proteins 65:490–498.

Monod, J., Wyman, J., and Changeux, J.-P. 1965. On the nature of allosteric transitions: a plausible model. J. Mol. Biol. 12:88–118.

Mylvaganam, S. E., Bonaventura, C., Bonaventura, J., and Getzoff, E. D. 1996. Structural basis for the root effect in haemoglobin. Nat. Struct. Biol. 3:275–283.

Perutz, M. F., and Brunori, M. 1982. Stereochemistry of cooperative effects in fish and amphibian haemoglobins. Nature 299:421–426.

Root, R.W. 1931. The respiratory function of the blood of marine fishes. Biol. Bull. Mar. Biol. Lab. Woods Hole 61:427–456.

Root, R.W., and Irving, L. 1941. The equilibrium between haemoglobin and oxygen in whole and hemolysed blood of the Tautog and a theory of the Haldane effect. Biol. Bull. Mar. Biol. Lab. Woods Hole 81:307–323.

Root, R. W., and Irving, L. 1943. The effect of carbon dioxide and lactic acid of the oxygen combining power of whole and hemolysed blood of the marine fish Tautog onitis. Biol. Bull. Mar. Biol. Lab. Woods Hole 84:207–212.

Wells, R. M. G. 1999. Evolution of haemoglobin function: molecular adaptation to environment. Clin. Exp. Pharm. Physiol. 26:591–595.

Wittenberg, J. B. 1958. The secretion of inert gas into the swim-bladder of fish. J. Gen. Physiol. 41:783–804.

Wittenberg, J. B. 1961. The secretion of oxygen into the swim-bladder of fish I. The transport of molecular oxygen. J. Gen. Physiol. 44:521–526.

Wittenberg, J. B., and Haedrich, R. L. 1974. The choroid rete mirabile of the fish eye II. Distribution and relation to the pseudobranch and to the swim bladder rete mirabile. Biol. Bull. 146:137–156.

Wittenberg, J. B., and Wittenberg, B. A. 1961. The secretion of oxygen into the swim-bladder of fish. II. The simultaneous transport of carbon monoxide and oxygen. J. Gen. Physiol. 44:527–542.

Wittenberg, J. B., and Wittenberg, B. A. 1962. Active secretion of oxygen into the eye of fish. Nature 194:107–108.

Wittenberg, J. B., and Wittenberg, B. A. 1974. The choroid rete mirabile of the fish eye I. Oxygen secretion and structure: comparison with the swim-bladder rete mirabile. Biol. Bull. 146:116–136.

Wittenberg, J. B., Schwend, M. J., and Wittenberg, B. A. 1964. The secretion of oxygen into the swim-bladder of fish III The role of carbon dioxide. J. Gen. Physiol. 48:337–355.

Woodland, W. N. F. 1911. On the structure and function of the gas glands and retina mirabilia associated with the gas bladder of some teleostean fishes, with notes on the teleost pancreas. Proc. Zool. Soc. London 11:183.

Wyman, J. 1964. Linked functions and reciprocal effects in hemoglobin: A second look. Adv. Prot. Chem. 19:223–286.

Yokoyama, T., Chong, K. T., Miyazaki, G., Morimoto, H., Shih, D. T., Unzai, S., Tame, J. R. H., and Park, S. Y. 2004. Novel mechanisms of pH sensitivity in tuna hemoglobin: a structural explanation of the Root effect. J. Biol. Chem. 279:28632–28640.

# 9

# Redox Reactions of Cross-linked Haemoglobins with Oxygen and Nitrite

Celia Bonaventura, Robert Henkens, Katherine D. Weaver, Abdu I. Alayash and Alvin L. Crumbliss

## Abstract

Redox reactions of haemoglobin (Hb) with oxygen can initiate a cascade of oxidative reactions that appear to underlie the adverse side reactions observed when cell-free Hbs are introduced into the circulation to enhance oxygen delivery to respiring tissues. Redox reactions of cell-free Hbs with nitrite may also be of significance *in vivo*, as these reactions can lead to formation of nitrosylated Hb (NO-Hb) along with oxidised Hb (MetHb). To clarify the factors governing these redox reactions we measured the kinetics of nitrite-induced and oxygen-induced heme oxidation and obtained oxygen binding and oxidation curves for unmodified human Hb and four cross-linked Hbs. The four cross-linked Hbs studied were generated by cross-linking Hb with glutaraldehyde, dextran, *O*-raffinose or bis(3,5-dibromosalicyl)fumarate. Oxygen binding by the cross-linked Hbs occurred with reduced oxygen affinity, reduced cooperativity and reduced responses to organic phosphate effectors. The redox potentials of the cross-linked Hbs were shifted to higher potentials relative to unmodified Hb in the absence of allosteric effectors, indicating a reduced thermodynamic driving force for oxidation. In spite of this, these Hbs showed increased rates in oxidative reactions. Elevated rates of heme oxidation were observed for their oxy derivatives under aerobic conditions, and upon exposure to nitrite under both aerobic and anaerobic conditions. These results show that heme accessibility rather than heme redox potential is the major determinant of the kinetics of redox reactions of Hb with both oxygen and nitrite.

## Introduction

Haemoglobin (Hb) within red blood cells is subject to homotropic and heterotropic allosteric control mechanisms that regulate oxygen affinity, which provides for efficient oxygen uptake in the lungs and effective oxygen delivery to respiring tissues (Perutz 1970, 1978). In contrast, cell-free Hbs are more readily lost to the circulation by renal filtration, and lack mechanisms for maintenance of reduced heme, and fine-tuning of oxygen uptake and delivery. It has more recently been shown that packaging Hbs within red blood cells also helps avoid nitric oxide (NO) scavenging and concomitant increases in blood pressure (Lancaster 1994). To substitute for effectors normally found in red blood cells and avoid loss by renal filtration, cross-linkers that stabilise Hb in the low-affinity (T-state) conformation have been used in the preparation of blood substitute candidates (Reiss 2001). Here we report on our studies of oxygen binding and of oxidation kinetics and thermodynamics for adult human Hb (Hb $A_0$) and four cross-linked Hbs that are in varying stages of clinical evaluation as blood substitute candidates.

Hb-DBBF, generated by reaction of deoxy Hb $A_0$ with bis(3,5-dibromosalicyl)fumarate, has a single intra-tetrameric cross-link between the $\alpha$ chains at 99Lys. It is an analogue of the commercially developed DCLHb™ and has been extensively evaluated as a potential blood substitute (D'Agnillo and Alayash 2000). PolyHbBv (Oxyglobin) is a highly purified bovine Hb that is intra- and inter-molecularly cross-linked with glutaraldehyde that has only recently been made commercially available for use as a blood substitute in dogs with anaemia. It contains a heterogeneous mixture of tetramers (~5%) and larger polymeric aggregates (~95%) ranging in size up to 500 kDa (Pearce and Gawryl 1998). HbDex is a human Hb polymer with intra- and inter-tetrameric cross-links formed by reaction with dextran. Dextrans of average molecular weight of approximately 10 kDa were used in the reaction and the resultant conjugate has a molecular weight average of 300 kDa (Prouchayret et al. 1992). $O$-R-PolyHbA$_0$ (Hemolink™) is a human Hb polymer with intra- and inter-tetrameric crosslinks formed by reaction of highly purified human Hb $A_0$ with $O$-raffinose. It has been reported to contain 32-kDa dimers (5%), 64-kDa tetramers (33%) and larger aggregates ranging in size up to 600 kDa (Adamson and Moore 1997). These cross-linked Hbs were compared to unmodified human Hb in the presence and absence of effectors. Our characterisation of these cross-linked Hbs includes determinations of their oxygen-binding curves, redox potentials, rates of MetHb and Hb-NO formation from nitrite, and rates of autoxidation.

# Materials and methods

## Haemoglobins

Hb $A_0$ was prepared by ammonium sulphate precipitation, stripped of endogenous organic phosphates and FPLC purified as previously described (Bonaventura et al. 1991). All Hbs used in this study were maintained at –80°C prior to their use in the studies described here. Hb-DBBF was a kind gift from the Walter Reed Army Institute of Research (Washington D.C). PolyHbBv, an FDA approved blood substitute for use in dogs with anaemia, is a glutaraldehyde-polymerised bovine Hb and was purchased from Biopure Inc. (Cambridge, MA). HbDex was a kind gift from Dr. Patrick Menu, Department of Hematology and Physiology, School of Pharmacy, University of Henri Poincaré-Nancy, 54001 Nancy, France, and was prepared as previously reported (Prouchayret et al. 1992) by cross-linking human Hb with benzene tetracarboxylate dextran. *O*-R-PolyHbA$_0$ was a kind gift of Hemosol, Inc. Canada. It was prepared by reaction of purified Hb A$_0$ with *O*-raffinose, a hexaldehyde obtained by oxidation of the trisaccharide raffinose (Adamson and Moore 1997). The air-equilibrated cross-linked Hbs had high initial percentages of MetHb when received. For use in the studies described here, they were reduced by addition of sodium dithionite (British Drug Houses) and put through a Sephadex G-25 column to remove dithionite byproducts immediately prior to functional studies. The spectra of the dithionite-treated Hbs showed no evidence of irreversible metHb, ferrylHb or hemichromes.

## Oxygen Binding

Oxygen equilibrium curves were determined tonometrically (Riggs and Wolbach 1956). Deoxygenation of the Hb samples before air addition was achieved within 15–20 min by three cycles of exposure of ~4 ml of Hb in large volume tonometers to $N_2$ and vacuum. A gastight syringe was used to inject measured volumes of room air through the rubber septum of the tonometer containing the Hb sample, ending with an equilibration with 100% oxygen. After each addition the tonometers were rotated in a water bath for 10 min before an absorbance spectrum was measured. At each equilibration step the PO$_2$ was calculated and changes in the visible absorption spectrum were used to calculate the corresponding fractional O$_2$ saturation.

## Spectroelectrochemistry

Experiments were carried out in an anaerobic OTTLE cell with $Ru(NH_3)_6Cl_3$ as a cationic electrochemical mediator. This mediator does not bind at Hb's anion binding sites and therefore allows anion effects on Hb redox potentials to be determined (Taboy, Bonaventura and Crumbliss 2002). Initial and final spectra were recorded from 350 nm to 750 nm. Absorbance changes were monitored at 406 nm (MetHb peak) and 430 nm (deoxy Hb peak) as a function of the applied potential. In a typical experiment, spectra were recorded starting at +250 mV, and proceeding to –100 mV (vs. NHE) in 20-mV increments. At each applied potential, the absorbance was monitored until no change was detected (5–30 min). Nernst plots of Hb's oxidation state *vs.* applied potential were derived from the observed changes in absorbance as previously described (Taboy et al. 2000; Taboy, Bonaventura and Crumbliss 2002). Although most experiments were performed going from fully oxidised to fully reduced protein, the system was shown to be reversible under our experimental conditions (i.e. the Nernst plot can be generated in either the oxidation or reduction direction).

## Nitrite-induced Oxidation

Deoxygenated or air-equilibrated Hb samples were exposed to 10–100-fold excess of $NaNO_2$ over heme. Changes in the visible spectrum were recorded to determine the rate of nitrite-induced heme oxidation. The rate and extent of nitrite-induced Hb-NO formation that accompanied formation of oxidised (MetHb) for the deoxygenated Hb samples was determined by spectral analysis.

# Results

## Oxygen Binding

We measured oxygen-binding equilibria for the cross-linked Hbs and unmodified Hb $A_0$. Figure 1 shows representative Hill plots for samples at 20°C at pH 7.4. Under these conditions the cross-linked Hbs have lower oxygen affinities and greatly reduced cooperativity in oxygen binding. We purposefully omitted use of a MetHb reducing system so that the intrinsic (anion-free) properties of the samples could be determined. This experimental detail merits mention, as the cross-linked Hbs were found to be very susceptible to oxidation. In spite of

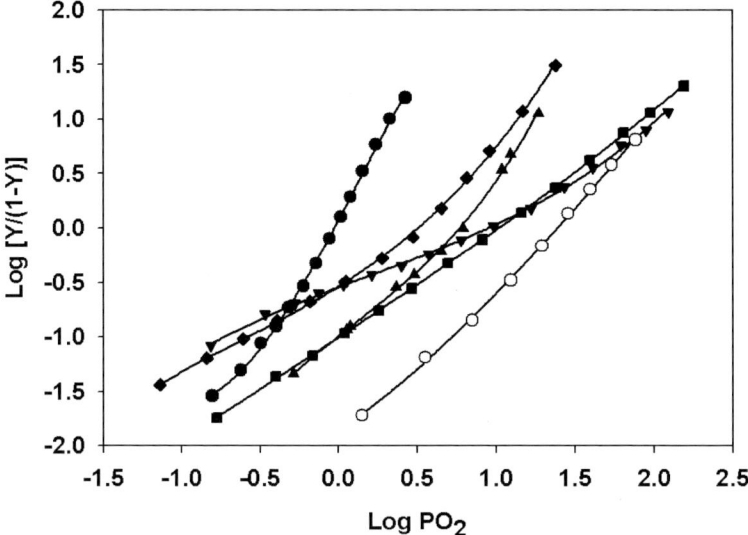

FIG. 1. Oxygen-binding curves for Hb $A_0$ and cross-linked Hbs. Hill plots are shown for Hb $A_0$ (●), $O$-R-PolyHbA$_0$ (□), PolyHbBv (■), HbDex (◆) and HbDBBF (□) without anionic effectors and for Hb $A_0$ with 0.15 mM IHP (▲). The $x$ axis represents log of oxygen pressure in mmHg and the $y$ axis represents the log of $Y$, the oxygen-bound species, divided by $(1-Y)$. Hbs were 0.06 mM (in heme) in 0.05 M bis-Tris/HCl at pH 7.4±0.1

this tendency, their $P_{50}$ values at 20°C were still fairly reproducible. Both $O$-R-PolyHbA$_0$ and HbDex contain subpopulations of material with elevated oxygen affinity, as evidenced by left-shifted Hill plots at low levels of oxygen saturation. The heterogeneous binding curves are consistent with the heterogeneous character of these cross-linked Hbs, which are known to contain subpopulations of unpolymerised material. While the presence of appreciable amounts of MetHb can also result in left-shifted Hill plots, the initial levels of MetHb were low (<5%) and the heterogeneity observed in the Hill Plots was paralleled by heterogeneity in the Nernst Plots (see below) where all MetHb was electrochemically reduced to the ferrous state.

Figure 1 shows that oxygen binding to Hb A$_0$ occurs with much higher affinity in the absence (filled circles) than in the presence (open circles) of the organic polyphosphate effector inositol hexaphosphate (IHP). In striking contrast, as shown in Table 1 (columns 4–6), the cross-linked Hbs show little or no alteration in oxygen binding associated with addition of inorganic or organic anions. Their $P_{50}$ values in the presence and absence of effectors are intermediate between the oxygen affinity of stripped (effector-free) Hb A$_0$ and the oxygen affinity of Hb A$_0$ in the presence of IHP. Table 1 summarises oxygen-binding data on the cross-linked Hbs along with data on their autoxidation and redox properties, determined as described in the following sections.

TABLE 1. Oxidation and oxygenation properties of Hb A$_0$ and cross-linked Hbs. Anaerobic redox potentials ($E_{1/2}$ in mV $vs.$ NHE) and slopes ($n_{50}$) of Nernst plots were determined from spectral changes at 430 nm for 0.031–0.183 mM (heme) samples in 0.05 M bis-Tris with 0.1 M Cl$^-$ at pH 7.4, 25°C. Autoxidation rate constants ($k$) were determined for 20 µM solutions of HbDex (Jia et al. 2004b) and for 0.05 mM solutions of other Hbs (Jia et al. 2004a) at 37°C in 0.05 M phosphate buffer, pH 7.4. Oxygenation properties (Log $P_{50}$ in mmHg) and slopes ($n_{50}$) of Hill plots were determined for 0.060 mM (heme) samples in 0.05 M bis-Tris/HCl at pH 7.4 (±0.1), 20°C, with the indicated anionic effectors.

| Haemoglobin | Oxidation | Autoxidation | Oxygenation | | |
|---|---|---|---|---|---|
| | $E_{1/2}$ ($n_{50}$) | $k$ (h$^{-1}$) | Log $P_{50}$ ($n_{50}$) | | |
| | 0.1 M Cl$^-$ | 0.05 M PO$_4^{-3}$ | 0.007 M Cl$^-$ | 0.1 M Cl$^-$ | 150 µM IHP |
| Hb A$_0$ | 82.9 (1.3) | 0.011 | −0.22 (2.4) | 0.3 (2.4) | 1.64 (2.0) |
| Hb-DBBF | 124.7 (1.0) | 0.024 | 0.77 (2.1) | – | – |
| PolyHbBv | 106.1 (0.9) | 0.036 | 1.02 (1.0) | 1.03 (1.1) | 1.08 (1.1) |
| HbDex | 93.6 (0.9) | 0.042 | 0.53 (1.5) | 0.62 (1.5) | 0.55 (1.3) |
| $O$-R-polyHbA$_0$ | 96.8 (0.7) | 0.039 | 0.96 (0.6) | 0.99 (0.7) | 1.03 (0.7) |

## Autoxidation of Cross-linked Hbs in Aerobic Solutions

As summarised in Table 1, the cross-linked Hbs studied have enhanced rates of autoxidation relative to Hb $A_0$. Their enhanced autoxidation tendencies were evident in appreciable MetHb formation during the determination of oxygen binding curves at 20°C, consistent with previous reports of the enhanced rates of autoxidation of these cross-linked Hbs (Nagababu et al. 2002; Alayash 2004; Jia et al. 2004). Dithionite reduction of MetHb followed by Sephadex G25 chromatography made it possible to start the oxygen-binding experiments with <5% MetHb in the cross-linked Hb solutions.

## Redox Potentials

Spectroelectrochemical determinations of anaerobic redox curves for the cross-linked Hbs studied established their redox potentials under conditions similar to those used for determinations of their oxygen equilibria. Representative Nernst plots for Hb $A_0$ and the cross-linked Hbs of this study, determined at physiological pH, are shown in Fig. 2. This figure illustrates the decreased thermodynamic driving force for oxidation and the reduced cooperativity in the oxidation process exhibited by the cross-linked Hbs relative to Hb $A_0$. Table 1 presents the redox data on Hb $A_0$ and the cross-linked Hbs along with oxygen binding and autoxidation data.

The Nernst plots of Fig. 2 are similar in many regards to the Hill plots shown in Fig. 1. The right-shifted curves of the cross-linked Hbs Nernst plots, relative to effector-free Hb $A_0$, parallel the right-shifted curves of the Hill plots of Fig. 1. The right-shifted Nernst plots indicate that these cross-linked Hbs are less easily oxidised than unmodified Hb $A_0$ in the absence of effectors, but more easily oxidised than Hb $A_0$ in the presence of IHP. While all of the cross-linked Hbs examined have higher $E_{1/2}$ values than unmodified Hb $A_0$ in the absence of anionic effectors, both $O$-R-PolyHbA$_0$ and HbDex contain some sites that are more easily oxidised, as evidenced by left-shifted curves at low levels of applied potential. These aspects of their Nernst plots parallel their heterogeneous Hill plots of oxygen binding; both reflect the fact that these cross-linked Hbs contain subpopulations of unpolymerised material.

The $E_{1/2}$ values obtained for the T-state stabilised cross-linked Hbs are in accord with previous studies showing that the redox potential of Hb is sensitive to its conformation (Taboy, Bonaventura and Crumbliss 2002). Higher values of $E_{1/2}$ are typical for T-state stabilised forms relative to unmodified and R-state stabilised forms. It is the T to R shift in Hb conformation that underlies the cooperativity of the oxidation process shown in the Nernst plots, in much the same fashion as this shift underlies the cooperativity of oxygen binding shown in Hill plots (Taboy, Bonaventura and Crumbliss 2002). The plot of $E_{1/2}$ vs. Log $P_{50}$ shown in Fig. 3 illustrates the trend in altered $E_{1/2}$ values associated with T-state stabilisation. Hb-DBBF does not fall on the $E_{1/2}$ vs Log $P_{50}$ trend line,

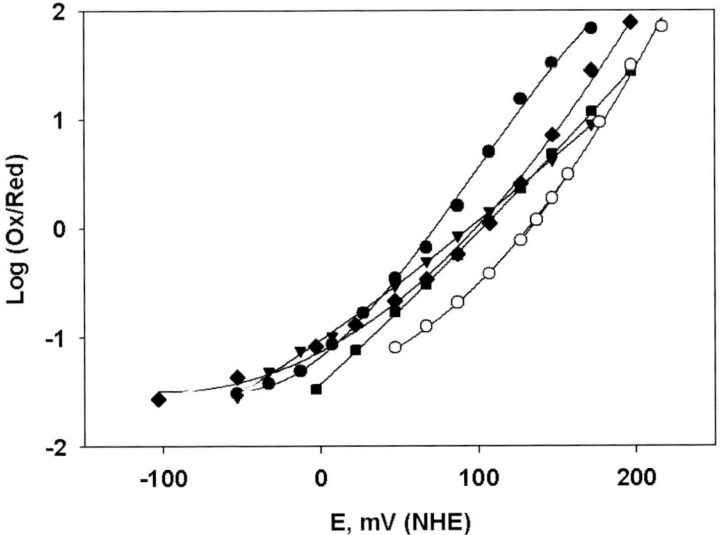

FIG. 2. Anaerobic oxidation curves for Hb A$_0$ and cross-linked Hbs. Nernst plots are shown for Hb A$_0$ (●), *O*-R-PolyHbA$_0$ (□), PolyHbBv (■) and HbDex (◆) in the absence of effectors, and for Hb A$_0$ with IHP (▲). The *x* axis represents applied potential *vs.* NHE and the *y* axis represents the log of the oxidised species divided by the reduced species. Hbs were 0.031–0.183 mM (in heme) in buffers containing a cationic mediator (0.975–2.57 mM Ru(NH$_3$)$_6$Cl$_3$). Studies were done at 25°C with Hbs at pH 7.4 in 0.05 M bis-Tris, 0.1 M KCl, except for HbA$_0$+IHP (from Taboy et al. 2000), which was done in 0.05 M HEPES, 0.2 M KCl, pH 7.5, 20°C

having a significantly higher redox potential than would be expected based on its oxygen affinity. As will be discussed, this departure from the trend line is indicative of a distinctive conformation brought about by the single cross-link at α99Lys in Hb-DBBF.

## Nitrite-induced Oxidation of Hbs

The reactions of the cross-linked Hbs with excess nitrite were first studied under oxygenated conditions where MetHb formation is coupled to nitrate formation via Hb's nitrite oxidase activity (Kosaka et al. 1979; Spagnuolo et al. 1987). As shown in Fig. 4a, the oxy forms of the cross-linked Hbs have enhanced rates of reaction with nitrite. Figure 4a shows the initial stages of nitrite-induced heme oxidation when the oxy-Hbs were exposed to a 100-fold excess of nitrite over heme. The oxy forms of the cross-linked Hbs show faster reactions with nitrite with shortened lag times relative to Hb A$_0$. Their reaction half-times with nitrite are reduced by 30% or more compared to those of unmodified Hb (see Table 2).

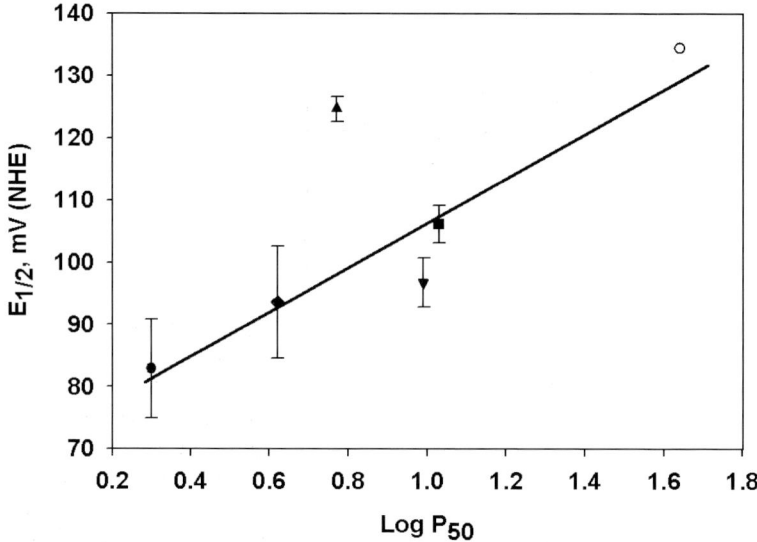

FIG. 3. Trend line for oxidation ($E_{1/2}$) *vs.* oxygenation ($P_{50}$) parameters for Hb $A_0$ and cross-linked Hbs. Data are shown for Hb $A_0$ (●), *O*-R-PolyHbA$_0$ (▢), PolyHbBv (■), HbDex (◆) and HbDBBF (▢) in the absence of effectors and for Hb $A_0$ with IHP (▲). The *x* axis represents the Log $P_{50}$ and the *y* axis represents the $E_{1/2}$ in mV *vs.* NHE. Log $P_{50}$ conditions as in Fig. 1. $E_{1/2}$ conditions as in Fig. 2

TABLE 2. Half-times for nitrite-induced oxidation. Deoxygenated and air-equilibrated 0.06 mM (heme) solutions of Hb $A_0$ and cross-linked Hbs in 0.05 M bis-Tris/HCl at pH 7.4 (±0.1), 20°C, were exposed to 6.0 mM nitrite. Changes in the visible spectrum were monitored to determine the half-times for formation of MetHb, which in experiments with deoxy Hb was coupled to a 1:1 formation of Hb-NO

| Haemoglobin | Half-time ($t_{50}$) in minutes | |
| --- | --- | --- |
| | Deoxy Hb | Oxy Hb |
| Hb $A_0$ | 8.6±0.2 | 28±1.5 |
| Hb-DBBF | – | 18±1 |
| PolyHbBv | 5.7±0.2 | 18±1 |
| HbDex | 7.5±0.2 | 17±1 |
| *O*-R-PolyHbA$_0$ | 5.6±0.2 | 18±1 |

The reactions of Hb with excess nitrite were also studied under deoxygenated conditions where MetHb and Hb-NO formation are coupled via Hb's NO-synthase activity. NO-Hb and MetHb were formed in a 1:1 ratio under our experimental conditions. The time-dependent changes in Hb absorbance following addition of a 10-fold excess of nitrite are shown in Fig. 4b. As shown,

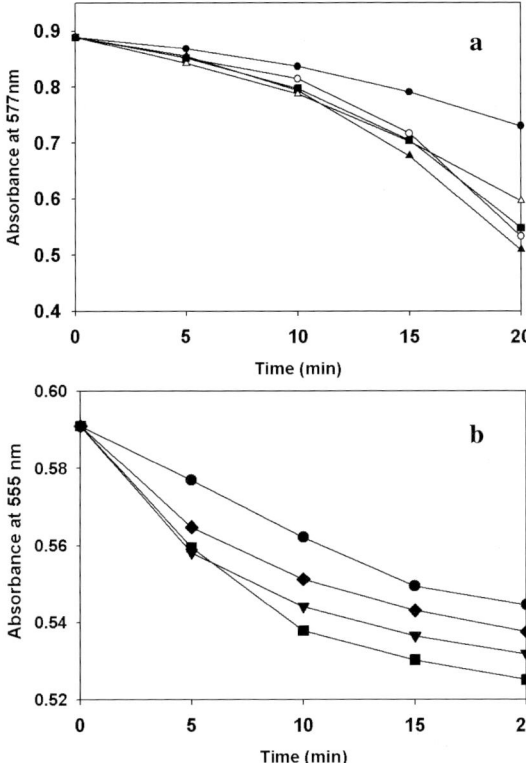

Fig. 4. Nitrite-induced oxidation of oxy and deoxy forms of Hb $A_0$ and cross-linked Hbs. Data are shown for the initial absorbance changes as a function of time following addition of 0.15 mM nitrite to 0.06 mM (in heme) samples of air-equilibrated (a) or deoxygenated (b) Hb samples in 0.05 M bis-Tris, pH 7.5 at 20°C. Hb $A_0$ (●), O-R-PolyHb$A_0$ (❑), PolyHbBv (■), HbDex (◆) and HbDBBF (❑). In b, MetHb and NO-Hb were formed in equimolar amounts

the deoxy forms of the cross-linked Hbs have faster initial rates of reaction than unmodified Hb in the absence of effectors. Their reaction half times were reduced by 30% or more, as was found for reaction of their oxy forms with nitrite (Table 2). As will be further discussed, the faster rates of their reactions with nitrite are not due to their altered redox potentials, but instead reflect the more facile entry of nitrite into their heme pockets.

## Discussion

Chemically and/or genetically altered cell-free Hbs have so far failed to demonstrate efficacy, let alone safety, outside the protective environment provided by the red blood cell. As shown in Table 1, the cross-linked Hbs of this study all have reduced cooperativity in oxygen binding and reduced sensitivity to pH and anionic effectors relative to Hb $A_0$. It remains to be seen if the reduced allosteric properties of these cross-linked Hbs make them ineffective as oxygen carriers *in vivo*. The physiological consequence of the reduced cooperativity, and reduced Bohr and chloride effects, of the cross-linked Hbs could be significant in affecting tissue acid–base balance, oxygen delivery to tissues, and $CO_2$ transport to the lungs.

While the altered allosteric responses and enhanced rates of heme oxidation exhibited by the cross-linked Hbs of this study make them problematic for use in an extracellular environment, their altered functional properties make them good model systems for analysis of the molecular controls of Hb function. Unlike the case of immobilised or gel-stabilised Hbs that are also conformationally constrained (Roche et al. 2006), the solution properties of these cross-linked Hbs can be directly compared to those of normal Hb in the presence and absence of allosteric effectors.

We determined oxygen-binding and oxidation curves for the cross-linked Hbs under comparable experimental conditions. Relative to unmodified Hb, the cross-linked forms studied all showed lowered oxygen affinity and cooperativity, and decreased responses to heterotropic allosteric effectors in both oxygen binding and heme oxidation processes. Their $P_{50}$ and $E_{1/2}$ values were found to be intermediate between those of Hb in the absence of effectors and Hb in the presence of IHP (Table 1 and Figs. 1 and 2). The cross-linked Hbs of this study have significantly enhanced rates of autoxidation, with appreciable loss of oxygen-carrying capacity after an $O_2$ binding experiment lasting ~2 h. The autoxidation process is enhanced at higher temperatures, such as those found *in vivo*, which suggests that the oxygen-binding functionality of these cross-linked Hbs could be rapidly compromised after infusion *in vivo*. This underscores the disadvantage of having rapid uncontrolled autoxidation processes as a functional characteristic for cross-linked Hbs intended for use as oxygen therapeutics.

Previous studies have shown that autoxidation of Hb is a source for non-functional MetHb as well as reactive oxygen species ($O_2^-$ and $H_2O_2$). The formation of reactive oxygen species can enhance production of highly reactive ferryl heme and heme degradation products (Nagababu et al. 2002). Cellular toxicity attributed to ferryl heme includes promotion of lipid peroxidation, lactate dehydrogenase release and DNA fragmentation; and phenomena recognised as markers of cell injury and death by apoptosis and necrosis (D'Agnillo and Alayash 2001).

The redox potentials for the cross-linked Hbs (given in Table 1) show them to be intrinsically *less* prone to oxidation than unmodified Hb $A_0$. As their equilibrium properties do not favour oxidation, it follows that their rapid rates of autoxidation are due to steric factors that influence the kinetics of the oxidation reaction, rather than the thermodynamic potential driving the reaction. The enhanced rates of autoxidation of the cross-linked Hbs appear to be due to heme pockets that are more open to the aqueous environment. In a previous study of six heme proteins for which crystallographic data was available, the heme redox potential was found to be inversely dependent upon the exposure of heme to aqueous solvent (Stellwagen 1978). As Hb's redox potential is dependent on its quaternary conformation, we can speculate that the redox potentials of the cross-linked Hbs of this study reflect a combination of their variable degrees of T-state stabilisation (promoting higher $E_{1/2}$ values) and increased solvent accessibility (promoting lower $E_{1/2}$ values).

The interpretation that more solvent-accessible hemes promote the more rapid redox reactions of the cross-linked Hbs is supported by structural data on glutaraldehyde-polymerised Hb (Chevalier et al. 1989; Guillochon et al. 1989). Its enhanced rate of autoxidation is attributed to changes in the heme environment that decrease proximal and distal side steric hindrance and open the heme pocket toward the solvent (Chevalier et al. 1989). This explanation was qualitatively supported by the observation that our spectroelectrochemical titrations came to equilibrium at each applied potential faster for the cross-linked Hbs than for Hb $A_0$.

As is evident from Fig. 3, the redox potential of Hb-DBBF ($\alpha 99$ cross-linked human Hb) is higher than would be expected based on its oxygen affinity. The fact that this singly cross-linked Hb did not fall on the trend line of Fig. 3 is an indication of specific structural constraints that differentiate this Hb from the other cross-linked Hbs studied. Alternatively, one may view the data in Fig. 3 as illustrating that the oxygen affinity of Hb-DBBF is higher than expected based on its $E_{1/2}$, which may be due to the presence of more exposed heme group, consistent with its rapid autoxidation. Hb-DBBF exhibits enhanced autoxidation rates and oxidative side reactions *in vitro* (Nagababu et al. 2002), in cell culture (D'Agnillo and Alayash 2001) and in mesenteric micro-circulation in small animals (Baldwin, Wiley and Alayash 2002; Baldwin et al. 2003; Baldwin, Wiley and Alayash 2004), in addition to more pronounced vasoactivity (Abassi et al. 1997) when compared with other cross-linked Hbs under the same experimental conditions. *These findings illustrate an important principle: that the nature and extent of cross-linking reactions can have variable consequences regarding the opening of the heme pocket (reflected in altered kinetics of heme oxidation) and regarding the thermodynamic driving force for oxidation of the heme iron (reflecting the Hb's redox potential).*

The cross-linked Hbs of this study showed faster initial rates of reaction with excess nitrite than unmodified Hb for both deoxy and air-equilibrated samples. The nitrite reaction with oxy-Hb is very complex. The slow lag-phase of the reaction has previously been shown to be associated with formation of $H_2O_2$ and MetHb, whose interaction leads to formation of ferryl heme and a progressive increase in rate of MetHb formation (Spagnuolo et al. 1987). The enhanced rates of reaction of the oxy derivatives of the cross-linked Hbs with nitrite parallel their faster rates of autoxidation. The observed rate enhancement is also consistent with prior studies indicating that the air-equilibrated cross-linked Hbs have enhanced rates of formation and transition to MetHb and ferryl heme upon exposure to $H_2O_2$ (Alayash et al. 1992; Nagababu et al. 2002).

Interaction of deoxy Hb with nitrite results in a net increase in NO, although the NO generated by this NO synthase activity of Hb can be readily scavenged by deoxy Hb with formation of Hb-NO. In the experiments reported here, changes in the visible spectrum showed that MetHb formation was coupled to a 1:1 formation of Hb-NO. This NO synthase function of Hb is under intensive investigation due to its potential importance for generation of NO from nitrite *in vivo* (Lancaster 1994; Cosby et al. 2003; Kim-Shapiro et al. 2005; Angelo, Singel and Stamler 2006; Roche et al. 2006). The faster initial rates of NO synthase reactions of the T-state stabilised cross-linked Hbs with nitrite under deoxygenated conditions were unexpected. This result presents a challenge to recent postulates concerning the basis of allosteric control of this reaction (Cosby et al. 2003; Kim-Shapiro et al. 2005). While there is considerable evidence that there is a redox shift toward greater ease of oxidation as Hbs assume the R-state conformation, the behaviour of the deoxy cross-linked Hbs indicates that the redox potential of Hb is less important for determining the kinetics of the reaction than the accessibility of the active site to nitrite. The faster rates of the initial stages of reaction of the deoxy cross-linked Hbs with nitrite parallels their enhanced rates of reaction with nitrite under oxy conditions, and their enhanced rates of autoxidation. The altered kinetics exhibited by the cross-linked Hbs in all these oxidative processes appears to have a similar basis, that of enhanced active site accessibility to oxidants.

## Disclaimer

The opinions and assertions contained herein are the scientific views of the author (AIA) and are not to be construed as policy of the United States Food and Drug Administration.

*Acknowledgements*
We thank NIH for support through grants #5PO1-HL-071064-04 to CB, #HL61411 to LLP and #GM-077387 to MPH. ALC thanks the National Science Foundation for partial support through grant CHE 0418006. We express our appreciation to Giulia Ferruzi for excellent technical assistance and to Maria Santabene, whose undergraduate research project at the Duke University Marine Laboratory helped lay the foundation for some of the results reported here.

## References

Abassi, Z., Kotob, S., Pieruzzi, F., Abouassali, M., Keiser, H. R., Fratantoni, J. C., and Alayash, A. I. 1997. Effects of polymerization on the hypertensive action of diaspirin cross-linked hemoglobin in rats. J. Lab. Clin. Med. 129:603–610.

Adamson, J. G., and Moore, C. 1997. Hemolink, an O-raffinose cross-linked hemoglobin-based oxygen carrier. In Blood Substitutes: Principals, Methods, Products and Clinical Trials, ed. T.M.S. Chang. pp. 62–79. Karger Landes Systems.

Alayash, A. I. 2004. Oxygen therapeutics: Can we tame haemoglobin? Nat. Rev. Drug. Disc. 3:152–159.

Alayash, A. I., Fratantoni, J. C., Bonaventura, C., Bonaventura, J., and Bucci, E. 1992. Consequences of chemical modifications on the free radical reactions of human hemoglobin. Arch. Biochem. Biophys. 298:114–120.

Angelo, M., Singel, D. J., and Stamler, J.S. 2006. An S-nitrosothiol SNO. synthase function of hemoglobin that utilizes nitrite as a substrate. Proc. Nat. Acad. Sci. 103:8366–8371.

Baldwin, A. L., Wiley, E. B., and Alayash, A. I. 2002. Comparison of effects of two hemoglobin-based oxygen carriers on intestinal integrity and microvascular leakage. Am. J. Physiol. 283:H1292–1301.

Baldwin, A. L., Wiley, E. B., and Alayash, A. I. 2004. Differential effects of sodium selenite in reducing tissue damage caused by three hemoglobin-based oxygen carriers. J. App. Physiol. 96:893–903.

Baldwin, A. L., Wiley, E. B., Summers, A. G., and Alayash, A. I. 2003. Sodium selenite reduces hemoglobin-induced venular leakage in the rat mesentery. Am. J. Physiol. 284:H81–91.

Bonaventura, C., Cashon, R., Bonaventura, J., Perutz, M., Fermi, G., and Shih, D. T. B. 1991. Involvement of the distal histidine in the low affinity exhibited by Hb Chico Lys beta 66 to Thr. and its isolated beta chains. J. Biol. Chem. 266:23033–23040.

Chevalier, A., Guillochon, D., Nadjar, N., Piot, J., Vijayakshmi, M. W., and Thomas, D. 1989. Effect of glutaraldehyde on haemoglobin: oxidation-reduction potentials and stability. Biochem. Cell. Biol. 68:813–818.

Cosby, K., Partovi, K. S., Crawford, J. H., Patel, R. P., Reiter, C. D., Martyr, S., Yang, R. K., Waclawiw, M. A., Zalos, G., Xu, S. L., Huang, K. T., Shields, H., Kim-Shapiro, D. B., Schechter, A. N., Cannon, R. O., and Gladwin, M. T. 2003. Nitrite reduction to nitric oxide by deoxyhemoglobin vasodilates the human circulation. Nat. Med. 9:1498–1505.

D'Agnillo, F., and Alayash, A. I. 2000. Site-specific modifications and toxicity of blood substitutes. The case of diaspirin cross-linked hemoglobin. Adv. Drug Deliv. Rev. 28:199–212.

D'Agnillo, F. and Alayash, A. I. 2001. Redox cycling of diaspirin cross-linked hemoglobin induces G2/M arrest and apoptosis in cultured endothelial cells. Blood 98:3315–3323.

Guillochon, D., Vijayakshmi, M. W., Thion-Sow, A., and Thomas, D. 1989. Effect of glutaraldehyde on hemoglobin: functional aspects and Mossbauer parameters. Biochem. Cell. Biol. 64:29–37.

Jia, Y., Ramasamy, S., Wood, F., Alayash, A. I., and Rifkind, J. M. 2004a. Cross-linking with O-raffinose lowers oxygen affinity and stabilizes haemoglobin in a non-cooperative T-state conformation. Biochem. J. 384:367–375.

Jia, Y., Wood, F., Menu, P., Faivre, B., Caron, A., and Alayash, A. I. 2004b. Oxygen binding and oxidation reactions of human hemoglobin conjugated to carboxylate dextran. Biochim. Biophys. Acta 1672:164–173.

Kim-Shapiro, D. B., Gladwin, M. T., Patel, R. P., and Hogg, N. 2005. The reaction between nitrite and hemoglobin: the role of nitrite in hemoglobin-mediated hypoxic vasodilation. J. Inorg. Biochem. 99:237–246.

Kosaka, H., Imaizumi, K., Imai, K. and Tyuma, I. 1979. Stoichiometry of the reaction of oxyhemoglobin with nitrite. Biochim. Biophys. Acta 581:184–188.

Lancaster, J. R. 1994. Simulation of the diffusion and reaction of endogeneously produced nitric oxide. Proc. Natl. Acad. Sci. U S A 91:8137–8141.

Nagababu, E., Ramasamy, S., Rifkind, J. M., Jia, Y., and Alayash, A. I. 2002. Site-specific cross-linking of human and bovine hemoglobins differentially alters oxygen binding and redox side reactions producing rhombic heme and heme degradation. Biochemistry 41:7407–7415.

Pearce, B. J. and Gawryl, M. S. 1998. The pharmacology of tissue oxygenation by Biopure's hemoglobin-based oxygen carrier, Hemopure. In Blood Substitutes: Principles, Methods, Products and Clinical Trials, ed. T. M. S. Chang, pp. 82–100. Vol II. Basel, Switzerland: Karger Landes Systems.

Perutz, M. F. 1970. Stereochemistry of cooperative effects in haemoglobin. Nature 228:726–734.

Perutz, M. F. 1978. Hemoglobin structure and respiratory transport. Sci. Am. 239:68–86.

Prouchayret, F., Fasan, G., Grandgeorge, M., Vigneron, C., Menu, P., and Dellacherie, E. 1992. A potential blood substitute from carboxylic dextran and oxyhemoglobin. I. Preparation, purification and characterization. Biomater. Artif. Cells Immobiliz. Biotechnol. 20:319–22.

Reiss, J. G. 2001. Oxygen carriers "blood substitutes" – raison d'etre, chemistry, and some physiology. Chem. Rev. 101:2797–2919.

Riggs, A. F. and Wolbach, R. A. 1956. Sulfhydryl groups and the structure of hemoglobin. J. Gen. Physiol. 39:585–605.

Roche, C. J., Dantsker, D., Samuni, U., and Friedman, J. M. 2006. Nitrite reductase activity of sol-gel encapsulated deoxy hemoglobin: Influence of quaternary and tertiary structure. J. Biol. Chem. 281:36874–36882.

Spagnuolo, C., Rinelli, P., Coletta, M., Chiancone, E., and Ascoli, F. 1987. Oxidation reaction of human oxyhemoglobin with nitrite: a reexamination. Biochim. Biophys. Acta 911:59–65.

Stellwagen, E. 1978. Haem exposure as the determinate of oxidation-reduction potential of haem proteins. Nature 275:73–74.

Taboy, C. H., Bonaventura, C., and Crumbliss, A. L. 2002. Anaerobic oxidations of myoglobin and hemoglobin by spectroelectrochemistry. In Methods in Enzymology, ed. M. I. Simon, pp. 187–209. New York: Academic Press.

Taboy, C. H., Faulkner, K. M., Kraiter, D., Bonaventura, C., and Crumbliss, A. L. 2000. Concentration-dependent effects of anions on the anaerobic oxidation of hemoglobin and myoglobin. J. Biol. Chem. 275: 39048–39054.

# 10

# Bis-histidyl Ferric Adducts in Tetrameric Haemoglobins

Alessandro Vergara, Cinzia Verde, Guido di Prisco and
Lelio Mazzarella

## Abstract

In the last decade crystallographic evidence for endogenous coordination at the sixth coordination site of the heme iron has been reported for monomeric haemoglobins (Hbs) in both the ferrous (haemochrome) and ferric (haemichrome) oxidation state. Usually, the sixth ligand is provided by the imidazole side chain of a His, the only putative ligand normally present in the distal site of the heme pocket. More recently, structural and spectroscopic evidence has been reported, which show that the bis-histidyl adduct in the ferric state represents a common accessible ordered state also for several tetrameric Hbs isolated from Antarctic fish, both in the solid and solution state. Bis-histidyl coordination was also discovered in the crystals of horse met-Hb exposed to acidic pH. All these crystal structures are characterised by a different binding state of $\alpha$ and $\beta$ chains. Tetrameric Hbs with all chains in the bis-histidyl coordination state have not yet been found. Herein we review the structural details of the recent results in this field, together with solution studies on the pathway of haemichrome formation.

## Historical Introduction

Haemoglobins (Hbs) carry out their function as oxygen carriers by keeping the iron atom, which binds the oxygen molecule, in the reduced Fe(II) state. Nonetheless, it is well known that Hbs, even under physiological conditions, frequently undergo spontaneous oxidation producing a variety of ferric species. The role of these species and their impact in different biological contexts has been highly debated in the last few decades. In the Hb superfamily, and in particular in the more restricted class of vertebrate Hbs that are assembled as $\alpha_2\beta_2$ tetramers in the native organisation, the exogenous hexa-coordinate (6C) aquo-met species, in equilibrium with the hydroxy-met form, is by far the most common species formed in the auto-oxidation process. In the presence of ligands such as $F^-$, $CN^-$ and $N_3^-$, many other 6C species can readily form. However, other coordination types in the ferric state have also been detected and characterised; in particular both penta-coordinate (5C) and endogenous hexa-coordinate (haemichrome) species have been observed. The distribution and relevance of these products are highly variable.

The 5C heme state is relatively rare in all Hbs. It has been reported for some mutants of myoglobin, in which distal His had been replaced by hydrophobic residue (Ikeda-Saito et al. 1992; Quillin et al. 1993). For instance, the mutations His→Val and His→Leu destabilise the heme-bound water molecule, producing a penta-coordinate heme iron at neutral pH. The EPR (Electronic Paramagnetic Resonance) spectra of these mutants at neutral pH have been interpreted as a mixture of intermediate ($S=3/2$) and high ($S=5/2$) spin states, which turns to high spin states at alkaline pH. A 5C coordination was also shown for a monomeric flavoHb, where the Leu-E11 side chain shields the heme iron from a nearby Tyr-B10 and Gln-E7, and stabilises a penta-coordinate ferric iron species (Ilari et al. 2002) and for the dimeric *Scapharca inaequivalvis* Hb at neutral and slightly alkaline pH (Boffi et al. 1994). More recently, Hbs with higher structural complexity, such as tetrameric Hbs from polar fish (Giordano et al. 2007; Vergara et al. 2007) and giant Hbs (Marmo Moreira et al. 2006) have also been found to have a 5C state.

With respect to the penta-coordinate ferric state, the occurrence of an endogenous hexa-coordination is more widespread. The residue that can occupy the sixth axial coordination site under native conditions depends on the residues which line the distal cavity, although in the great majority of the cases there is a His residue (bis-histidyl adduct or His-haemichrome). More rarely, in Hbs with lower structural complexity (Das et al. 1999a; Milani et al. 2005) or in mutants of human Hb (HbA) (Nagai et al. 2000), a Tyr has been found to act as the sixth ligand at the iron site. Since in wild-type tetrameric Hbs both the proximal and distal residues are His, in this chapter only the His-haemichrome derivatives will be described.

# State of the Art on Bis-histidyl Structure

The studies of bis-histidyl hexa-coordination in monomeric Hbs, such as legHb (Appleby et al. 1976), barley Hb (Das et al. 1999b) and cyanobacterium synechocystis PCC6803 Hb (Couture et al. 2000), were pioneered by the Wittenbergs' group. The monomeric *Caudina arenicola* was the first Hb with a haemichrome structure studied by X-ray analysis in 1995 by Mitchell, Kitto and Hackert (1995). Other monomeric and dimeric Hbs have also been studied by X-ray crystallography, and some of them at very high resolution.

For tetrameric Hbs, the initial findings of haemichrome formation were associated with a partial unfolding of the native structure (Rifkind et al. 1994). However, it was soon recognised that the haemichrome state had to be considered as an accessible structural sub-state of the native-like Hb population. These states can be distinguished from those associated with denatured conditions by reversibility and absence of precipitation and heme release (Rifkind et al. 1994). A detailed description of the haemichrome formation and denaturation in solution for HbA in the tetrameric state, as well as for the separated chains, has been previously reported (Rifkind et al. 1994). These results also indicated that the hexa-coordination is favoured under conditions where the chains are separated, suggesting a possible role of the quaternary structure in the destabilisation of the haemichrome species.

A first picture of a tetrameric Hb in partial haemichrome state was obtained by our team while investigating air-oxidised crystals of Hb1Tn, the principal Hb component isolated from the Antarctic fish *Trematomus newnesi* (Riccio et al. 2002). Successively, a detailed analysis of the haemichrome species of several Antarctic fish Hbs, their formation and stability in solution as well as their structural features in the solid state, was obtained by complementing X-ray crystallography with optical and Raman spectroscopy in solution and in the solid state (Vitagliano et al. 2004). EPR studies were also conducted in collaboration with Peisach's team at the Albert Einstein College of Medicine (Giordano et al. 2007; Vergara et al. 2007). In particular the crystal structure of HbTb from the Antarctic fish *Trematomus bernacchii*, in two different crystal forms, has been elucidated (Vitagliano et al. 2004). Crystallographic detection of the haemichrome has also been reported for horse Hb (Robinson, Smith and Arnone 2003) and for the α chain of HbA complexed with the α-Hb-stabilising protein (AHSP) (Feng et al. 2005).

Currently, only about a dozen Hb 3D structures, which show the bis-histidyl endogenous coordination, have been deposited in PDB (Protein Data Bank) (Mitchell, Kitto and Hackert 1995; Hargrove et al. 2000; Riccio et al. 2002; Robinson, Smith and Arnone 2003; de Sanctis et al. 2004a; Hoy et al. 2004; Pesce et al. 2004b; Vallone et al. 2004; Vitagliano et al. 2004; de Sanctis et al. 2005; Feng et al. 2005; Vergara et al. 2007) and they are summarised in Table 1. In all cases, as well as for other haemoproteins, the histidyl coordination has always been interpreted as due to $N\epsilon_{2proxymal}$–Fe–$N\epsilon_{2distal}$ binding (Fig. 1). His bonding through a ring carbon has recently been suggested to be a possible coordination mode, particularly in a local slightly acidic environment (Mercs et al. 2006); however, a carbene–Fe binding mode of His to heme in haemoproteins is not supported by experimental observations.

# Structural Features

## *The Haeme Environment*

Since in the usual penta- or exogenous hexa-coordinate state the distal histidyl $N\epsilon_2$ is located about 4 Å away from the iron position, the bis-histidyl coordination requires significant rearrangement of the distal site. Moreover, as both the proximal and distal His are embedded in a α-helical environment (helices E and F, respectively), it is likely that the perturbation produced by the establishment of the internal covalent linkage is not confined to the distal pocket.

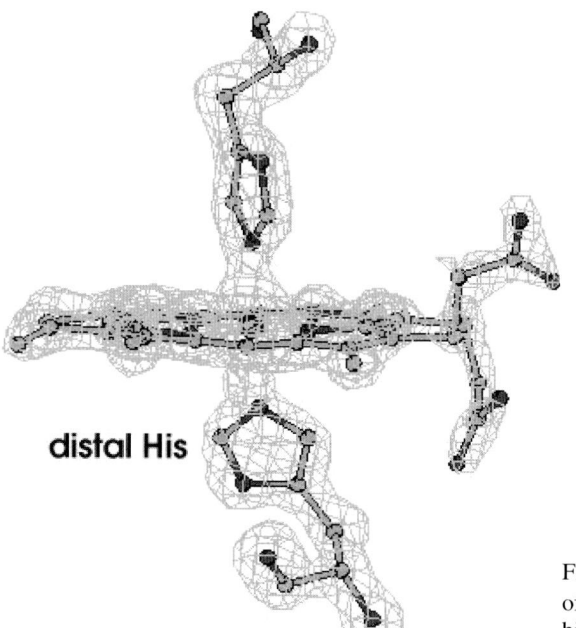

**distal His**

FIG. 1. Electron density maps of ferric-HbTb β-haeme in a bis-histidyl coordination

However, how far this conformational readjustment will be sensed by distant regions of the protein is a matter of the protein matrix rigidity and might depend on the aggregation state of Hb. In turn protein rigidity may cause a more or less pronounced distortion from ideality of the coordination geometry at the haeme site. In the following, we shall first examine the local geometrical features of the bis-histidyl coordination in Hbs, keeping in mind that a meaningful comparison should be based on a more populated set of structures. In this case, high-resolution diffraction data are particularly important, because the presence of the electron-rich iron ion increases the positional errors of the bonded lighter atoms. Some relevant parameters for the bis-histidyl complex are collected in Table 1, where, for the sake of comparison, the data for three structures of the closely related cytochrome $b_5$ are also shown. Table 1 summarised the tilt angle between the haeme plane and the distal ($\vartheta_d$) and proximal ($\vartheta_p$) imidazole plane, the dihedral angle between the two imidazole planes ($\omega$), the bond length of iron with the $N\varepsilon_2$ of the two His, the bond angle $N\varepsilon_{2prox}$–Fe–$N\varepsilon_{2dist}$ and the angle $\varphi$ that the $sp^2$ direction of the $N\varepsilon_2$ atom makes with the $N\varepsilon_2$–Fe bond for both His (Vergara et al. 2007). Despite the low number of structures, some indications emerge from the figures given in Table 1. The deviations from ideality are generally more pronounced for Hbs with respect to cytochrome $b_5$, and for tetrameric *vs.* monomeric Hbs (consider for instance the $N\varepsilon_{2prox}$–Fe–$N\varepsilon_{2dist}$ angle, which measures the deviation from lin-

TABLE 1. Relevant stereochemical parameters (see text) of the bis-histidyl coordinated haeme group

| Haemoprotein (PDB code) | Res. (Å) | $\vartheta_p$ (°) | $\vartheta_d$ (°) | $\omega$ (°) | $N\varepsilon_{2im}$–Fe–$N\varepsilon_{2im}$ (°) | $\varphi$ (°) Prox. | $\varphi$ (°) Dist. | $C\alpha_{prox}$–$C\alpha_{dist}$ (Å) | Fe–$N\varepsilon_{2im}$ (Å) Prox. | Fe–$N\varepsilon_{2im}$ (Å) Dist. |
|---|---|---|---|---|---|---|---|---|---|---|
| T. bernacchii (Hb1Tb β chain)* (2 PEG) | 1.4 | 88.5 | 68.8 | 57.7 | 173 | 12 | 16 | 12.6 | 2.02 | 1.98 |
| T. newnesi (Hb1Tn β chain)* (1LA6) | 2.0 | 76.8 | 73.0 | 62.7 | 165 1 | 5 | 11 | 12.4 | 2.00 | 2.00 |
| E. caballus (Hb α chain)* (1NS6) | 2.1 | 83.1 | 67.6 | 22.7 | 173 | 4 | 16 | 12.5 | 2.11 | 2.18 |
| H. sapiens [(AHSP) HbA α chain] (1Z8U) | 2.4 | 83.2 | 85.1 | 89.1 | 174 | 5 | 3 | 12.8 | 2.13 | 2.10 |
| | | 89.8 | 85.7 | 68.6 | 175 | | | 12.8 | | |
| O. sativa† (1D8U) | 2.3 | 87.1 | 88.4 | 64.4 | 178 | 3 | 4 | 12.1 | 2.08 | 2.08 |
| | | 87.9 | 87.2 | 63.8 | 177 | 4 | 4 | 12.1 | 2.11 | 2.04 |
| H. sapiens cytoglobin† (1URV) | 2.0 | 83.4 | 82.4 | 67.4 | 171 | 13 | 12 | 12.3 | 1.96 | 2.20 |
| | | 84.3 | 81.3 | 67.4 | 174 | 3 | 5 | 12.6 | 2.21 | 2.26 |
| H. sapiens neuroglobin† (1OJ6) | 1.9 | 88.4 | 81.8 | 65.5 | 175 | 8 | 2 | 12.3 | 2.09 | 2.15 |
| | | 89.3 | 78.1 | 67.3 | 175 | 8 | 3 | 12.2 | 2.02 | 2.17 |
| | | 88.5 | 79.1 | 66.1 | 175 | 10 | 9 | 12.1 | 2.10 | 2.06 |
| | | 88.8 | 79.4 | 60.9 | 176 | 6 | 5 | 12.3 | 2.02 | 2.08 |
| Synechocystis s.‡ (1RTX) | 1.8 | 79.0 | 72.7 | 77.0 | 175 | 17 | 15 | 11.5 | 2.11 | 1.99 |
| C. arenicola‡ (1HLB) | 2.5 | 89.3 | 77.0 | 65.0 | 177 | 14 | 9 | 11.6 | 2.06 | 2.07 |
| Mus musculus neuroglobin‡ (1Q1F) | 1.5 | 89.5 | 79.6 | 61.1 | 177 | 4 | 11 | 12.1 | 2.12 | 1.91 |
| D. melanogaster‡ (2BK9) | 1.2 | 86.9 | 88.5 | 89.0 | 179 | 1 | 2 | 12.2 | 1.97 | 2.02 |
| Bovis taurus Cyt b5 (1CYO) | 1.5 | 85.9 | 86.3 | 21.2 | 178 | 4 | 5 | 11.7 | 2.00 | 2.07 |
| R. norvegicus Cyt b5 (1B5M) | 2.7 | 89.1 | 81.3 | 11.8 | 166 | 9 | 10 | 11.5 | 1.82 | 1.92 |
| H. sapiens Cyt b5 (sulphite oxidase domain) (1MJ4) | 1.2 | 87.0 | 86.3 | 0.7 | 178 | 3 | 2 | 11.6 | 2.01 | 2.01 |

The first two entries refer to the β chains of tetrameric Antarctic fish Hbs; the third entry refers to the α chains in the tetramers of the horse met-Hb at acidic pH and the fourth to the ferric α chains of *Homo sapiens* HbA complexed with the α-Hb stabilising protein (AHSP). Sources of cytochrome (Cyt) $b_5$ are reported for comparative analysis.

*Tetrameric

†Dimeric

‡Monomeric

earity of the apical bonds at the iron atom). However, there is no clear indication whether the approach of distal His to the iron, as monitored by $\varphi$, deviates from the ideality more than proximal His, although this may appear to be the case for tetrameric Hbs. In contrast, a geometry closer to ideality is observed only for Hb isolated from *Drosophila melanogaster* and for HbA $\alpha$-chain complexed with AHSP.

The elucidation of the crystal structure of Hbs in a bis-histidyl state and the measurement of the stereochemical parameters of the heme coordination found in the solid state has provided an important guide for the interpretation of the spectroscopic features observed in solution, especially for tetrameric Hbs. EPR data have been collected with several Hbs and they satisfactorily correlate with the X-ray data of these Hbs, as well as with the data observed for model systems (Walker 2004; Vergara et al. 2007). With the exception of human cytoglobin, the g anisotropy parameter is found to increase with the deviation of $\vartheta_d$ from 90°, that is with the tilt of the distal imidazole plane with respect to the heme plane (Vergara et al. 2007). This parameter also increases as the dihedral angle $\omega$ between the two imidazole planes deviates from 0° or 90°. Further evidence of haemichrome form in solution was provided by resonance Raman spectra, which allow unambiguous detection of the haemichrome in solution and were used to monitor its formation and stability as a function of pH and other environmental parameters (Vergara et al. 2008; Vitagliano et al. unpublished).

## Modification of the Polypeptide Chain

At the chain level, interesting modifications are produced by the endogenous hexa-coordination of the iron. The formation of the covalent bond between distal His and the iron necessarily requires a scissor-like closure of the fork formed by the helices E and F: the two helices hold the heme moiety between them and include distal and proximal His, respectively. The scissor-like movement is highlighted by the distance between the C$\alpha$ carbon atoms of the two residues. Depending on the protein, this distance is about 2.5–3.5 Å smaller (Table 1) with respect to non-endogenous coordinated forms; the movement of the two helices also causes sliding of the heme group toward the solvent. However, in most structures, the largest modifications of the polypeptide chain are observed in the nearby CD region, in line with the greater mobility of this segment. This is also of great interest because this region is directly involved in the allosteric interface of tetrameric Hbs.

## Quaternary Modifications

Particularly intriguing are the structural results obtained in tetrameric Hbs (Riccio et al. 2002; Robinson, Smith and Arnone 2003; Vitagliano et al. 2004;

Vergara et al. 2007). In this case, the endogenous hexa-coordination may also perturb the quaternary association. Moreover, within the framework of the tetrameric structure, chain heterogeneity may play a significant role. Indeed, crystallographic studies (Riccio et al. 2002; Robinson, Smith and Arnone 2003; Vitagliano et al. 2004; Vergara et al. 2007) have so far shown that these Hbs form only partial haemichrome states either in the α or in the β chain: these results dramatically underline the intrinsic differences between the two chains in terms of backbone flexibility. Interestingly, in horse Hb haemichrome is formed at the α haeme (Robinson, Smith and Arnone 2003), whereas in Antarctic fish Hbs (Hb1Tb and Hb1Tn) only the β haeme is involved (Riccio et al. 2002; Vitagliano et al. 2004; Vergara et al. 2007). As the quaternary assembly is very similar, these findings suggest a change in the relative flexibility of the two chains. As for monomeric and dimeric Hbs, haemichrome formation requires similar sliding of the heme toward the exterior of the haeme pocket (Riccio et al. 2002; Vitagliano et al. 2004; Vergara et al. 2007). There are, however, considerable differences between horse Hb and Antarctic fish Hbs. In the former, haemichrome was obtained as a result of a pH-induced transition starting from a met-Hb single crystal; the solid state transformation is coupled with extensive modification of the packing contacts and tertiary modifications. Also in this case the largest modifications are observed in the CD α region. It is converted from an extended loop at pH 7.1 to a helix at pH 5.4 (Fig. 2a) and packs against the helix β C of a symmetry related molecule. In particular, Phe46α moves more than 6 Å away from the position in the hydrophobic distal cavity to the surface, forming a π-stacking interaction with the side chain of symmetry related Arg40β, and the α-heme propionate groups of two adjacent tetramers form an intertetramer hydrogen bond. Thus the conformational change at the CD corner is intimately linked to the packing interactions, which help to stabilise the transition to haemichrome. Despite this remarkable change, which affects a region directly involved in the $\alpha_1\beta_2$ ($\alpha_2\beta_1$) interface, the quaternary-structure modifications are negligible and the molecule fully maintains the standard R-state organisation of the starting met-Hb form (Robinson, Smith and Arnone 2003).

In comparison to horse Hb, bis-histidyl adducts in Antarctic fish Hbs exhibit both similarities and differences. First of all haemichrome is formed at the β subunits, where the closer approach of the two His is favoured by Tyr70 in F2, which replaces Phe in HbA (Fig. 2b). The hydroxyl group of Tyr causes a kink in helix E by forming two hydrogen bonds (Riccio et al. 2002). In addition, the CD corner of the β chain becomes highly disordered, whereas the same region in the α chain of horse Hb is well ordered; however, in the latter, the position of the CD corner away from the body of the molecule (Robinson, Smith and Arnone 2003) may suggest that its ordered structure is merely the result of the packing interactions, and that this region could be disordered in solution.

In HbTb and HbTn, the closure movement of the EF fragment produces further modifications to other structural elements of the tertiary structure of the β subunit, directly involved in the quaternary assembly of the molecule. In par-

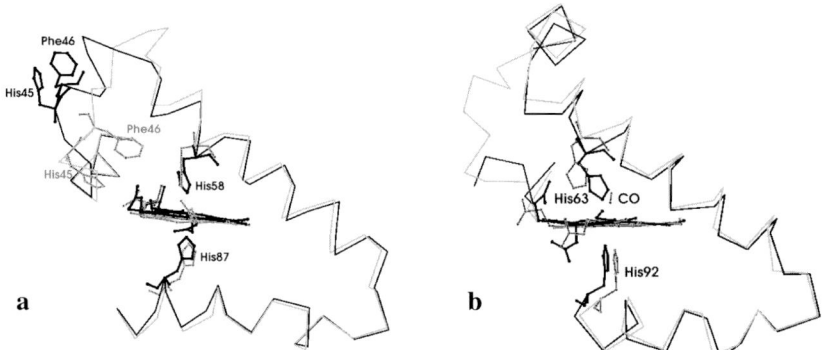

FIG. 2. Superimposition of 6C hexogenous state (grey) and 6C endogenous state (black) for horse Hb (a) and HbTb (b)

ticular, the FG corner takes a conformation that is more similar to that found in the T state rather than that in the R state. The displacement of the FG corner produces variations in the position of His97β whose location is crucial in stabilising the quaternary structure of liganded and unliganded forms of tetrameric Hbs. Indeed, the Antarctic Hbs adopt a quaternary assembly which is intermediate between the classic R and T states. After superimposition of the $\alpha_1\beta_1$ dimer of ferric HbTb to that of the deoxy (T) structure (Riccio et al. 2002), a rotation of 6.7° is required to superimpose the $\alpha_2\beta_2$ dimers. A similar analysis performed between ferric HbTb and the carbomonoxy (R) structure (Camardella et al. 1992) provides an angle of 4.7° in the opposite direction. The combined structural variations induced by haemichrome formation on the β chain and on the quaternary assembly cause significant changes of the FG region of α subunits that extend up to the FG region in contact with the α heme. Therefore, the structural data on Antarctic fish Hbs indicate that the scissoring-like motion of the EF pocket may initiate a cascade of events that favour the switch from the R to the T state. The hypothesis that reversible bis-histidyl complexes, such as haemochrome, could populate the landscape of native Hbs state in solution is particularly attractive as it may suggest a functional role of this species in tetrameric Hbs.

A mechanistic study of the mixed aquo-met/haemichrome formation in tetrameric Hbs would shed light on the origin of the heterogeneity between α and β chains and on the distinct behaviour of mammalian and Antarctic fish Hbs (AFHbs). Preliminary studies on the auto-oxidation of AFHbs upon air exposure show that distinct but interdependent chemical processes on α and β subunits are strictly related to quaternary-structure modification (Vitagliano et al. unpublished). In the early steps of auto-oxidation only the β iron is oxidised, coexisting with an α ($O_2$) form (Vitagliano et al. unpublished). Furthermore, if bis-histidyl coordination is thought to form through aquo-met in HbA, it is formed through a 5C ferric state in AFHbs .

Bis-histidyl coordination is strongly affected by mutations at the haeme distal pocket. Indeed, EPR analysis of both the cathodic component of *T. newnesi* (HbCTn) (Vergara et al. 2007) and Hb of the Arctic fish *Liparis tunicatus* (Giordano et al. 2007) revealed a penta-coordinated ferric state presumably preferred to the bis-histidyl adduct at the β haeme. Probably, the replacement of Val67β with bulkier Ile generates unfavourable interactions in the haeme pocket, which negatively affects haemichrome formation (Vergara et al. 2007).

# State of the Art of Bis-Histidyl Function

This section describes the relevance of bis-histidyl adducts in biomedical applications and in physiological media. In this framework, we summarise the current hypotheses on the role of bis-histidyl coordination on the organism physiology.

The ferric forms of Hb are physiologically inert to further oxygenation, but several subsequent side reactions in the Hb auto-oxidation may interfere with or merge into other biochemical pathways, including formation of haemichrome. The relevance of haemichrome spans from biomedical to physiological aspects (Vergara et al. 2008). For example, auto-oxidation is a serious problem that limits the storage time of cellular Hb-based blood substitutes (Ray, Friedman and Friedman 2002). Also, haemichrome detection has been suggested as a valuable tool for tumour diagnosis (Croci et al. 2001); the reaction of acetylphenylhydrazine (APH) with erythrocytes leads to haemichrome formation in healthy people and not in breast-cancer patients. Some of these hypotheses were reviewed in Rifkind et al. (1994), and others followed more recently. The bis-histidyl complex can be involved in ligand binding (de Sanctis et al. 2004b; Pesce et al. 2004a), in the *in vivo* reduction of met-Hb, in Heinz body formation (Rifkind et al. 1994) and in nitric oxide scavenging.

The Wittenbergs extensively contributed to investigating the function of members of the Hb superfamily, both in the context of peroxidase activity (Ouellet et al. 2007) and in NO scavenging (Wittenberg et al. 1986; Ouellet et al. 2002). Recently, it has also been suggested that haemichrome can be involved in Hb protection from peroxidation attack (Feng et al. 2005). Haemichrome in isolated human α subunits complexed with the AHSP does not exhibit peroxidase activity (Feng et al. 2005). Interestingly, preliminary results show that AFHbs exhibit a peroxidase activity higher not only than that of HbA but also of temperate fish Hbs (Vergara et al. unpublished).

The redox properties are well known for bis-histidyl adducts of haemoproteins different from tetrameric Hbs. A useful summary of their redox potential has been reported (Walker 2004). No data as yet exist in the literature on the hybrid forms of the AFHbs. Further work on the distinct electrochemical behaviour of α and β chains in AFHbs would be beneficial. The accessibility of an oxidation hybrid (α ferrous/β ferric) in AFHb (Vitagliano et al. unpublished) suggests possible dismutase activity currently under investigation.

# Perspectives

The knowledge of how bis-histidyl adducts are formed in the members of the Hb superfamily is a target of future research. Quantum-dynamics computation, currently performed in our team, would help to understand the different occurrence of ferrous and ferric bis-histidyl coordinations. The computational and experimental structural data will drive the design of mutants in which the bis-histidyl coordination is stable. This project would allow the bis-histidyl functions to be transferred to mutants of mammalian Hbs. To understand the functional role of bis-histidyl adducts, monitoring of reactions with NO, peroxidase activity and other redox properties of tetrameric Hbs in a bis-histidyl coordination will be a target of further work. These topics have been treated in a pioneering manner in the long and brilliant carrier of the Wittenbergs.

Furthermore, by unambiguously unveiling the binding state of the individual chains within the tetramer, crystal analysis has also provided invaluable information for the interpretation of spectroscopic features in solution. Another common goal, pertaining to the available structures of tetrameric Hbs in a bis-histidyl coordination, is related to the consideration that up to now no tetrameric Hb has been observed in a complete ($\alpha$ and $\beta$) haemichrome state. The finding that only partial ($\alpha$ or $\beta$) haemichrome states have been observed so far suggests that a complete ($\alpha$ and $\beta$) haemichrome state might not be reached within a folded tetramer. Possibly, full coordination would require a significant shrink of the helices E-F scissors, not compatible with the folded structure.

# *References*

Appleby, C. A., Blumberg, W. E., Peisach, J., Wittenberg, B. A., and Wittenberg, J. B. 1976. Leghemoglobin. 3. An electron paramagnetic resonance and optical spectral study of the free protein and its complexes with nicotinate and acetate. J. Biol. Chem. 251:6090–6096.

Boffi, A., Takahashi, S., Spagnuolo, C., Rousseau, D. L., and Chiancone E. 1994. Structural characterization of oxidized dimeric *Scapharca inaequivalvis* hemoglobin by resonance Raman spectroscopy. J. Biol. Chem. 269:20437–20440.

Camardella, L., Caruso, C., D'Avino, R., di Prisco, G., Rutigliano, B., Tamburrini, M., Fermi, G., and Perutz, M. F. 1992. Hemoglobin of the Antarctic fish *Pagothenia bernacchii*. Amino acid sequence, oxygen equilibria and crystal structure of its carbonmonoxy derivative. J. Mol. Biol. 224:449–460.

Couture, M., Das, T. K., Savard, P.-Y., Ouellet, Y., Wittenberg, J. B., Wittenberg, B. A., Rousseau D. L., and Guertin, M. 2000. Structural investigations of the hemoglobin of the cyanobacterium *Synechocystis* PCC6803 reveal a unique distal heme pocket. Eur. J. Biochem. 267:4770–4780.

Croci, S., Pedrazzi, G., Passeri, G., Piccolo, P., and Ortalli, I. 2001. Acetylphenylhydrazine induced haemoglobin oxidation in erythrocytes studied by Mossbauer spectroscopy. Biochim. Biophys. Acta 1568:99–104.

Das, T. K., Couture, M., Lee, H. C., Peisach, J., Rousseau, D. L., Wittenberg, B. A., Wittenberg, J. B., and Guertin, M. 1999a. Identification of the ligands to the ferric heme of *Chlamydomonas* Chloroplast hemoglobin: evidence for ligation of Tyrosine-63 (B10) to the heme. Biochemistry 38:15360–15368.

Das, T. K., Lee, H. C., Duff, S. M. G., Hill, R. D., Peisach, J., Rousseau, D. L., Wittenberg, B. A., and Wittenberg, J.B. 1999b. The heme environment in barley hemoglobin. J. Biol. Chem. 274:4207–4212.

de Sanctis, D., Dewilde, S., Pesce, A., Moens, L., Ascenzi, P., Hankeln, T., Burmester, T., and Bolognesi, M. 2004a. Crystal structure of cytoglobin: The fourth globin type discovered in man displays heme hexa-coordination. J. Mol. Biol. 336:917–927.

de Sanctis, D., Dewilde, S., Vonrhein, C., Pesce, A., Moens, L., Ascenzi, P., Hankeln, T., Burmester, T., Ponassi, M., Nardini, M., et al. 2005. Bishistidyl heme hexacoordination, a key structural property in *Drosophila melanogaster* hemoglobin. J. Biol. Chem. 280:27222–27229.

de Sanctis, D., Pesce, A., Nardini, M., Bolognesi, M., Bocedi, A., and Ascenzi, P. 2004b. Structure–function relationships in the growing hexa-coordinate hemoglobin sub-family. IUBMB Life 56:643–651.

Feng, L., Zhou, S., Gu, L., Gell, D., Mackay, J., Weiss, M., Gow, A., and Shi, Y. 2005. Structure of oxidized a-haemoglobin bound to AHSP reveals a protective mechanism for haem. Nature 435:697–701.

Giordano, D., Vergara, A., Lee, H., Peisach, J., Balestrieri, M., Mazzarella, L., Parisi, E., di Prisco, G., and Verde, C. 2007. Hemoglobin structure/function and globin-gene evolution in the Arctic fish *Liparis tunicatus*. Gene 406:58–68.

Hargrove, M. S., Brucker, E. A., Stec, B., Sarath, G., Arredondo-Peter, R., Klucas, R. V., Olson, J. S., and Phillips, G.N. 2000. Crystal structure of a nonsymbiotic plant hemoglobin. Structure 8:1005–1014.

Hoy, J. A., Kundu, S., Trent, J. T., III, Ramaswamy, S., and Hargrove, M. S. 2004. The crystal structure of *Synechocystis* hemoglobin with a covalent heme linkage. J. Biol. Chem. 279:16535–16542.

Ikeda-Saito, M., Hori, H., Andersson, L. A., Prince, R. C., Pickering, I. J., George, G. N., Sanders, C. R., Lutz, R. S., McKelvey, E. J., and Mattera, R. 1992. Coordination structure of the ferric heme iron in engineered distal histidine myoglobin mutants. J. Biol. Chem. 267:22843–22852.

Ilari, A., Bonamore, A., Farina, A., Johnson, K., and Boffi, A. 2002. The X-ray structure of ferric *Escherichia coli* flavohemoglobin reveals an unexpected geometry of the distal heme pocket. J. Biol. Chem. 26:23725–23732.

Marmo Moreira, L., Lima Poli, A., Costa-Filho, A. J., and Imasato, H. 2006. Pentacoordinate and hexacoordinate ferric hemes in acid medium: EPR, UV-Vis and CD studies of the giant extracellular hemoglobin of *Glossoscolex paulistus*. Biophys. Chem. 124:62–72.

Mercs, L., Labat, G., Neels, A., Ehlers, A., and Albrecht, M. 2006. Piano-stool iron(II) complexes as probes for the bonding of N-heterocyclic carbenes: indications for π-acceptor ability. Organometallics 25:5648–5656.

Milani, M., Pesce, A., Nardini, M., Ouellet, H., Ouellet, Y., Dewilde, S., Bocedi, A., Ascenzi, P., Guertin, M., Moens, L. et al. 2005. Structural bases for heme binding and diatomic ligand recognition in truncated hemoglobins. J. Inorg. Biochem. 99:97–109.

Mitchell, D. T., Kitto, G. B., and Hackert, M. L. 1995. Structural analysis of monomeric hemichrome and dimeric cyanomet hemoglobins from *Caudina arenicola*. J. Mol. Biol. 251:421–431.

Nagai, M., Aki, M., Li, R., Jin, Y., Sakai, H., Nagatomo, S., and Kitagawa, T. 2000. Heme structure of hemoglobin M Iwate [R87(F8)HisfTyr]: a UV and Visible Resonance Raman study. Biochemistry 39:13083–13105.

Ouellet, H., Ouellet, Y., Richard, C., Labarre, M., Wittenberg, B., Wittenberg, J., and Guertin, M. 2002. Truncated hemoglobin HbN protects *Mycobacterium bovis* from nitric oxide. Proc. Natl. Acad. Sci. U.S.A. 277:5902–5907.

Ouellet, H., Ranguelova, K., Labarre, M., Wittenberg, J., Wittenberg, B., Magliozzo, R., and Guertin, M. 2007. Reaction of *Mycobacterium tuberculosis* truncated hemoglobin O with hydrogen peroxide: evidence for peroxidatic activity and formation of protein-based radicals. J. Biol. Chem. 282:7491–7503.

Pesce, A., De Sanctis, D., Nardini, M., Dewilde, S., Moens, L., Hankeln, T., Burmester, T., Ascenzi, P., and Bolognesi, M. 2004a. Reversible hexa- to penta-coordination of the heme Fe atom modulates ligand binding properties of neuroglobin and cytoglobin. IUBMB Life 56:657–664.

Pesce, A., Dewilde, S., Nardini, M., Moens, L., Ascenzi, P., Hankeln, T., Burmester, T., and Bolognesi, M. 2004b. The human brain hexacoordinated neuroglobin three-dimensional structure. Micron 35:63–65.

Quillin, M., Arduini, R., Olson, J., and Phillips, G. J. 1993. High-resolution crystal structures of distal histidine mutants of sperm whale myoglobin. J. Mol. Biol. 234:140–155.

Ray, A., Friedman, B. A., and Friedman, J. M. 2002. Trehalose glass-facilitated thermal reduction of metmyoglobin and methemoglobin. J. Am. Chem. Soc. 124:7270–7271.

Riccio, A., Vitagliano, L., di Prisco, G., Zagari, A., and Mazzarella, L. 2002. The crystal structure of a tetrameric hemoglobin in a partial hemichrome state. Proc. Natl. Acad. Sci. U.S.A. 99:9801–9806.

Rifkind, J. M., Abugo, O., Levy, A., and Heim, J. M. 1994. Detection, formation, and relevance of hemichrome and hemochrome. Meth. Enzymol. 231:449–480.

Robinson, V. L., Smith, B. B., and Arnone, A. 2003. A pH-dependent aquomet-to-hemichrome transition in crystalline horse methemoglobin. Biochemistry 42:10113–10125.

Vallone, B., Nienhaus, K., Matthes, K., Brunori, M., and Nienhaus, G. 2004. The structure of murine neuroglobin: Novel pathways for ligand migration and binding. Proteins: Struct. Funct. Bioinf. 56:85–92.

Vergara, A. Franzese, M., Merlino, A., Vitagliano, L., di Prisco, G., Verde, C., Lee, H. C., Peisach, J., and Mazzarella, L. 2007. Structural characterization of ferric hemoglobins from three Antarctic fish species of the suborder *Notothenioidei*. Biophys. J. 93:2822–2829.

Vergara, A., Vitagliano, L., di Prisco, G., Verde, C., and Mazzarella, L. 2008. Spectroscopic and crystallographic characterization of hemichromes in tetrameric hemoglobins. Meth. Enzymol. 436A:421–440.

Vitagliano, L., Bonomi, G., Riccio, A., di Prisco, G., Smulevich, G., and Mazzarella, L. 2004. The oxidation process of Antarctic fish hemoglobins. Eur. J. Biochem. 271:1651–1659.

Walker, F.A. 2004. Models of the bis-histidine-ligated electron-transferring cytochromes. Comparative geometric and electronic structure of low-spin ferro- and ferrihemes. Chem. Rev. 104:589–615.

Wittenberg, J., Wittenberg, B., Gibson, Q., Trinick, M., and Appleby, C. 1986. The kinetics of the reactions of *Parasponia andersonii* hemoglobin with oxygen, carbon monoxide, and nitric oxide. J. Biol. Chem. 261:13624–13631.

# 11

# Cooperativity and Ligand-linked Polymerisation in *Scapharca* Tetrameric Haemoglobin

Gianni Colotti, Alberto Boffi and Emilia Chiancone

## Abstract

The assembly of two heterodimers into the $A_2B_2$ tetrameric haemoglobin from *Scapharca inaequivalvis* (HbII) confers to the molecule additional properties relative to the dimeric component (HbI), namely the capacity to undergo an oxygen- and anion-linked polymerisation process. This manifests itself functionally in an increase in cooperativity and a decrease in oxygen affinity at high protein concentrations. The functional parameters of the HbII tetramer as distinct from the effect of ligand-linked polymerisation, i.e. from the so-called polysteric effect, were evaluated in the present work. In conditions where polymerisation is abolished, the $A_2B_2$ tetramer is characterised by a significantly higher Hill coefficient than the HbI dimer ($n=1.8$ *vs.* 1.5), indicating that in *Scapharca* HbII heme–heme communication takes place also over long-range pathways that differ with respect to the direct pathway operative in HbI and by inference in the AB dimer of HbII. At high HbII concentration, where polymerisation of deoxygenated HbII is at a maximum, an additional increase in cooperativity is observed due to the association of the $A_2B_2$ tetramer into the $(A_2B_2)_4$ and $(A_2B_2)_8$ species. Thus, the consequent increase in Hill coefficient from 1.8 to 3.0 can be attributed to polysteric linkage.

## Introduction

The nucleated erythrocytes of the clam *Scapharca inaequivalvis* contain two cooperative haemoglobins, a homodimer (HbI) and a heterotetramer (HbII), which are constructed from three types of polypeptide chain (Petruzzelli et al. 1985, 1989; Piro, Gambacurta and Ascoli 1996) (Fig. 1). The subunit arrangement and hence the molecular basis of cooperativity are unique among haemoglobins (Royer, Love and Fenderson 1985). The HbI chains are assembled back to front with respect to vertebrate haemoglobins, such that the E and F helices do not face the solvent as in the vertebrate proteins, but form the subunit interface and thereby bring the two heme groups in contact through a hydrogen bond network comprising several structured water molecules

```
        |-preA--|  |-----A------|    |------B------|  |-C--|
HbIIA   1  ---VDAAVAKVCGSEAIKANLRRSWGVLSADIEATGLMLMSNLFTLRPDTKTYFT 52
HbIIB   1  -SKVAELANAVVSNADQKDLLRMSWGVLSVDMEGTGLMLMANLFKTSPSAKGKFA 54
HbI     1  MPSVYDAAAQL--TADVKKDLRDSWKVIGSDKKGNGVALMTTLFADNQETIGYFK 53
            *    .  :      *  ** ** *:. * :..*: **:.**    .:   *

        |--------E--------|    |-------F-------|
HbIIA  53  RLGDVQKGKANSKLRGHAITLTYALNNFVDSLDDPSRLKCVVEKFAVNHINRKIS 107
HbIIB  55  RLGDVSAGKDNSKLRGHSITLMYALQNFVDALDDVERLKCVVEKFAVNHINRQIS 109
HbI    54  RLGDVSQGMANDKLRGHSITLMYALQNFIDQLDNPDDLVCVVEKFAVNHITRKIS 108
            *****. *  *.*****:*** ***:**:* **:  . * ***********.*:**

        |--------G-------|      |------H------|
HbIIA 108  GDAFGAIVEPMKETLKARMGNYYSDDVAGAWAALVGVVQAAL 149
HbIIB 110  ADEFGEIVGPLRQTLKARMGNYFDEDTVAAWASLVAVVQASL 151
HbI   109  AAEFGKINGPNKKVLASKN---FGDKYANAWAKLVAVVQAAL 147
            .  ** *  *  ::.*  ::      :.:. . *** **.****:*
```

FIG. 1. Alignment of the *Scapharca* haemoglobin amino acid sequences. The two chains of heterotetrameric haemoglobin (HbIIA and HbIIB) and of the homodimeric HbI are shown; the helices are indicated; identical residues are marked with an asterisk, while similar residue are marked with one or two dots according to the similarity matrices of ClustalW

(Royer, Love and Fenderson 1985; Royer, Hendrickson and Chiancone 1990; Royer 1994; Royer et al. 1996). Oxygen binding induces important changes in the tertiary structure within the heme pocket and at the subunit interface. In particular, the heme becomes more planar and sinks into its pocket, while Phe97 moves from the heme pocket itself into the interface, whose hydrophobicity increases significantly. The consequent changes at the heme–heme contacts allow the information on the ligand-binding event to be transferred from one heme to the opposite subunit such that the affinity for the second ligand is increased (Royer, Love and Fenderson 1985; Royer, Hendrickson and Chiancone 1990; Royer 1994; Royer et al. 1996; Pardanani et al. 1997; Zhou, Zhou and Karplus 2003; Knapp and Royer 2003; Knapp et al. 2006). In contrast to these marked ligand-linked tertiary changes, the quaternary structural movements are very limited (<1° rotation of a chain with respect to the other, with a translation of only 0.1 Å), an observation which finds its functional counterpart in the absence of heterotropic effectors (Chiancone et al. 1981). Small ligand-linked effects on the quaternary structure are apparent in the dissociation of HbI into monomers, which is detectable in the liganded derivatives, but not in the deoxygenated protein (Royer et al. 1997). In *Scapharca* HbI therefore cooperativity, which can be described by a Hill coefficient, *n*, of 1.5, is based essentially on direct heme–heme communication with a small contribution of quaternary constraints as opposed to vertebrate haemoglobins where the phenomenon is dominated by a long range, protein matrix-mediated information transfer accompanied by large quaternary structural changes

(Perutz 1970; Mills, Johnson and Ackers 1976; Perutz 1989; Doyle and Ackers 1992).

The HbII tetramer ($A_2B_2$) is a dimer of identical heterodimers (AB), whose chains interact through the E and F helices, in a manner very similar to that of HbI (Royer et al. 1995). The assemblage of dimers into tetramers confers to HbII additional properties with respect to HbI, namely a higher degree of cooperativity in ligand binding ($n\sim2.0$ at a protein concentration of 1 mg/ml, in 0.1 M phosphate buffer at pH 7.0 and 20°C) and the sensitivity to heterotropic effectors, as shown by the presence of an acid Bohr effect (Chiancone et al. 1981). However, the increase in cooperativity cannot be ascribed in a straightforward fashion to the assembly of two AB dimers into the $A_2B_2$ HbII structure due to the occurrence of ligand- and anion-linked polymerisation. In fact, in a sedimentation velocity study Boffi, Vecchini and Chiancone (1990) established that oxy-HbII undergoes a rapidly reversible dimerisation, whereas deoxy-HbII polymerises into tetramers and octamers to an extent that depends on type and concentration of the anion, pH and temperature.

In the present work, the functional parameters of the HbII tetramer have been assessed distinctly from the effect of ligand-linked polymerisation, namely from the so-called polysteric effect. The use of a much wider range of protein concentrations than in Boffi, Vecchini and Chiancone (1990) allowed us to either abolish HbII polymerisation or shift it towards formation of tetramers and octamers. Oxygen equilibrium experiments performed in parallel indicate that the Hill coefficient pertaining to the $A_2B_2$ tetramer is 1.8, as opposed to $n=1.5$ for HbI, and that any further increase in cooperativity (up to $n=3.0$) is due to ligand-linked polymerisation. In HbII therefore two additional levels of heme–heme communication act over long-range pathways in concert with the simple and direct mechanism operative in HbI and by inference in the AB dimer.

## Materials and Methods

HbII was purified from the nucleated red cells of *S. inaequivalvis* as described previously (Chiancone et al. 1981). Protein concentration was determined using $\varepsilon_M$ (heme basis)=14,200 $M^{-1}$ $cm^{-1}$ at 578 nm and $\varepsilon_M$=105,000 $M^{-1}$ $cm^{-1}$ at 416 nm for the oxygenated derivative, and $\varepsilon_M$=12,000 $M^{-1}$ $cm^{-1}$ at 557 nm and $\varepsilon_M$=143,000 $M^{-1}$ $cm^{-1}$ at 434 nm for the deoxygenated derivative.

Oxygen equilibria were measured at 20°C on a Cary 3 spectrophotometer (Varian, Mulgrave, Victoria, Australia), using a tonometer (Rossi Fanelli and Antonini 1958). In the experiments at high protein concentration the optical path was reduced with a spacer.

Sedimentation velocity experiments were carried out at 20°C on a Beckman-Coulter Optima XL-A ultracentrifuge at 40,000 rpm, at protein concentrations ranging between 0.04 and 8 mg/ml, in 100 mM sodium phosphate at pH 7.0, and BisTris-Cl⁻ (50 mM) at pH 6.5. In the experiments performed with deoxy-HbII, the solutions were degassed in a tonometer and were transferred in nitrogen-flushed ultracentrifuge cells with a syringe that had been washed with deoxygenated buffer containing sodium dithionite (at concentrations below $10^{-4}$ M). The small amount of dithionite-containing buffer that remains in the needle ensured complete deoxygenation for the duration of the run as indicated by the absorption spectra measured in the cell. The sedimentation coefficients were calculated using the programmes provided by the manufacturers.

The oxygen equilibrium data were fitted according to a two-state model, while polymerisation was taken into account by using a scheme based on the principles of polysteric linkage (Colosimo, Brunori and Wyman 1974, 1976). The fitting of the data was carried out with a least-squares method by using the Matlab program (The Math Works Inc., Natick, MA, USA).

# Results

Previous sedimentation velocity studies (Chiancone et al. 1981; Ikeda-Saito et al. 1983; Boffi, Vecchini and Chiancone 1990) revealed that at pH values near neutrality and low protein concentrations, HbII is a stable $A_2B_2$ tetramer with no tendency to dissociate into AB dimers. Upon increase in protein concentration, oxy-HbII dimerises whereas deoxy-HbII forms tetramers and octamers. Accordingly, sedimentation velocity experiments on the deoxygenated protein at concentrations lower than about 4 mg/ml display only one peak, while an additional fast peak appears in the experiments carried out at higher concentrations. Polymerisation increases with increase in temperature from 8°C to 20°C, decreases at ionic strengths above 0.25 M, and is anion-linked. In fact, the distinct dependence of polymer formation on anion concentration at different pH values indicates that the binding of anions is linked to proton binding. In the case of chloride, at pH 6.2–6.5 polymerisation goes through a maximum at around 50 mM $Cl^-$. When polymer formation is significant, the lower asymptote of the Hill plots shifts to the right while the upper asymptote is unaltered, indicating that the HbII polymers are characterised by a lower oxygen affinity and a higher cooperativity than the tetrameric structure.

In the present work, to assess the functional parameters of the HbII tetramer distinctly from the effect of ligand-linked polymerisation, parallel oxygen equilibrium and sedimentation velocity experiments were carried out at protein concentrations between 0.01 and 16 mg/ml (corresponding to $6\times10^{-7}$–$1\times10^{-3}$ M in heme) in sodium phosphate (100 mM) at pH 7.0, and BisTris-HCl ([$Cl^-$]=50 or 100 mM) at pH 6.2–6.5. Boffi, Vecchini and Chiancone (1990) showed that in these buffer conditions polymerisation of HbII is at a maximum; thus, any observed difference in cooperativity and oxygen affinity can be attributed with confidence to the effect of protein concentration on the extent of polymer formed.

In full accordance with previous data, the sedimentation pattern of the oxy- and CO-HbII displays a single, symmetrical peak with the typical velocity of the tetrameric protein ($s_{20,w}$=4.4±0.2 S) even at the highest concentration analysed (8 mg/ml, data not shown). In contrast, deoxy-HbII displays a single peak of sedimentation velocity around 4.3 S at concentrations below 3 mg/ml, although small differences are apparent in different buffer conditions (Fig. 2). At higher protein concentrations, the progressive polymerisation of the $A_2B_2$ tetramer into $(A_2B_2)_4$ tetramers and $(A_2B_2)_8$ octamers leads to the appearance of a faster peak whose velocity depends on protein concentration (Gilbert and Gilbert 1973). Thus, at 5–6 mg/ml, the sedimentation coefficient of the faster component is 10–11 S in 50 mM Cl⁻ when BisTris is the cation, at pH 6.2–6.5, and 8.4 S in 100 mM phosphate at pH 7; in the latter buffer condition, the peak velocity increases to 10 S at HbII concentrations around 8 mg/ml (Fig. 2).

Oxygen equilibria measured in parallel at 20°C show that both cooperativity and oxygen affinity change as a function of protein concentration over the range 0.01–16 mg/ml, in line with the varying extent of polymerisation of deoxy-HbII in the buffer systems utilised (100 mM phosphate at pH 7.0, and 50 mM or 100 mM BisTris at pH 6.2–6.5). The data in 50 mM Cl⁻ at pH 6.2 exemplify the changes in cooperativity and oxygen affinity. The Hill coefficient varies between 1.8 at very low protein concentration and 2.7 at the upper limit of the range examined, with a sharp and sudden increase above 4 mg/ml (Fig. 3a). Oxygen affinity decreases correspondingly: the $p_{1/2}$ value increases

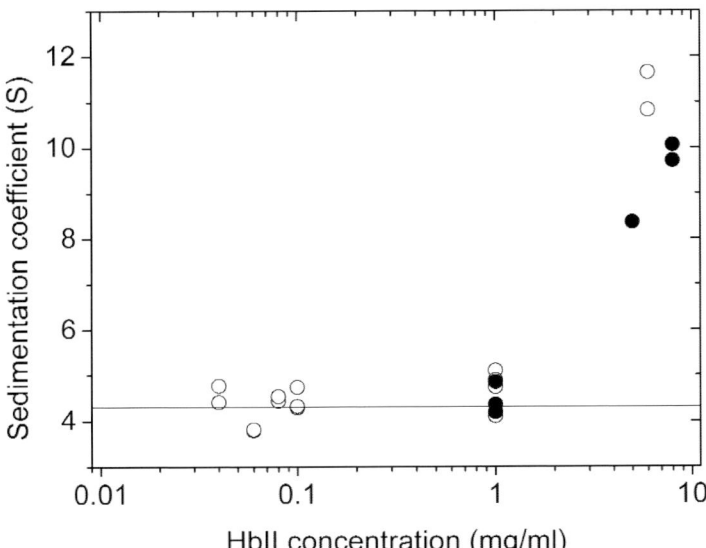

FIG. 2. Sedimentation velocity of the faster component of deoxygenated *Scapharca* HbII as a function of protein concentration. The sedimentation velocity of the slower component corresponds to 4.3 S (horizontal line); the data below 3 mg/ml refer to the only component present, the slower one. The experiments were carried out at 20°C in 50 mM BisTris-HCl at pH 6.5 (O) or in 0.1 M phosphate buffer at pH 7.0 (■)

slowly from 6.5 to 9.8 torr when HbII concentration increases from 0.02 to 4 mg/ml, and quite sharply at higher concentrations (Fig. 3b). The abrupt changes in $n$ and $p_{1/2}$ therefore take place in correspondence with the appearance of the fast sedimenting peak in the sedimentation velocity patterns (Fig. 2). The situation is similar in the other buffers examined: in 0.1 M phosphate at pH 7.0, for example, the Hill coefficient increases slowly from $n=1.8$ at [HbII]=0.01 mg/ml to $n=2.1$ at [HbII]=2–3 mg/ml, with a burst above 4 mg/ml reaching values of 2.7–3.0 at [HbII]=10–15 mg/ml (Fig. 3a, 3b) while oxygen affinity decreases. The shift in the lower asymptote of the Hill plot that accompanies these changes indicates that they can be ascribed to the deoxygenated T form (Fig. 4).

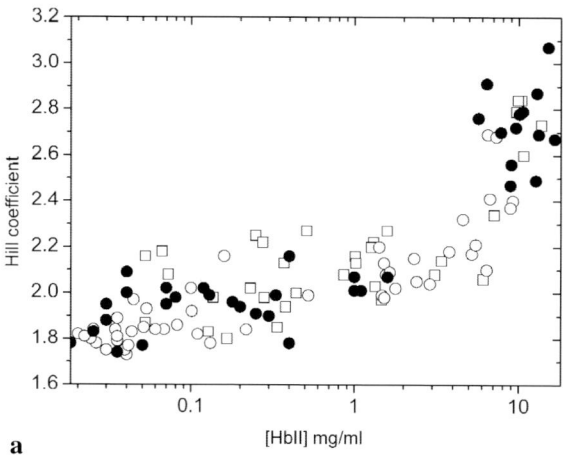

**a**

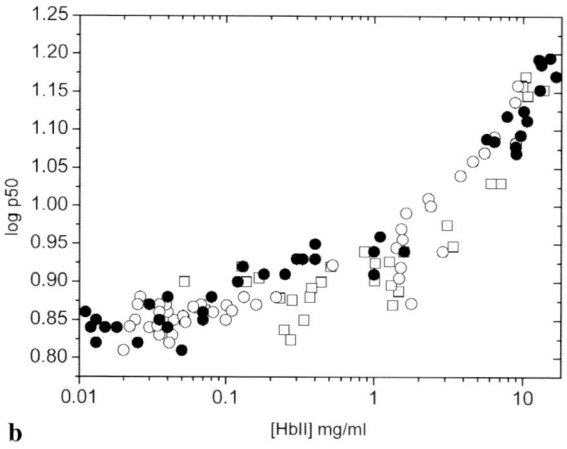

**b**

FIG. 3. Hill coefficient (a) and oxygen affinity (b) as a function of *Scapharca* HbII concentration. The experiments were carried out at 20°C in 0.1 M phosphate buffer at pH 7.0 (●), in 5 mM BisTris-HCl at pH 6.5 (□) and in 50 mM BisTris-HCl at pH 6.5 (○)

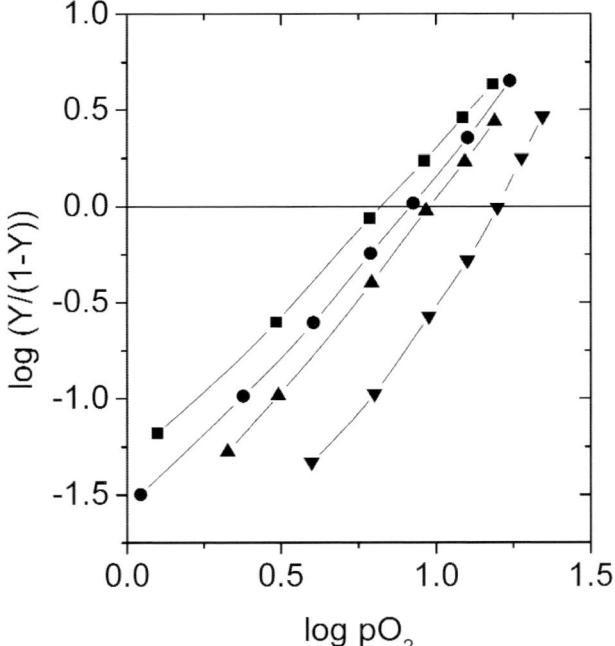

FIG. 4. Hill plots of *Scapharca* HbII as a function of protein concentration. The protein concentrations in 100 mM phosphate buffer, at pH 7.0, at 20°C are (from left to right) 0.01 mg/ml (■), 1.0 mg/ml (●), 4.5 mg/ml (▲) and 14 mg/ml (▼). The data were fitted to the polysteric linkage scheme (Colosimo, Brunori and Wyman 1976) using the following values: $K_R$=0.55 mmHg$^{-1}$; $L_0$=800 (from Ikeda-Saito et al. 1983), and the experimentally determined parameters $K_{T1}$=0.067 mmHg$^{-1}$; $K_{T2}$=0.013 mmHg$^{-1}$ for the $A_2B_2$ and the $(A_2B_2)_4$ species respectively

## Discussion

*S. inaequivalvis* HbII offers the unique opportunity of studying a haemoglobin that is assembled from cooperative dimers and is characterised by a strong tendency to form higher molecular mass polymers, especially when deoxygenated, as is usually the case with oxygen-binding respiratory pigments (Antonini and Chiancone 1977). In addition, HbII polymerisation is proton- and anion-linked. Due to this complex linkage pattern, the functional parameters pertaining to the HbII tetramer have never been assessed separately from the effects of polymerisation and it is not known whether association of the AB dimers into the $A_2B_2$ tetramer leads to an increase in cooperativity.

   To address this specific point, in the analysis of the oxygen equilibrium and sedimentation velocity data, various levels of association have been considered for oxy- and deoxy-HbII, namely the hypothetical AB heterodimer, the $A_2B_2$ tetramer, tetramers $(A_2B_2)_4$ and octamers $(A_2B_2)_8$ of the tetrameric structure, according to the following scheme:

$A+B \rightarrow AB \rightarrow A_2B_2 \leftrightarrow (A_2B_2)_2$                                   oxy-HbII

$A+B \rightarrow AB \rightarrow A_2B_2 \leftrightarrow (A_2B_2)_4 \leftrightarrow (A_2B_2)_8$                deoxy-HbII

The amounts of all species in the oxygenated and deoxygenated derivatives at each HbII concentration were calculated from the sedimentation velocity experiments presented in Fig. 2, using the association constants obtained by Boffi, Vecchini and Chiancone (1990) for similar buffer conditions. In deoxy-HbII, the presence of the $(A_2B_2)_8$ octamer is negligible at 1 mg/ml, but rises to about 5–10% of the total protein content at 10 mg/ml, while the $(A_2B_2)_4$ tetramer amounts to about 0.1% and 2–5% of the total at 0.1 and 1 mg/ml, respectively, and reaches 50–60% at a protein concentration of 10 mg/ml. In oxy-HbII, polymerisation is much less evident: $(A_2B_2)_2$ dimers represent only 0.1% of the protein at 1 mg/ml and about 5% at 10 mg/ml. It cannot be excluded that polymerisation may reach the tetramer or octamer stage when protein concentration is at the levels found in the red cell. In fact, $(A_2B_2)_4$ superstructures are observed in the liganded HbII crystals (Royer et al. 1995).

A simple analysis of the sedimentation velocity and oxygen equilibrium experiments indicates that each association step adds cooperativity to the protein and lowers oxygen affinity. Thus, as the amounts of $(A_2B_2)_4$ and $(A_2B_2)_8$ species become significant in the deoxygenated protein with increase in protein concentration, cooperativity increases from $n=1.8$ (at 0.01 mg/ml) to $n=2.1$ (at 2 mg/ml), up to $n=2.7–3.0$ (at 10 mg/ml). The $p_{1/2}$ value increases in parallel from 6.5, to 9.8 and 14 torr (Fig. 3) due to the progressive shift of the lower asymptote in the Hill plot (Fig. 4). The corresponding change in $K_T$ was evaluated by fitting the oxygen equilibrium data with an algorithm based on the scheme in Fig. 5, derived in part from the studies on polysteric linkage of Colosimo, Brunori and Wyman (1974, 1976). The association constants were taken from Boffi, Vecchini and Chiancone (1990) and the $K_R$ and $L_0$ values of the $A_2B_2$ structure from Ikeda-Saito et al. (1983). On this basis, the decrease in oxygen affinity that accompanies polymerisation of HbII can be accounted for by a $K_T$ value of 0.067 mmHg$^{-1}$ for the $A_2B_2$ tetramer and of 0.013 mmHg$^{-1}$ for the $(A_2B_2)_4$ species.

FIG. 5. Linkage scheme of the association equilibria of HbII

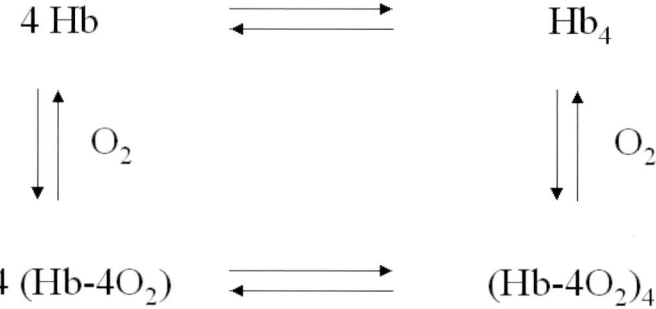

One can only speculate on the molecular basis of the increase in cooperativity pertaining to the different levels of association as only the carbonmonoxy HbII structure has been determined (Royer et al. 1995). However, the AB heterodimer, the first level of association in HbII, and HbI share a high sequence homology (53% identity and 66% homology in the case of HbIIA, 54% identity and 67% homology in the case of HbIIB, Fig. 1) and a strong structural similarity (Royer, Hendrickson and Chiancone 1989; Royer, Hendrickson and Chiancone 1990; Royer 1994) that can be utilised. Thus, the three-dimensional structures of the HbI chain and the A and B chains of HbII are practically superimposable, and the correspondence of the HbI crystal structure with that of the AB dimer in HbII is striking, especially around the oxygen-binding site (Petruzzelli et al. 1985; Petruzzelli et al. 1989; Royer, Hendrickson and Chiancone 1990; Royer 1994; Royer et al. 1995). The haeme-carrying E and F helices that form the intersubunit contact and the heme pocket are almost identical in the AB dimer and in HbI and the area buried at the interface is similar (about 2100 Å$^2$). In particular, the relative positions of the heme groups and of most residues located at the haeme-E-F region of the interface, which are responsible for the direct communication between the hemes (e.g., the heme propionate groups, Lys F10, Phe F11, Asn F14 and His F15), are almost superimposable in HbI-CO and in the carbonmonoxy AB dimer. Even the structure of the water network appears to be very similar given the conservation of at least 18 water molecules in the core of the interface. In HbI and in the HbII AB dimer, therefore, communication between the hemes and hence cooperativity

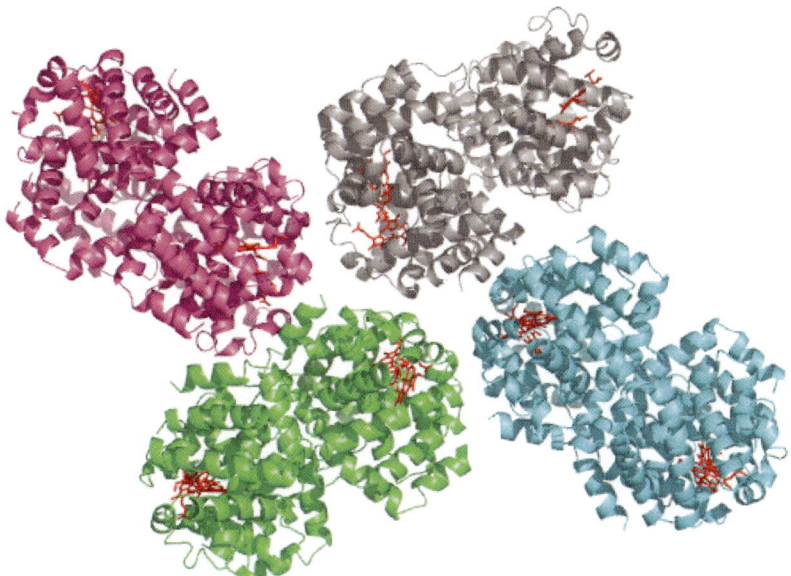

FIG. 6. Packing of four *Scapharca* HbII-CO molecules in the crystal lattice. The four molecules are coloured of grey, magenta, green and cyan; the heme group are in ruby.

can be taken with confidence to arise from the same mechanism and to follow the same pathway(s). It follows that, although the Hill coefficient of the HbII AB dimer cannot be measured directly, as the tetramer does not dissociate into subunits even at very low protein concentrations, it is reasonably ~1.5 (at 20°C) as in HbI.

The second level of association of HbII is represented by the $A_2B_2$ tetramer. The interface between AB dimers involves the A helices, and the AB and GH corners of both subunits (Royer et al. 1995), such that the distance between the heme groups belonging to two different AB dimers in the $A_2B_2$ structure is 35 Å. HbII is tetrameric both in the liganded and unliganded state at low protein concentrations as the amount of polymer formation is negligible up to about 0.01 mg/ml. The $n=1.83\pm0.03$ and $\log(p_{1/2})=0.864\pm0.008$ mmHg values measured at these low concentrations (Fig. 3a, 3b) therefore can be assigned to the $A_2B_2$ tetramer. On this basis, cooperativity in oxygen binding in the $A_2B_2$ tetramer increases by about 0.3 with respect to the (putative) value pertaining to the AB dimer. In terms of structure, this increase can be ascribed to a long-range communication among the heme groups belonging to different AB dimers in the $A_2B_2$ tetramer. Whereas the distance between the two heme iron atoms is about 18 Å within the AB dimer (the edges of the hemes are practically in contact), it is 45 Å when two AB dimers are associated in the tetramer (Fig. 6). As mentioned above, only the structure of ligand-bound HbII has been solved. It must differ from that of the unliganded species in view of the preferential polymerisation of the deoxy protein. It may be envisaged that the rigidity of the AB dimer structure allows the movement of one dimer relative to the other (e.g. a twist at the level of the dimer–dimer interface) and hence communication of structural changes over long distances. In turn, such structural changes would provide the basis for the change in the allosteric properties of the protein.

It has been suggested (Royer et al. 1995) that the increase in cooperativity pertaining to the $A_2B_2$ structure is not due to communication between dimers within a tetramer, but to communication between dimers on different tetramers. This hypothesis appears to be unlikely, as the Hill coefficient of HbII is 1.8 at very low protein concentration where polymerisation of $A_2B_2$ does not occur: e.g., at 0.01 mg/ml at pH 6.3, the association equilibrium of $A_2B_2$ tetramers into HbII oligomers is fully shifted towards the $A_2B_2$ species also in the deoxygenated derivative as the ratio $[A_2B_2]/[(A_2B_2)_4]$ is about $10^7$ and the ratio $[A_2B_2]/[(A_2B_2)_8]$ is about $10^{18}$, based on the association constants calculated by Boffi, Vecchini and Chiancone (1990); even at $[HbII]=0.1$ mg/ml, the ratio $[A_2B_2]/[(A_2B_2)_4]$ is about $10^4$ and the ratio $[A_2B_2]/[(A_2B_2)_8]$ is about $10^{11}$.

The $(A_2B_2)_4$ and $(A_2B_2)_8$ polymers of the tetrameric structure represent the third level of association. Formation of these species accounts in terms of polysteric linkage for the increase in cooperativity from $n=1.8$ up to 3.0 and for the concomitant decrease in oxygen affinity that take place upon increase in HbII concentration (Colosimo, Brunori and Wyman 1974, 1976). A possible structural explanation for the increased value of the Hill coefficient can be found in the HbII-CO crystal structure (Royer et al. 1995). A superquaternary assem-

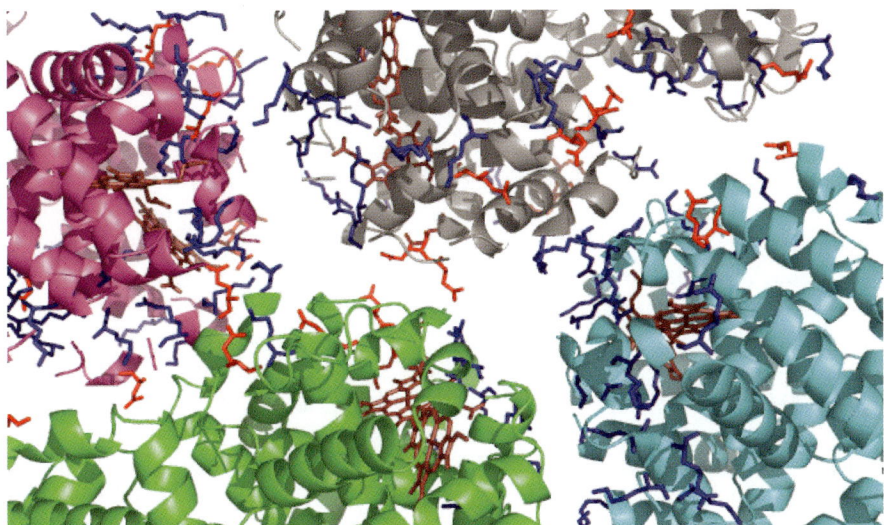

FIG. 7. Residues participating in the contact regions between four *Scapharca* HbII-CO molecules in the crystal lattice. Positively and negatively charged residues are shown (in blue and red, respectively). The picture, adapted from Royer et al. (1995), has been drawn with PyMol

blage, namely a tetramer of tetramers, can be observed within the crystal lattice. In this structure, as apparent from Fig. 6, the ligand-binding sites belonging to different $A_2B_2$ tetramers are much closer (20–30 Å) than those between dimers within a tetramer (45 Å). The tetramer–tetramer interactions involve residues in the pre-A, A, H helices and, more noticeably, in the CD corner and in the FG corner, which are expected to be sensitive to binding of heme ligands due to their proximity to the proximal histidine and heme group. Eight hemes of the $(A_2B_2)_4$ tetramer are brought close to each other such that new pathways of heme–heme communication may be formed, presumably involving polar residues like Glu pre-A3, Lys C4, Lys C6, Arg FG1, Lys or Gln FG2, Asp G2 and Glu G3. Some of these residues, i.e., Arg FG1 and Lys/Gln FG2, may be part of the anion-binding site(s) responsible for anion-linked polymerisation (Boffi, Vecchini and Chiancone 1990). In particular Arg FG1 is not only involved in tetramer–tetramer contacts, but is most likely sensitive to oxygenbinding, as it undergoes significant structural changes upon oxygenation in HbI. Thus, it may participate in the stabilisation of the superquaternary $(A_2B_2)_4$ structure in deoxy-HbII. If one assumes that this superquaternary structure resembles that of the $(A_2B_2)_4$ tetramer in deoxy-HbII, tetramer polymerisation would bring the haemes on a more direct communication pathway than in the tetramer itself, leading to an increase in cooperativity.

Lastly, the difference in the tendency to form polymers displayed by liganded HbII in solution and in the crystal deserves a comment. In the crystal, HbII-CO may be forced to associate into structures that are not favoured in solution as the crystal lattice forces are of the same order of magnitude as the energies involved in quaternary changes (Mozzarelli et al. 1996).

In conclusion, the present study demonstrates that the $A_2B_2$ tetramer is characterised by a significantly higher Hill coefficient than the HbI dimer ($n=1.8$ *vs.* 1.5), indicating that haeme–haeme communication takes place in *Scapharca* HbII also over long-range pathways that differ with respect to the simple and direct mechanism operative in HbI and by inference in the AB dimer of HbII. The additional increase of cooperativity ($n=1.8{\to}3.0$) observed upon association of the $A_2B_2$ tetramer into the $(A_2B_2)_4$ and $(A_2B_2)_8$ species can be attributed to polysteric linkage.

## *References*

Antonini, E., and Chiancone, E. 1977. Assembly of multisubunit respiratory proteins. Annu. Rev. Biophys. Bioeng. 6:239–271.

Boffi, A., Vecchini, P., and Chiancone, E. 1990. Anion-linked polymerization of the tetrameric hemoglobin from *Scapharca inaequivalvis*. Characterization and functional relevance. J. Biol. Chem. 265:6203–6209.

Chiancone, E., Vecchini, P., Verzili, D., Ascoli, F., and Antonini, E. 1981. Dimeric and tetrameric hemoglobins from the mollusc *Scapharca inaequivalvis*. Structural and functional properties. J. Mol. Biol. 152:577–592.

Colosimo, A., Brunori, M., and Wyman, J. 1974. Concerted changes in an allosteric macromolecule. Biophys. Chem. 2:338–344.

Colosimo, A., Brunori, M., and Wyman, J. 1976. Polysteric linkage. J. Mol. Biol. 100:47–57.

Doyle, M. L., and Ackers, G. K. 1992. Cooperative oxygen binding, subunit assembly, and sulfhydryl reaction kinetics of the eight cyanomet intermediate ligation states of human hemoglobin. Biochemistry 31:11182–11195.

Gilbert, L. M., and Gilbert, G. A. 1973. Sedimentation velocity measurement of protein association. Methods Enzymol. 27:273–296.

Ikeda-Saito, M., Yonetani, T., Chiancone, E., Ascoli, F., Verzili, D., and Antonini, E. 1983. Thermodynamic properties of oxygen equilibria of dimeric and tetrameric hemoglobins from *Scapharca inaequivalvis*. J. Mol. Biol. 170:1009–1018.

Knapp, J. E., and Royer, W. E. Jr. 2003. Ligand-linked structural transitions in crystals of a cooperative dimeric hemoglobin. Biochemistry 42:4640–4647.

Knapp, J. E., Pahl, R., Srajer, V., and Royer, W. E. Jr. 2006. Allosteric action in real time: time-resolved crystallographic studies of a cooperative dimeric hemoglobin. Proc. Natl. Acad. Sci. U.S.A. 103:7649–7654.

Mills, F. C., Johnson, M. L., and Ackers, G. K. 1976. Oxygenation-linked subunit interactions in human hemoglobin: experimental studies on the concentration dependence of oxygenation curves. Biochemistry 15:5350–5362.

Mozzarelli, A., Bettati, S., Rivetti, C., Rossi, G. L., Colotti, G., and Chiancone, E. 1996. Cooperative oxygen binding to *Scapharca inaequivalvis* hemoglobin in the crystal. J. Biol. Chem. 271:3627–3632.

Pardanani, A., Gibson, Q. H., Colotti, G., and Royer, W. E. Jr. 1997. Mutation of residue Phe97 to Leu disrupts the central allosteric pathway in *Scapharca* dimeric hemoglobin. J. Biol. Chem. 272:13171–13179.

Perutz, M. F. 1970. Stereochemistry of cooperative effects in haemoglobin. Nature (London) 228:726–739.

Perutz, M. F. 1989. Mechanisms of cooperativity and allosteric regulation in proteins. Q. Rev. Biophys., Cambridge University Press, New York, U.S.A., 1–101.

Petruzzelli, R., Goffredo, M., Barra, D., Bossa, F., Boffi, A., Verzili, D., Ascoli, F., and Chiancone, E. 1985. Amino acid sequence of the cooperative homodimeric hemoglobin from the mollusc *Scapharca inaequivalvis* and topology of the intersubunit contacts. FEBS Lett. 184:328–332.

Petruzzelli, R., Boffi, A., Barra, D., Bossa, F., Ascoli, F., and Chiancone, E. 1989. Scapharca hemoglobins, type cases of a novel mode of chain assembly and heme-heme communication. Amino acid sequence and subunit interactions of the tetrameric component. FEBS Lett. 259:133–136.

Piro, M. C., Gambacurta, A., and Ascoli, F. 1996. Scapharca inaequivalvis tetrameric hemoglobin A and B chains: cDNA sequencing and genomic organization. J. Mol. Evol. 43:594–601.

Rossi Fanelli, A., and Antonini, E. 1958. Studies on the oxygen and carbon monoxide equilibria of human myoglobin. Arch. Biochem. Biophys. 77:478–492.

Royer, W. E. Jr. 1994. High-resolution crystallographic analysis of a co-operative dimeric hemoglobin. J. Mol. Biol. 235:657–681.

Royer, W. E. Jr., Love, W. E., and Fenderson, F. F. 1985. Cooperative dimeric and tetrameric clam haemoglobins are novel assemblages of myoglobin folds. Nature 316:277–280.

Royer, W. E. Jr., Hendrickson, W. A., and Chiancone, E. 1989. The 2.4-Å crystal structure of *Scapharca* dimeric hemoglobin. Cooperativity based on directly communicating hemes at a novel subunit interface. J. Biol. Chem. 264:21052–21061.

Royer W. E. Jr., Hendrickson, W. A., and Chiancone, E. 1990. Structural transitions upon ligand binding in a cooperative dimeric hemoglobin. Science 249:518–521.

Royer, W. E. Jr., Heard, K. S., Harrington, D. J., and Chiancone, E. 1995. The 2.0 Å crystal structure of *Scapharca* tetrameric hemoglobin: cooperative dimers within an allosteric tetramer. J. Mol. Biol. 253:168–186.

Royer, W. E. Jr., Pardanani, A., Gibson, Q. H., Peterson, E. S., and Friedman, J. M. 1996. Ordered water molecules as key allosteric mediators in a cooperative dimeric hemoglobin. Proc. Natl. Acad. Sci. U.S.A. 93:14526–14531.

Royer, W. E. Jr., Fox, R. A., Smith, F. R., Zhu, D., and Braswell, E. H. 1997. Ligand linked assembly of *Scapharca* dimeric hemoglobin. J. Biol. Chem. 272:5689–5694.

Zhou, Y., Zhou, H., and Karplus, M. 2003. Cooperativity in Scapharca dimeric hemoglobin: simulation of binding intermediates and elucidation of the role of interfacial water. J. Mol. Biol. 326:593–606.

# 12
# Human Serum Haeme-albumin: An Allosteric 'Chronosteric' Protein

Mauro Fasano, Gabriella Fanali, Riccardo Fesce
and Paolo Ascenzi

## Abstract

Serum albumin (SA) participates in plasma haeme scavenging. Sequestering
the haeme, SA accounts for most of the antioxidant capacity of plasma. In
turn, serum haeme albumin (SA-haeme) displays ligand-binding and pseudo-
enzymatic properties. Recently, engineered SA-haeme has been proposed as
an $O_2$ carrier not only for red blood cell substitutes but also as an $O_2$-therapeu-
tic agent. Eventually, SA-haeme could be considered an allosteric 'chronosteric' haeme-protein, the transient haeme-based reactivity being modulated by
third components.

## Introduction

The long known and widely characterised haemoglobin superfamily has grown
in the last 15 years to include flavo-haemoglobins, truncated-haemoglobins,
mini-haemoglobin, neuroglobin, cytoglobin, plant (non)symbiotic haemoglobins, protoglobins and globin coupled sensors (see Vinogradov et al. 2006;
Vuletich and Lecomte 2006). Remarkably, Jonathan B. Wittenberg's studies
were pivotal in highlighting myoglobin (Mb)-facilitated oxygen diffusion (see
Wittenberg 2007) and truncated-haemoglobin structure–function relationships
(see Wittenberg et al. 2002).

Recently, proteins involved in haeme scavenging attracted considerable
attention because of their haeme-based reactivity (see Ascenzi et al. 2005). In
this context, serum haeme-albumin (SA-haeme) displays ligand-binding and
pseudo-enzymatic properties (see Peters 1996; Fasano et al. 2005; Ascenzi et
al. 2006a; Ascenzi and Fasano 2007). Engineered SA-haeme has been proposed as an $O_2$ carrier not only for red blood cell substitute but also as an $O_2$-
therapeutic agent (see Carter, Ho and Rüker 1999; Komatsu, Matsukawa and
Tsuchida 2000; Komatsu et al. 2005; Nakagawa et al. 2006). SA-haeme also
has therapeutic relevance, being used to deliver the haeme in the treatment of
acute intermittent porphyria (Bonkovsky et al. 1991). Eventually, SA could be
considered an allosteric 'chronosteric' haeme-protein, the transient haeme-

based reactivity being modulated by third components (see Fasano et al. 2005; Ascenzi et al. 2006a; Ascenzi and Fasano 2007).

# Haeme Binding to Serum Albumin

SA is constituted by a single non-glycosylated all-$\alpha$ chain of 65 kDa containing three homologous domains (labelled I, II and III). Each domain is known to be made up of two separate helical subdomains (named A and B), connected by random coils. Terminal regions of sequential domains contribute to the formation of interdomain helices linking domain IB to IIA, and IIB to IIIA, respectively (Fig. 1) (see Peters 1996; Curry 2002; Fasano et al. 2005).

SA is able to bind up to seven equivalents of long-chain fatty acids (FA) at multiple binding sites (labelled FA1 to FA7, see Fig. 1) with different affinity. FA1, located in subdomain IB, hosts the haeme, with the tetrapyrrole ring arranged in a D-shaped cavity limited by Tyr138 and Tyr161 residues that provide $\pi$–$\pi$ stacking interaction with the porphyrin and supply a donor oxygen (from Tyr161) for the haeme-Fe(III)-iron atom. Ferric haeme is secured by the long IA–IB connecting loop that fits into the cleft opening. Haeme propionates point toward the interface between domains I and III and are stabilised by salt bridges with Arg114 and Lys190 residues (Wardell et al. 2002; Zunszain et al. 2003). Recent reports suggest that FA1 structure has evolved to specifically bind the haeme, FAs being secondary ligands (see Simard et al. 2006; Fasano et al. 2007).

Haeme binds to SA by a simple equilibrium (see Peters 1996). Values of the second-order rate constant (i.e., $k_{on}$) and of the association equilibrium constant (i.e., $K (=k_{on}/k_{off})$) for haeme binding to SA ($7.4\times10^5$ M$^{-1}$ s$^{-1}$ and $7.7\times10^7$ M$^{-1}$ s$^{-1}$, respectively) decrease by about one order of magnitude upon ligand binding to Sudlow's site I. According to linked functions, haeme inhibits ligand binding to Sudlow's site I. In contrast, values of the first-order rate constant for the dissociation of the haeme (i.e., $k_{off}=9.6\times10^{-3}$ s$^{-1}$) and of Sudlow's site I ligands are insensitive to third component(s) (i.e., Sudlow's site I ligands and haeme, respectively). This allosteric regulation may be mediated by the rearrangement of the Phe149-Tyr150 dyad, Phe149 contacting the haeme ring and Tyr150 protruding into Sudlow's site I (i.e., FA7) (see Ghuman et al. 2005; Ascenzi et al. 2006a; Simard et al. 2006).

FAs are effective in the regulation of haeme binding, both by direct competition and allosteric mechanisms. Indeed, FA binding to SA promotes a conformational rearrangement at the I–II and II–III interfaces. Sudlow's site I (FA7) ligands (e.g., warfarin) have a higher affinity for the FA-free conformation, whereas FA1 ligands (e.g., haeme) have a higher affinity for the FA-bound conformation, FA1 and FA7 being functionally coupled (see Ascenzi et al. 2006a; Simard et al. 2006; Fanali et al. 2007). A structural explanation would involve Tyr150 again; actually, binding of myristate to FA2 attracts Tyr150 and Arg252 towards the carboxylate moiety (Simard et al. 2006). Therefore, Arg252 is no

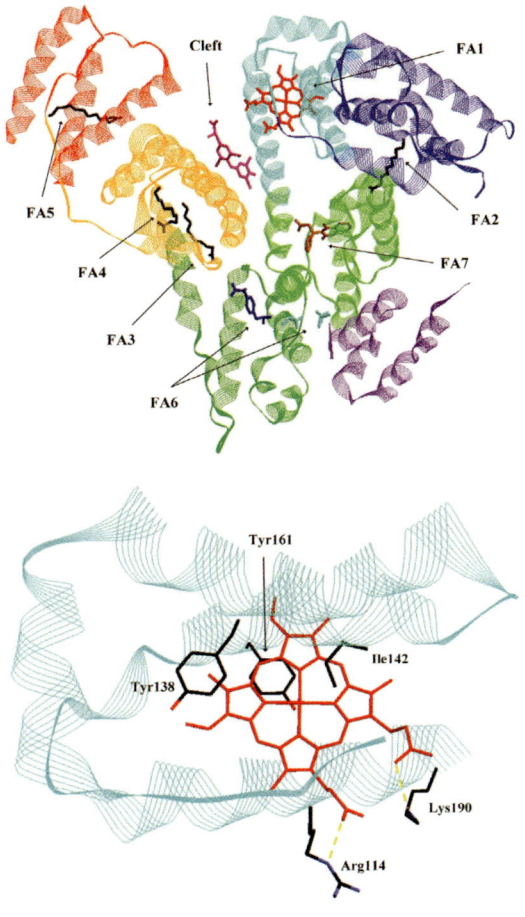

FIG. 1. The human SA structure. (Top panel) SA is rendered with semi-transparent ribbons, coloured as follows: subdomain IA, blue; subdomain IB, cyan; subdomain IIA, pale green; subdomain IIB, dark green; subdomain IIIA, orange; subdomain IIIB, red. The haeme (in red) fits the primary cleft in subdomain IB, corresponding to FA1. Sudlow's site I (in subdomain IIA, corresponding to FA7) is occupied by warfarin (in brown). Sudlow's site II (in subdomain IIIA, corresponding to FA3-FA4), FA2 and FA5 are occupied by myristate anions (in black). FA6 is occupied by two halothane molecules (in cyan) and an additional ibuprofen molecule (in blue) fitting its secondary binding site. Thyroxine (in magenta) fits the cleft between domains I and III. The bacterial protein GA is in purple. (Bottom panel) The SA-haeme environment. Residues establishing main contacts with the haeme moiety are shown. Salt bridges with haeme propionates are shown as dashed dark-yellow lines. Atomic coordinates were taken from PDB entries 1E7C, 1H9Z, 1HK4, 1O9X, 1TF0 and 2BXG (Berman et al. 2000). The picture was drawn with Swiss-PDB-Viewer (Guex and Peitsch 1997)

longer available to stabilise FA7 ligands; on the other hand, the reorientation of Tyr150 may stabilise the interaction of Phe149 with haeme, thus explaining the allosteric modulation observed in solution studies (see Ascenzi et al. 2006a; Fanali et al. 2007). This allosteric regulation is not observed for short FAs (e.g., octanoate) that preferably bind to Sudlow's site II displacing the specific ligands (e.g., ibuprofen) without inducing SA allosteric rearrangement(s) (see Ghuman et al. 2005; Ascenzi et al. 2006a; Simard et al. 2006).

The modulation of haeme binding to SA by drugs and metabolites is relevant in pharmacological therapy management. Indeed, the increase of the haeme plasma level under pathological conditions may increase metabolite plasma concentration and induce a release of SA-bound drugs with the concomitant intoxication of the patient. As expected, the toxic plasma haeme concentration may increase in patients after drug administration (see Ascenzi et al. 2005; Ascenzi et al. 2006a).

# Ligand Binding Properties of Serum Haeme-albumin

## $O_2$ Binding to SA-haeme(II)

SA appears of considerable interest in preparing artificial $O_2$ carriers; however, even if one reduces SA-haeme(III) to SA-haeme(II), it is immediately oxidised by $O_2$. This is due to the fact that SA lacks a proximal His residue, which enables the haeme group to bind $O_2$ reversibly (Wardell et al. 2002; Zunszain et al. 2003). Artificial SA-haeme has been built complexing haeme(II) with SA mutants bearing a proximal His residue at position 142 or 185 (see Komatsu et al. 2004, 2005). Moreover, SA-tetraphenylporphinatoiron(II) (SA-FeP(II)) complexes bearing a covalently linked proximal imidazole have been prepared (Komatsu, Matsukawa and Tsuchida 2000; Nakagawa et al. 2006).

The affinity of $O_2$ for mutant SA-haeme(II) is lower than that of sperm whale Mb(II) (Table 1), but similar to that of human Hb(II). Biphasic dissociation kinetics for $O_2$ binding to mutant SA-haeme(II) (Table 1) reflects the two different geometries of the axial His142 and His185 residues arising from the two orientations of the porphyrin plane in the haeme pocket (Komatsu et al. 2004, 2005).

Monomeric, dimeric, trimeric and tetrameric SA-FeP(II) species display similar $O_2$-binding properties. The affinity of $O_2$ for SA-FeP(II) complexes is lower than that of sperm whale Mb(II) (Table 1), but similar to that of human Hb(II). Multiexponential kinetics for oxygenation and deoxygenation of SA-FeP(II) complexes reflect the different protein environment of FeP(II) molecules bound to multiple SA clefts (Komatsu, Matsukawa and Tsuchida 2000; Nakagawa et al. 2006).

## CO Binding to SA-haeme(II)

The affinity of CO for mutant SA-haeme(II) and sperm whale Mb(II) is grossly similar (Table 1). Biphasic kinetics for CO dissociation from mutant human SA-haeme(II) may reflect the different geometries of the axial His142 and His185 residues arising from the two orientations of the porphyrin plane in the haeme pocket (Zunszain et al. 2003; Komatsu et al. 2005). Biphasic kinetics for CO binding to wild-type bovine SA-haeme(II) (Marden et al. 1989) may also reflect the different orientation of the haeme(II) as observed in mutant human

TABLE 1. Values of kinetic and thermodynamic parameters for ligand binding to SA-haeme(II), SA-FeP(II) and sperm whale Mb(II)[a]

| Haeme-protein | Ligand | $h_{on}$ (M$^{-1}$ s$^{-1}$) | | $h_{off}$ (s$^{-1}$) | | $H$ (M$^{-1}$) | |
|---|---|---|---|---|---|---|---|
| | | I | II | I | II | I | II |
| Bovine SA-haeme[b] | CO[c] | $1.5\times10^6$ | $\sim1\times10^5$ | | n.d. | | n.d. |
| Human SA-haeme[b] | •NO[d] | n.d. | | $\sim5\times10^{-5}$ | | n.d. | |
| Mutant human SA-haeme[e,f] | O$_2$ | $2.0\times10^7$ | | $1.0\times10^2$ | $9.9\times10^2$ | $2.0\times10^5$ | $2.0\times10^4$ |
| | CO | $6.8\times10^6$ | $7.2\times10^5$ | $9.0\times10^{-3}$ | $6.1\times10^{-2}$ | $7.6\times10^8$ | $1.2\times10^7$ |
| Human SA-FeP[g,h] | O$_2$ | $3.4\times10^7$ | $9.5\times10^6$ | $7.5\times10^2$ | $2.0\times10^2$ | $4.5\times10^4$ | $4.7\times10^4$ |
| | CO | $4.9\times10^6$ | $6.5\times10^5$ | n.d. | | n.d. | |
| | •NO | $1.5\times10^7$ | | $6.7\times10^{-5}$ | | $2.2\times10^{11}$ | |
| Sperm whale Mb | O$_2$[j] | $1.4\times10^7$ | | $1.2\times10^1$ | | $1.2\times10^6$ | |
| | CO[i] | $5.1\times10^5$ | | $1.9\times10^{-2}$ | | $2.7\times10^7$ | |
| | •NO[i] | $1.7\times10^7$ | | $1.2\times10^{-4}$ | | $1.4\times10^{11}$ | |

[a]$H=h_{on}/h_{off}$; for details, see text. [b]SA incorporates one haeme. From Marden et al. 1989; Kharitonov et al. 1997. [c]pH=7.0 and 20.0°C. From Marden et al. 1989. [d]pH=7.4 and 20.0°C. From Kharitonov et al. 1997. [e]The SA mutant Ile142His/Tyr161Phe incorporates one haeme. From Komatsu et al. 2005. [f]pH=7.0 and 22.0°C. From Komatsu et al. 2005. [g]SA incorporates a maximum of eight tetraphenylporphinatoiron(II) molecules bearing a covalently linked axial base. From Komatsu, Matsukawa and Tsuchida 2000, 2001. [h]pH=7.3 and 25.0°C. From Komatsu et al. 2005. [i]pH=7.0 and 20.0°C. From Rohlfs et al. 1990; Komatsu et al. 2005. [j]pH=7.0 and 20.0°C. From Moore and Gibson 1976; Komatsu, Matsukawa and Tsuchida 2001. n.d., not determined

SA-haeme(II) (Komatsu et al. 2004, 2005). On the other hand, biphasic kinetics for CO binding to human SA-FeP(II) may reflect the different protein environment of FeP(II) molecules bound to multiple SA clefts (Komatsu, Matsukawa and Tsuchida 2000; Komatsu et al. 2005; Nakagawa et al. 2006).

## •NO Binding to SA-haeme(II)

•NO binds to SA(-haeme) clefts, such as Sudlow's site I, and the haeme-Fe atom. Furthermore, Cys34 undergoes •NO-mediated S-nitrosylation. Therefore, •NO can modify SA(-haeme) structure and function(s) by different mechanisms (see Peters 1996; Fasano et al. 2005; Ascenzi et al. 2006a).

In the absence of allosteric effectors, SA-haeme(II)-NO is a five-coordinate haeme-iron species, characterised by the three-line splitting observed in the high magnetic field region of the X-band EPR spectrum. On the other hand, in the presence of drugs that stabilise the B-form, SA-haeme(II)-NO is a six-coordinate haeme-iron species, characterised by an X-band EPR spectrum with an axial geometry. This represents further evidence for allosteric coupling of the haeme cleft with drug binding sites (e.g., FA2, FA6, FA7, and Sudlow's site II) (see Fasano et al. 2005; Ascenzi et al. 2006a).

Kinetic and thermodynamic parameters for •NO binding to SA-haeme(II) and SA-FeP(II) are similar to those reported for sperm whale Mb(II) nitrosylation (Table 1) (Komatsu, Matsukawa and Tsuchida 2001) Unexpectedly, the $k_{off}$ value for •NO dissociation from penta-coordinate SA-haeme(II)-NO is closely similar to those for denitrosylation of hexa-coordinate SA-FeP(II), sperm whale Mb(II)-NO (Table 1) and of the R-state of human Hb(II)-NO. This may reflect different protein structure-dependent 'cage effects'; following •NO dissociation, there is a significant likelihood that •NO undergoes different geminate recombination with Fe(II) before it escapes to the surrounding solvent (Kharitonov et al. 1997).

## Azide, Cyanide, Fluoride and Imidazole Binding to SA-haeme(III)

Haeme-Fe(III) ligands bind to SA-haeme(III) by a simple equilibrium. Values of the association equilibrium constant for azide, cyanide, fluoride and imidazole to SA-haeme(III) are lower than those reported for ligand binding to sperm whale Mb(III) (Fasano et al. 2001; Monzani et al. 2001). Actually, this may reflect steric clash of the exogenous ligand with the SA-haeme(III) distal residues and/or the scarce stabilisation of the SA-haeme(III)-bound ligand by hydrogen bonding to the protein matrix (Monzani et al. 2001).

# Pseudo-Enzymatic Properties of Serum Haeme-albumin

## $^{\bullet}$NO/Peroxynitrite Scavenging by SA-haeme(II)

Mixing of SA-haeme(II)-NO and peroxynitrite solutions leads to SA-haeme(III) by way of SA-haeme(III)-NO according to Scheme 1 (Ascenzi and Fasano 2007):

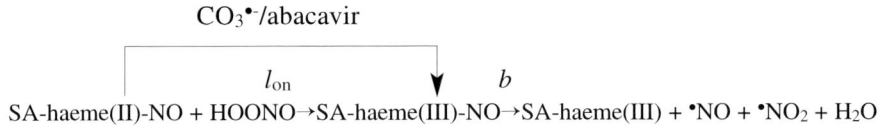

$$\text{SA-haeme(II)-NO + HOONO} \rightarrow \text{SA-haeme(III)-NO} \rightarrow \text{SA-haeme(III)} + {}^{\bullet}\text{NO} + {}^{\bullet}\text{NO}_2 + \text{H}_2\text{O}$$

Values of $l_{on}$ and $b$ for peroxynitrite-mediated oxidation of SA-haeme(II)-NO and horse heart Mb(II)-NO, in the absence and presence of $CO_2$ and/or abacavir, are shown in Table 2 (see Herold and Boccini 2006; Ascenzi and Fasano 2007).

These data represent the first evidence for the allosteric modulation of SA-haeme reactivity by heterotropic interaction(s). In fact, abacavir enhances the value of the second-order rate constant $l_{on}$ for peroxynitrite-mediated oxidation of SA-haeme(II)-NO, in the absence and presence of $CO_2$ (Table 2) (Ascenzi and Fasano 2007). This may reflect abacavir binding to Sudlow's site I and the concomitant drug-induced hexa-coordination of the haeme-Fe-atom of SA-haeme(II)-NO by Tyr161 (see Fasano et al. 2005; Ascenzi et al. 2006b; Fanali et al. 2007). Note that the formation of the ferric form of haeme-proteins is facilitated by binding of Tyr residue(s) and phenolate (derivatives) to the haeme-Fe atom (Adachi et al. 1993; Rydberg, Sigfridsson and Ryde 2004). Furthermore, $CO_2$ enhances the rate of peroxynitrite-mediated oxidation of haeme(II)-NO proteins (i.e., $l_{on}$) via rapid formation of the transient reactive species $CO_3^{\bullet-}$ and

TABLE 2. Kinetic parameters for peroxynitrite-mediated oxidation of human SA-haeme(II)-NO and horse heart Mb(II)-NO[a]

| Haeme-protein | [$CO_2$] (M) | [Abacavir] (M) | $l_{on}$ ($M^{-1}$ $s^{-1}$) | $b$ ($s^{-1}$) |
|---|---|---|---|---|
| Human SA-haeme[b] | – | – | $6.5 \times 10^3$ | $1.9 \times 10^{-1}$ |
| | $1.2 \times 10^{-3}$ | – | $1.3 \times 10^5$ | $1.7 \times 10^{-1}$ |
| | – | $5.0 \times 10^{-3}$ | $2.2 \times 10^4$ | $1.8 \times 10^{-1}$ |
| | $1.2 \times 10^{-3}$ | $5.0 \times 10^{-3}$ | $3.6 \times 10^5$ | $1.9 \times 10^{-1}$ |
| Horse heart Mb | –[c] | – | $3.1 \times 10^{4}$ [c] | $\sim 1.2 \times 10^{1}$ [c] |
| | $1.2 \times 10^{-3}$ [d] | – | $1.7 \times 10^5$ [d] | $1.1 \times 10^{1}$ [d] |

[a]For details, see Scheme 1 and text. [b]pH 7.0 and 10.0°C. From Ascenzi and Fasano 2007. [c]pH 7.5 and 20.0°C. From Herold and Boccini 2006. [d]pH 7.0 and 20.0°C. From Herold and Boccini 2006.

$^{\bullet}NO_2$; $CO_3^{\bullet-}$ is a stronger oxidant than $^{\bullet}NO_2$ and peroxynitrite (see Goldstein, Lind and Merényi 2005; Ascenzi et al. 2006b; Papina and Koppenol 2006).

## Catalase and Peroxidase Activity of SA-haeme

SA-haeme exhibits weak catalase and peroxidase activity in the oxidation of phenolic compounds related to tyrosine (i.e., p-cresol, 3-(p-hydrox-yphenyl)propionic acid, tyramine and tyrosine) (Monzani et al. 2001), and 2,2´-azinobis(3-ethylbenzthiazoline-6-sulfonate) (Kamal and Behere 2002). The catalytic mechanism of SA-haeme appears to be different from that of per-oxidases. In fact, the rate of formation of the active SA-haeme species is lower than its reaction with either the substrate (in a peroxidase reaction), the perox-ide (in a catalase reaction) or the protein itself to restore resting haeme(III) (Monzani et al. 2001). The main factors affecting the catalase and peroxidase activity of SA-haeme are the reduced accessibility of the haeme-Fe centre, and the lack of an Arg residue in the SA-haeme pocket that in peroxidases assists the cleavage of bound peroxide and accelerates the formation of the active species (Monzani et al. 2001). Lastly, a glutathione-linked thiol peroxidase activity of SA, based on protein sulphydryl(s), has been reported (Cha and Kim 1996).

## Conclusions

Haeme binding to SA basically serves as buffer to avoid the promotion of oxida-tive stress conditions. In turn, SA-haeme shares ligand-binding and pseudo-enzy-matic properties as haemoglobin and Mb. Although SA-haeme lacks the ability to bind molecular $O_2$, a number of point mutants have been proposed as blood substi-tutes and $O_2$-therapeutic agents with encouraging results. Due to the flexible mod-ular structure of SA, haeme binding and reactivity are modulated by heterotropic interaction(s); accordingly, haeme affects the ligand-binding properties of func-tionally linked sites. This aspect is fundamental for a better management of the patient. Lastly, data available for SA-haeme describe a curious situation where haeme binding to a non-classical haeme-protein (i.e., SA) confers (although tran-siently) ligand binding and pseudo-enzymatic properties. Furthermore, the effects arising from haeme binding to SA might have some role in the regulation of bio-logical functions. As these actions are dependent on the interaction of a ligand (e.g., haeme) with a carrier (e.g., SA), they have been called 'chronosteric' effects.

*Acknowledgements*
This work was partially supported by a grant from the Ministry for Health of Italy (National Institute for Infectious Diseases I.R.C.C.S. 'Lazzaro Spallanzani', Roma, Italy, *Ricerca corrente* 2007 to P.A.) and from the University of Insubria (*Fondo di Ateneo per la Ricerca* 2005-06 to M.F.).

# References

Adachi, S., Nagano, S., Ishimori, K., Watanabe, Y., Morishima, I., Egawa, T., Kitagawa, T., and Makino, R. 1993. Roles of proximal ligand in heme proteins: replacement of proximal histidine of human myoglobin with cysteine and tyrosine by site-directed mutagenesis as models for P-450, chloroperoxidase, and catalase. Biochemistry 32:241–252.

Ascenzi, P., and Fasano, M. 2007. Abacavir modulates peroxynitrite-mediated oxidation of ferrous nitrosylated human serum heme-albumin. Biochem. Biophys. Res. Commun. 353:469–474.

Ascenzi, P., Bocedi, A., Visca, P., Altruda, F., Tolosano, E., Beringhelli, T., and Fasano, M. 2005. Hemoglobin and heme scavenging. IUBMB Life 57:749–759.

Ascenzi, P., Bocedi, A., Notari, S., Fanali, G., Fesce, R., and Fasano, M. 2006a. Allosteric modulation of drug binding to human serum albumin. Mini Rev. Med. Chem. 6:483–489.

Ascenzi, P., Bocedi, A., Visca, P., Minetti, M., and Clementi, E. 2006b. Does $CO_2$ modulate peroxynitrite specificity? IUBMB Life 58:611–613.

Berman, H. M., Westbrook, J., Feng, Z., Gilliland, G., Bhat, T. N., Weissig, H., Shindyalov, I. N., and Bourne, P. E. (2000) The Protein Data Bank. Nucleic Acids Res. 28:235–242.

Bonkovsky, H. L., Healey, J. F., Lourie, A. N., and Gerron, G. G. 1991. Intravenous heme-albumin in acute intermittent porphyria: evidence for repletion of hepatic hemoproteins and regulatory heme pools. Am. J. Gastroenterol. 86:1050–1056.

Carter, D.C., Ho, J.X., and Rüker, F. 1999. Oxygen-transporting albumin: albumin-based blood replacement composition and blood volume expander. US Pat. No.5,948,609.

Cha, M. K., and Kim, I. H. 1996. Glutathione-linked thiol peroxidase activity of human serum albumin: a possible antioxidant role of serum albumin in blood plasma. Biochem. Biophys. Res. Commun. 222:619–625. Erratum in: (1996) Biochem. Biophys. Res. Commun. 225:695.

Curry, S. 2002. Beyond expansion: structural studies on the transport roles of human serum albumin. Vox Sang. 83[Suppl. 1]:315–319.

Fanali, G., Bocedi, A., Ascenzi, P., and Fasano, M. 2007. Modulation of heme and myristate binding to human serum albumin by anti-HIV drugs. An optical and NMR spectroscopic study. FEBS J. 274:4491–4502.

Fasano, M., Baroni, S., Vannini, A., Ascenzi, P., and Aime, S. 2001. Relaxometric characterization of human hemalbumin. J. Biol. Inorg. Chem. 6:650–658.

Fasano, M., Curry, S., Terreno, E., Galliano, M., Fanali, G., Narciso, P., Notari, S., and Ascenzi, P. 2005. The extraordinary ligand binding properties of human serum albumin. IUBMB Life 57:787–796.

Fasano, M., Fanali, G., Leboffe, L., and Ascenzi, P. 2007. Heme binding to albuminoid proteins is the result of recent evolution. IUBMB Life 59:436–440.

Ghuman, J., Zunszain, P.A., Petitpas, I., Bhattacharya, A.A., Otagiri, M., and Curry, S. 2005. Structural basis of the drug-binding specificity of human serum albumin. J. Mol. Biol. 353:38–52.

Goldstein, S., Lind, J., and Merényi, G. 2005. Chemistry of peroxynitrites and peroxynitrates. Chem. Rev. 105:2457–2470.

Guex, N., and Peitsch, M. C. 1997. SWISS-MODEL and the Swiss-PdbViewer: an environment for comparative protein modeling. Electrophoresis 18:2714–2723.

Herold, S., and Boccini, F. 2006. NO• release from MbFe(II)NO and HbFe(II)NO after oxidation by peroxynitrite. Inorg. Chem. 45:6933–6943.

Kamal, J. K. A., and Behere, D. V. 2002. Spectroscopic studies on human serum albumin and methemalbumin: optical, steady-state, and picosecond time-resolved fluorescence studies, and kinetics of substrate oxidation by methemalbumin. J. Biol. Inorg. Chem. 7:273–283.

Kharitonov, V. G., Sharma, V. S., Magde, D., and Koesling, D. 1997. Kinetics of nitric oxide dissociation from five- and six-coordinate nitrosyl hemes and heme proteins, including soluble guanylate cyclase. Biochemistry 36:6814–6818.

Komatsu, T., Matsukawa, Y., and Tsuchida, E. 2000. Kinetics of CO and $O_2$ binding to human serum albumin-heme hybrid. Bioconjug. Chem. 11:772–776.

Komatsu, T., Matsukawa, Y., and Tsuchida, E. 2001. Reaction of nitric oxide with synthetic hemoprotein, human serum albumin incorporating tetraphenylporphinatoiron(II) derivatives. Bioconjug. Chem. 12:71–75.

Komatsu, T., Ohmichi, N., Zunszain, P. A., Curry, S., and Tsuchida, E. 2004. Dioxygenation of human serum albumin having a prosthetic heme group in a tailor-made heme pocket. J. Am. Chem. Soc. 126:14304–14305.

Komatsu, T., Ohmichi, N., Nakagawa, A., Zunszain, P. A., Curry, S., and Tsuchida, E. 2005. $O_2$ and CO binding properties of artificial hemoproteins formed by complexing iron protoporphyrin IX with human serum albumin mutants. J. Am. Chem. Soc. 127:15933–15942.

Marden, M. C., Hazard, E. S., Leclerc, L., and Gibson, Q. H. 1989. Flash photolysis of the serum albumin-heme-CO complex. Biochemistry 28:4422–4426.

Monzani, E., Bonafè, B., Fallarini, A., Redaelli, C., Casella, L., Minchiotti, L., and Galliano, M. (2001) Enzymatic properties of hemalbumin. Biochim. Biophys. Acta 1547:302–312.

Moore, E. G., and Gibson, Q. H. 1976. Cooperativity in the dissociation of nitric oxide from hemoglobin. J. Biol. Chem. 251:2788–2794.

Nakagawa, A., Komatsu, T., Iizuka, M., and Tsuchida, E. 2006. Human serum albumin hybrid incorporating tailed porphyrinatoiron(II) in the a,a,a,b-conformer as an $O_2$-binding site. Bioconjug. Chem. 17:146–151.

Papina, A. A., and Koppenol, W. H. 2006. Two pathways of carbon dioxide catalysed oxidative coupling of phenol by peroxynitrite, Chem. Res. Toxicol. 19:382–391.

Peters, T. Jr. 1996. All about Albumin: Biochemistry, Genetics and Medical Applications. San Diego and London: Academic Press.

Rohlfs, R. J., Mathews, A. J., Carver, T. E., Olson, J. S., Springer, B. A., Egeberg, K. D., and Sligar, S. G. 1990. The effects of amino acid substitution at position E7 (residue 64) on the kinetics of ligand binding to sperm whale myoglobin. J. Biol. Chem. 265:3168–3176.

Rydberg, P., Sigfridsson, E., and Ryde, U. 2004. On the role of the axial ligand in heme proteins: a theoretical study. J. Biol. Inorg. Chem. 9:203–223.

Simard, J. R., Zunszain, P. A., Hamilton, J. A., and Curry, S. 2006. Location of high and low affinity fatty acid binding sites on human serum albumin revealed by NMR drug-competition analysis. J. Mol. Biol. 361:336–351.

Vinogradov, S. N., Hoogewijs, D., Bailly, X., Arredondo-Peter, R., Gough, J., Dewilde, S., Moens, L., and Vanfleteren, J. R. 2006. A phylogenomic profile of globins. BMC Evol. Biol. 6:31.

Vuletich, D. A., and Lecomte, J. T. 2006. A phylogenetic and structural analysis of truncated hemoglobins. J. Mol. Evol. 62:196–210.

Wardell, M., Wang, Z., Ho, J. X., Robert, J., Rüker, F., Ruble, J., and Carter, D. C. 2002. The atomic structure of human methemalbumin at 1.9 Å. Biochem. Biophys. Res. Commun. 291:813–819.

Wittenberg, J. B. 2007. On optima: the case of myoglobin-facilitated oxygen diffusion. Gene 398:156–161.

Wittenberg, J. B., Bolognesi, M., Wittenberg, B. A., and Guertin, M. 2002. Truncated hemoglobins: a new family of hemoglobins widely distributed in bacteria, unicellular eukaryotes, and plants. J. Biol. Chem. 277:871–874.

Zunszain, P. A., Ghuman, J., Komatsu, T., Tsuchida, E., and Curry, S. 2003. Crystal structural analysis of human serum albumin complexed with hemin and fatty acid. BMC Struct. Biol. 3:6.

# 13

# T- and R-state Tertiary Relaxations in Sol-gel Encapsulated Haemoglobin

Uri Samuni, Camille J. Roche, David Dantsker
and Joel M. Friedman

## Abstract

Tertiary relaxations within the T and R quaternary states of human adult haemoglobin (HbA) are compared for sol-gel encapsulated samples bathed in buffer with either 25% or 75% (v/v) glycerol. T-state tertiary relaxations are initiated by adding CO to an encapsulated T-state deoxyHbA sample, thus generating liganded T-state species. The conformational evolution of the liganded T-state samples is followed by monitoring the frequency of $v$(Fe-His), the conformation-sensitive iron-proximal histidine stretching mode observed in the resonance Raman spectra of either of the deoxy sample of the 7 ns photoproduct derived from the CO samples. In parallel, the functional properties are monitored by following the evolution of the kinetic traces associated with CO recombination subsequent to nanosecond photodissociation of the CO-heme unit. In contrast, the R-state relaxations are initiated by adding dithionite to encapsulated samples of either oxyHbA or cyanometHbA, thus generating deoxy hemes whose resonance Raman spectra reflect the influence of the relaxing tertiary structure within the R state. After the "deoxy" sample is allowed to relax for a defined time period, CO is introduced. The evolution of the relegated samples is now followed by monitoring the photoproduct frequency of the $v$(Fe-His) Raman band and the kinetic traces for the CO recombination.

The results reveal a hierarchy of R/T-dependent tertiary relaxation processes whose differences can be explained based on differences in solvent slaving properties of the different relaxations. The results also support models of HbA allostery in which there are multiple functionally distinct tertiary conformations with each quaternary state.

## Introduction

Many of the allosteric aspects of ligand binding to haemoglobin have been understood using the two-state model based on a ligation-dependent equilibrium between two quaternary structures: the low-affinity T-state structure and the high-affinity R-state structure (Perutz et al. 1987, 1998). Despite the seeming success of this model, there are many questions that require a more detailed molecular level model that can account for such phenomena as: (i) modulation of ligand reactivity within a given quaternary state by added effectors (Yonetani et al. 2002); (ii) the trajectories associated with ligand binding-induced T to R transitions and deligation-induced R to T transitions (Hofrichter et al. 1983; Scott and Friedman 1984; Rousseau and Friedman 1988; Jayaraman et al. 1995; Jayaraman and Spiro 1995; Bjorling et al. 1996; Ghelichkhani et al. 1996; Juszczak and Friedman 1999; Esquerra et al. 2000; Goldbeck, Paquette and Kliger 2001; Henry et al. 2002; Balakrishnan et al. 2004; Goldbeck et al. 2004; Samuni et al. 2004; Viappiani et al. 2004; Samuni et al. 2006); and (iii) the existence of and function of multiple T and R structures (Silva, Rogers and Arnone 1992; Mueser, Rogers and Arnone 2000; Kavanaugh, Rogers and Arnone 2005; Safo and Abraham 2005).

A major part of the problem in addressing these molecular-level issues is that for a given set of solution conditions the equilibrium populations are dominated by either the deoxy T-state or liganded R-state structures. The use of time-resolved techniques can be used to follow the evolution of structure; however, rapid mix ligand-binding experiments preclude observation of the initial intermediates due to the mixing times, and photolysis experiments are complicated by geminate recombination (Alpert et al. 1979; Duddell, Morris and Richards 1979; Friedman and Lyons 1980), which creates evolving populations containing different partially liganded intermediates.

The use of sol-gel encapsulation overcomes these difficulties. It has been shown that encapsulation of Hb in tetramethoxysilane (TMOS)-derived solgels greatly slows down the transitions between the T and R state structures (Shibayama and Saigo 1995; Bettati and Mozzarelli 1997; Das et al. 1999; Khan et al. 2000; Abbruzzetti et al. 2001; Bruno et al. 2001; Shibayama and Saigo 2001). In effect, the sol-gel environment allows the equivalent of a solution rapid mix experiment (done slowly) but with the advantage that mixing times (i.e. diffusion of substrates, ligands, effectors, osmolytes, etc.) typically occur on a timescale that is faster than those associated with many proteins' conformational relaxations. Upon the addition of CO or $O_2$ to an encapsulated deoxy human adult haemoglobin (HbA) sample, there is rapid binding of the ligand to the heme. The result is the formation of a ligand-saturated T-state species whose functional and spectroscopic properties can be followed as the protein population slowly evolves towards the equilibrium liganded R-state distribution of conformations. Similarly, by starting with an encapsulated oxy R-state population and then adding dithionite to deoxygenate the sample, it is possible to trap deoxy R-state species whose properties can be followed as it relaxes toward the deoxy T-state endpoint population (Das et al. 1999; Schiro and Cupane 2007).

Much of the encapsulated Hb work to date has focused on the properties and the evolution of deoxy samples that were converted into liganded (CO or $O_2$) T-state samples (Das et al. 1999; Khan et al. 2000; Bruno et al. 2001; Shibayama 2001; Samuni et al. 2003, 2004, 2006). These studies have revealed at least two liganded T-state populations: a low-affinity T-state and high-affinity T-state. The rate at which these populations are accessed upon ligand binding is a strong function of both the encapsulation protocol (including the addition of glycerol into the buffer bathing the thin sol-gel layer) and the presence or absence of allosteric effectors. The limited numbers of R-state experiments were based on starting with an encapsulated $O_2$Hb sample that is then exposed to dithionite in order to create a nonequilibrium population of R-state deoxy molecules. In an early study using resonance Raman (Das et al. 1999), it was shown that the relaxation pattern within the sol-gel mimicked the tertiary and quaternary relaxations seen in nanosecond time-resolved Raman studies (Scott and Friedman 1984) but with the timescale greatly expanded in a temperature-dependent manner. Near 0°C, the relaxation in the sol-gel occurs over days; whereas, progressing with temperatures above ambient, the relaxation times are observed to decrease to hours and then minutes (near 80°C). In a more recent study (Schiro and Cupane 2007), the R to T relaxation was followed using both the R/T-sensitive heme associated Band III in the near-IR absorption spectrum and the 291 nm R/T-sensitive near UV tryptophan absorption band. In that study, it was shown that the heme environment undergoes three distinct viscosity-dependent relaxations during the R to T relaxation and the R-T-sensitive Trpβ93 interactions evolve in parallel with one of these three relaxations.

In the present work, these initial type studies are extended for both T- and R-state samples. For the deoxy T-state plus ligand experiments, higher amounts of glycerol are added to the bathing buffer (75% by volume). For the R-state relaxation studies we build on the $O_2$Hb→deoxy protocol by including cyanometHb→deoxy as well (thus eliminating the complications associated with autooxidation for samples of encapsulated $O_2$Hb that require extending ageing). For both T- and R-state experiments the objectives include: (i) characterising the tertiary structures accessed upon ligand binding or ligand dissociation; (ii) probing the hierarchical nature of the different conformational relaxations; (iii) characterising the energy landscape of both the ligand-saturated and ligand-free (deoxy) derivatives within the T and R quaternary states; and (iv) exposing evidence for solvent slaving of functionally important relaxations within and between the T and R quaternary states. The intermediates are characterised functionally using CO recombination traces and structurally using the resonance Raman frequency of the iron-proximal histidine stretching mode ($v$(Fe-His)), which is highly responsive to functionally significant tertiary and quaternary conformational changes at the heme (Friedman et al. 1982; Rousseau and Friedman 1988; Peterson and Friedman 1998).

# Materials and Methods

All chemicals, including inositol hexaphosphate (IHP) and tetramethylorthosilicate (TMOS), were commercially obtained at the highest purity available. The potent allosteric effector 2,[4-([(3,5-dichlorophenyl)amino]carbonyl amino)phenoxy]-2-methylpropanoic acid (L35) (Lalezari et al. 1990) was obtained as a gift from Dr. I. Lalezari. Human haemoglobin (HbA) was purified as described previously (Doyle et al. 1992). Several mutant recombinant forms of HbA were used in this study. Purified and characterised Hb(Wβ37E) and Hb(Yα140F) were provided by Professor R. Noble.

Sol-gel-encapsulated Hbs were prepared as thin layers on the inner surface of 10 mm diameter NMR tubes as previously described (Samuni et al. 2002, 2003, 2004). After a suitable ageing period the tubes were filled with an excess of bathing buffers containing different concentrations of glycerol. Except where noted, the ageing occurs with a bathing buffer containing 25% (v/v) glycerol. Liganded T-state samples were prepared by starting with encapsulated deoxy HbA and then adding a CO-saturated bathing buffer. Deoxy R-state samples were prepared starting with either encapsulated $O_2$HbA or encapsulated cyanometHbA, to which is added a dithionite-containing bathing buffer under a nitrogen atmosphere. The dithionite-containing buffer is added after repeated buffer changes to remove excess cyanide or oxygen. The redox and ligation status of these samples were established using the visible and near IR absorption spectra. In several instances the global status of the encapsulated samples was confirmed using UV CD measurements (200–450 nm). In all cases the UV CD measurements were consistent with secondary structures for the encapsulated samples being comparable to that of the solution phase.

The Soret-enhanced resonance Raman spectra of deoxy heme species derived either from deoxy Hb samples or from the 7-ns photoproducts of CO or $O_2$ derivatives of Hb were generated as previously described (Samuni et al. 2002, 2003, 2004).

The kinetic traces associated with the recombination of CO to Hb subsequent to photodissociation using a 7-ns photolysis pulse were generated as previously described (Samuni et al. 2002, 2003, 2004; Dantsker et al. 2005a; Dantsker et al. 2005b). Measurements were made at 3.5°C.

Two basic protocols are used in this study. The first protocol utilises encapsulated deoxyHb as the starting sample. The addition of CO is then used to initiate the evolution of the liganded T-state population towards the liganded R state. In the second protocol, the starting samples are encapsulated forms of either $O_2$HbA or cyanometHbA. The addition of dithionite-containing buffer is used to initiate the evolution of the deliganded R-state populations towards the deoxy T state. In the second protocol, CO was added after different time intervals to provide a functional probe of the evolving populations via CO recombination kinetics as well as the time-resolved Raman spectrum from the 7 ns photoproduct. Thus an evolving as well as a stable liganded population derived from either protocol could be compared with respect to functional and spectroscopic properties. In the following text, as in our earlier work, brackets are used to signify the Hb derivative when initially encapsulated and the plus and minus signs following to the right of the brackets refer to the additions or removals made to the samples after the initial preparation. For example, [deoxyHbA]75%glycerol+CO indicates that CO is added to an encapsulated deoxyHbA sample bathed in a buffer containing 75% (v/v) of glycerol. All measurements, unless otherwise noted, were conducted on samples bathed in BisTris buffer at pH 6.5 with the stated amount of added glycerol. Samples were stored in a standard laboratory cold room at ~4°C.

## Results

### *Tertiary Relaxation within the T state: Ligand Binding-induced Evolution of the Resonance Raman Spectrum*

Figure 1 shows the ligand (CO)-binding induced evolution of the Soret-enhanced resonance Raman spectrum of [deoxyHbA]75%+CO (i.e., sol-gel encapsulated deoxy HbA bathed in an excess of buffer containing 75% glyc-

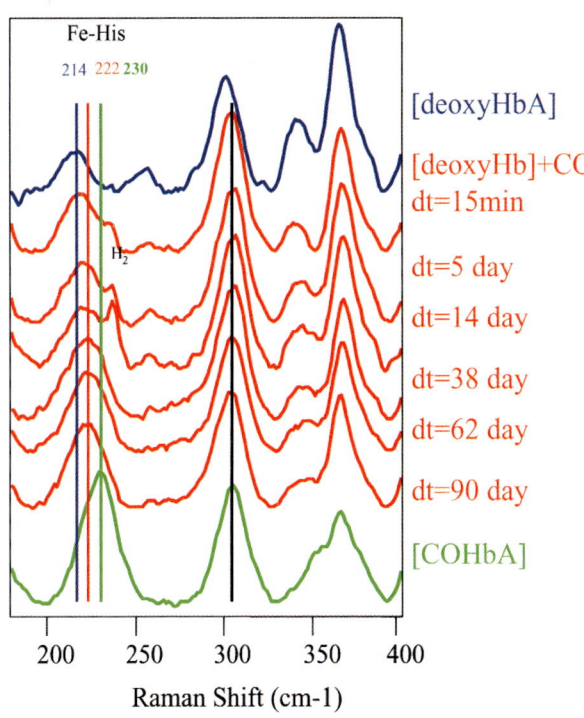

FIG. 1. Time evolution of the low-frequency Soret enhanced resonance Raman spectrum of sol-gel encapsulated deoxy HbA+IHP/L35 bathed in pH 6.5 buffer containing 75% v/v glycerol subsequent to the addition of CO. The top spectrum is of the starting deoxy sample and the bottom spectrum is of a similarly prepared sample of CO-HbA. The designation dt indicates number of days subsequent to the addition of CO. The vertical lines indicate the peak frequencies for the $\nu$(Fe-His) for the starting deoxy species (blue), the quasi-stable T-state COHbA species (red) and R-state COHbA (green). The Raman spectra of the CO derivatives in this and all subsequent figures are of the 7-ns photoproduct

erol to which CO is added). This sample also contains the allosteric effectors IHP and L35. The initial spectrum of [deoxyHbA]75% prior to adding CO is identical to that of solution-phase deoxyHbA samples. The subsequent spectra are of the five-coordinate ferrous "deoxy-like" photoproduct of the CO derivative generated within the 7-ns excitation pulse. It can be seen that the frequency of $v$(Fe-His), the iron-proximal histidine stretching mode, increases over a period of several days from the initial 214 cm$^{-1}$ to a relatively stable value of 222 cm$^{-1}$. The sample showed no significant additional shifting over the subsequent two-year period, when stored in the cold room. In the absence of either both effectors or just L35, a similar starting spectrum and rate of ligation-induced evolution are seen with the exception that the sample containing both effectors levelled off at 222 cm$^{-1}$, the other samples levelled off at ~225 cm$^{-1}$. In all three cases the stable spectrum (over the two-year period) is attained within 14 days of adding the CO. Also shown in the figure is the anticipated endpoint photoproduct spectrum derived from the equilibrium population of sol-gel-encapsulated COHbA generated under similar conditions. When the bathing buffer has 50% glycerol or less (Samuni et al. 2003, 2004, 2006), the evolution of the spectra exhibits similar patterns, but the rates become much faster such that the 222 cm$^{-1}$ peak position for $v$(Fe-His) is now observed within minutes of ligand binding. In addition, the relaxation rates for the samples bathed with 75% glycerol are greatly slowed if the deoxy sample is pre-treated with the 75% glycerol-containing buffer for several days prior to having the CO-saturated buffer introduced (vis à vis pretreating with buffer containing only 25% glycerol).

Figure 2 shows a similar evolution for the Raman spectrum of Hb(Nβ102A), a noncooperative HbA mutant that remains in the T state even when fully liganded (Samuni et al. 2006; Kwiatkowski et al. 2007). The figure shows the evolution for [deoxyHb(Nβ102A)]75%+CO. It can be seen that the evolution of the peak position of $v$(Fe-His) is similar to what is observed for wt HbA. However, unlike the case shown in Fig. 1 for HbA where for the time scale shown, the photoproduct spectrum of the liganded R structure ($v$(Fe-His) frequency of 230 cm$^{-1}$) is not attained, here the evolving spectrum ends up much closer to the photoproduct spectrum of the corresponding stable encapsulated CO-saturated T-state derivative i.e. [COHb(βN102A)+IHP].

## Tertiary Relaxation within the R State: Deoxygenation-induced Evolution of the Resonance Raman Spectrum

Figure 3 shows the evolution of the Raman spectrum of [oxyHbA]75%glycerol when the sample first undergoes dithionite-initiated deoxygenation and then after a period of 8 months is religanded with CO. The bottom spectrum is the 7-ns photoproduct spectrum derived from the initial [oxyHbA]75% glycerol samples. The next five spectra (starting at the bottom and moving up) show the slow evolution of $v$(Fe-His) when the sample undergoes deoxygenation. Within

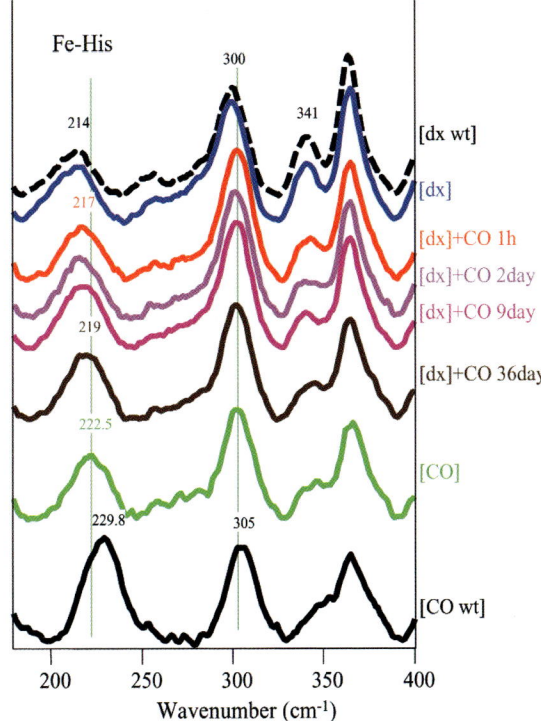

FIG. 2. Time evolution of the low-frequency Soret enhanced resonance Raman spectrum of sol-gel encapsulated deoxy Hb(β(N102A) +IHP bathed in 75% glycerol buffer subsequent to addition of CO, which is designated in the figure as [dx]+CO. The green spectrum is of the corresponding encapsulated CO derivative+IHP and the black trace is from encapsulated COHbA

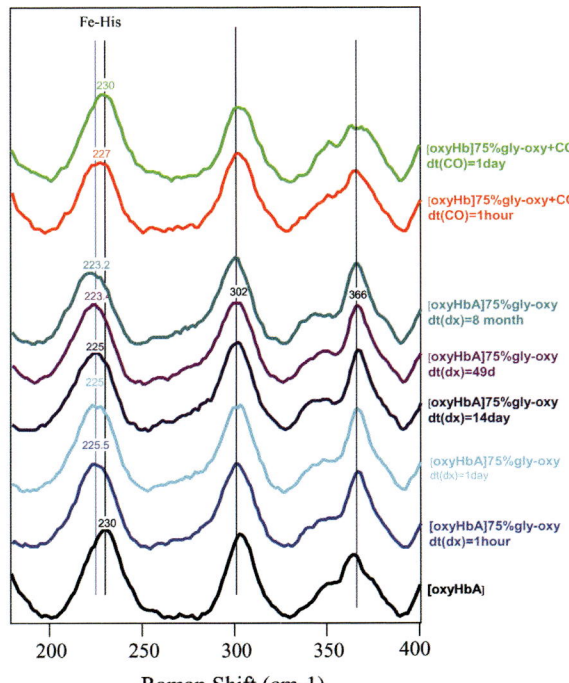

FIG. 3. The resonance Raman spectra of an encapsulated oxyHbA sample bathed in 75% glycerol as a function of time subsequent to deoxygenation and then 8 months after the addition of CO. The bottom spectrum is from the initial [oxyHbA] sample. The temporal progression is from the bottom to the top

an hour the frequency drops to 225 cm$^{-1}$ and only after a period of weeks does it relax to 223 cm$^{-1}$, where it remains stable for months. This frequency is in the range of what is observed for Hbs that can be stabilised in the R structure even as the fully deoxy derivative (Ondrias et al. 1982). The top two spectra in Fig. 3 show that in marked contrast to the slow T-state relaxations observed for the [deoxyHbA]+CO and [deoxyHb($\beta$N102A)]+CO samples (Figs. 1 and 2), in this case the exposure of the 8-month quasi-stable deoxy species to CO, i.e., [oxyHbA]75%glycerol-oxy→8 months+CO, results in a very rapid evolution of the photoproduct spectrum to the endpoint value associated with [COHbA]. It can be seen that the addition of CO to the partially relaxed deoxy species derived from the deoxygenated sample results in the frequency of $v$(Fe-His) rapidly (within 1 day) evolving from 223 cm$^{-1}$ to the endpoint value of 230 cm$^{-1}$ (vis à vis the weeks to months required for the ligation-induced evolution of the encapsulated T-state deoxy samples).

## *Tertiary Relaxation within the R State: Reduction of Cyanomet Hb-induced Evolution of the Resonance Raman Spectrum*

The use of oxyHbA as a starting material for relaxation studies is often complicated by the propensity of these samples to autoxidise. Although sol-gel encapsulation appears to reduce the rate of autoxidation, the need to age samples for varying lengths of time prior to ligand removal makes the issue of autoxidation a concern as formation of an aquomet heme introduces a much weaker ligand which may not create the same initial local tertiary structure as either oxygen or CO. For the results presented here we did not experience any autooxidation. Nonetheless, we also present a set of experiments where we use the low-spin ferric heme cyanomet derivative of HbA as our starting R-state liganded species. Figure 4 shows the evolution of the Raman spectrum subsequent to the addition of dithionite to the [cyanometHbA]75% glycerol sample. The top spectrum is that of [deoxyHbA] whereas the subsequent progression (top to bottom) shows the relaxation of $v$(Fe-His) starting within minutes of addition of dithionite out to 41 days subsequent to the reduction of the cyanomet heme sites. As with the oxy samples, the conversion to a derivative with deoxy hemes results in relaxation of $v$(Fe-His) to lower frequencies. In this case the peak frequency for $v$(Fe-His) appears to level off at ~220 cm$^{-1}$, which is a slightly lower frequency than seen in Fig. 3 for the oxy turned deoxy sample but still in the range of R-state deoxyHb derivatives. This difference in the levelling off value is attributable to a slightly different encapsulation protocol which allows for a faster rate and greater extent of relaxation over a given time window for the cyanomet sample. The bottom spectra show the evolution of the 7-ns photoproduct Raman spectrum when, after 41 days as a deoxy-like species, the [cyanometHbA]+dithionite sample is exposed to CO. As with the previous oxy turned deoxy encapsulated sample, the addition of CO results in

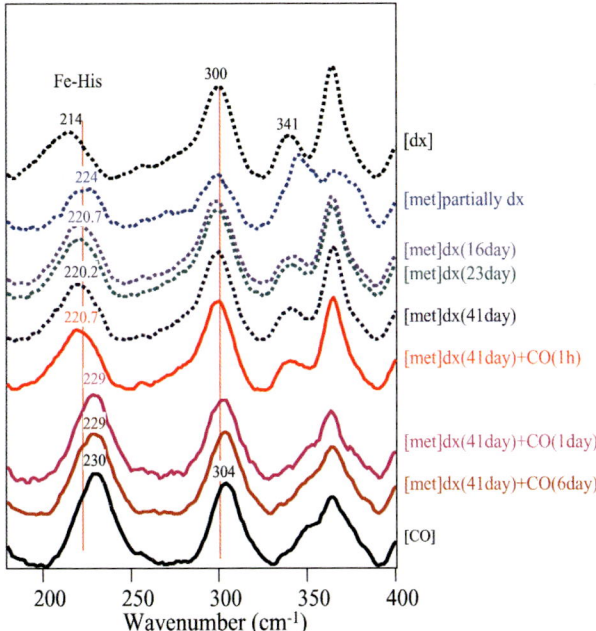

FIG. 4. The resonance Raman spectra of an encapsulated cyanomet HbA sample (designated as [met]) bathed in 75% glycerol as a function of time subsequent to reduction to a deoxy derivative (dotted spectra, designated as [met]dx) followed by conversion to the CO derivative after 41 days as a deoxy species (solid red lines). The number in parenthesis after the CO indicates the number of days after addition of CO. The black curve is from [COHbA]

a relatively rapid formation of a COHbA species whose 7-ns photoproduct yields a frequency for the $v$(Fe-His) that is essentially identical to that of the encapsulated COHbA under similar conditions.

When the bathing buffer for the cyanomet turned deoxy sample contains 25% glycerol instead of 75% glycerol as for the previously discussed sample, the evolution of the $v$(Fe-His) band is more extensive over the same time period. It can be seen in Fig. 5 that over the 41-day period the frequency of the $v$(Fe-His) band initially relaxes to values associated with R-state deoxy Hb species but then continues to relax to ~217 cm$^{-1}$ where the spectrum remains stable over the extensive timescale of the measurement. Thus when the bathing buffer contains 75% glycerol, the relaxation greatly slows after accessing the deoxy R-state value; whereas, for the sample bathed with 25% glycerol, the relaxation slows only after the $v$(Fe-His) attains a frequency associated with high-affinity deoxy T-state Hbs such as Hb($\beta$W37E) (Peterson and Friedman 1998; Samuni et al. 2006) and Hb($\alpha$F140Y) (Juszczak, Samuni and Friedman 2005). Upon addition of CO after 41 days, the sample rapidly reverts to a COHb sample with a photoproduct yielding a frequency for $v$(Fe-His) at ~225 cm$^{-1}$ vis à vis the 230 cm$^{-1}$ seen for the sample bathed in 75% glycerol. This photoproduct frequency is similar to that which is observed for the liganded derivatives of high-affinity T-state species (Peterson and Friedman 1998; Samuni et al. 2004, 2006).

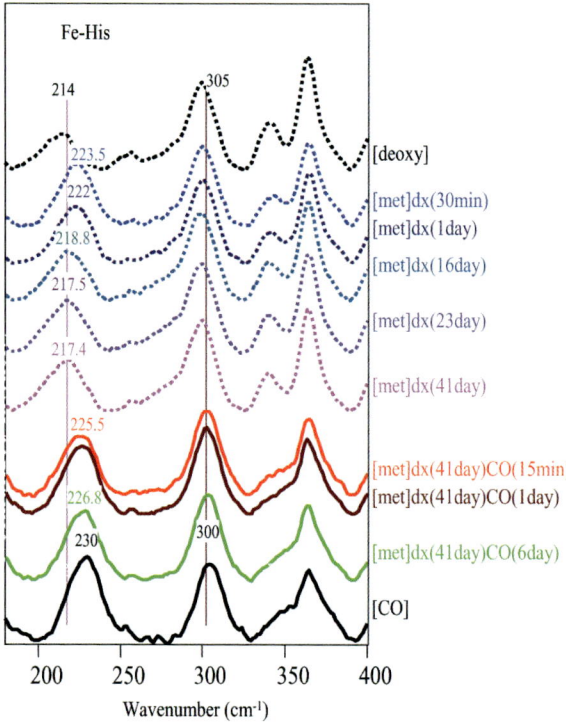

FIG. 5. The resonance Raman spectra of an encapsulated cyanomet HbA sample (designated as [met]) bathed in 25% glycerol as a function of time subsequent to reduction to a deoxy derivative (dotted spectra, designated as [met]dx) followed by conversion to the CO derivative after 41 days as a deoxy species (solid red, brown and green lines)

## CO Recombination Kinetics as a Function of Tertiary Structure

Figure 6 shows the evolution over a period of 6 months of traces for CO recombination kinetics from samples of encapsulated deoxy HbA samples bathed in 75% glycerol that are exposed to CO after allowing the [deoxyHb] gels to age for several days while being bathed in an excess of nitrogen-purged 75% glycerol buffer (same series as discussed for Fig. 1). This ageing protocol results in a sol-gel that is especially effective in slowing down relaxations. The three deoxy samples differ in that one sample has no added allosteric effectors, one sample has just IHP and the other has both IHP and L35. The bottom trace is derived from an encapsulated sample of COHbA bathed in 75% glycerol. The traces shown resemble what has been previously reported for similarly prepared samples that are bathed either in aqueous buffer or with 25% or 50% added glycerol (Samuni et al. 2004). The major difference is not in the appearance of the traces but in the rate of evolution. As in the earlier studies, the presence of added effectors slows the evolution with the combination of IHP and L35 having the largest effect. Also shown in the figure in parentheses are the

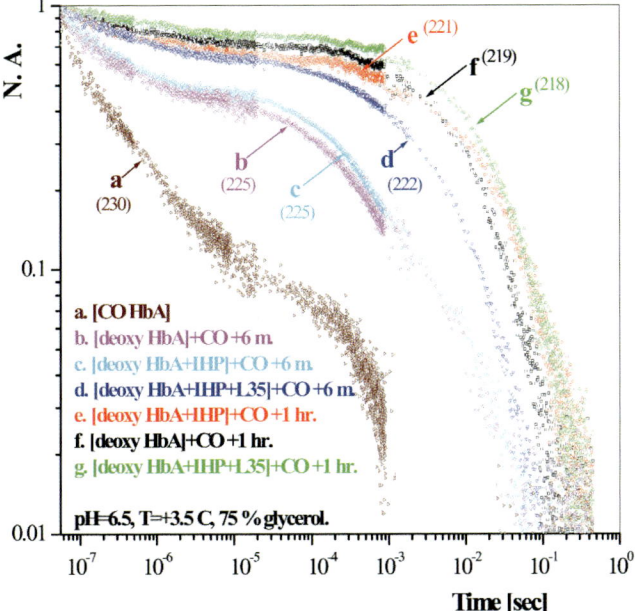

FIG. 6. Kinetic traces for the CO recombination from encapsulated samples bathed in 75% glycerol. Trace a is from R state [COHbA]. Traces b, c, d are from samples of [deoxyHbA] six months after addition of CO and traces e, f, g are from the same samples one hour after being converted to the CO derivative. The three different [deoxy HbA]+CO samples were prepared: without added effectors (traces b and f), with added IHP (traces c and e) and with added IHP and L35 (traces d and g). The ν(Fe-His) frequencies from the photoproducts of the different samples giving rise to the shown traces are given in the colour-coded parentheses. Traces comparable to those shown at six months were observed to develop within less than a month and exhibited minimal subsequent change to at least two years

frequencies of ν(Fe-His) for the 7-ns photoproduct derived from the samples that yield the corresponding kinetic trace. The frequencies start at 218 cm$^{-1}$ for the initial CO saturated [deoxyHbA]+CO sample and end up at 225 cm$^{-1}$ for the effector-free and IHP-containing samples 6 months after addition of CO. The [deoxyHbA]+CO sample containing both effectors ends up at 222 cm$^{-1}$ after the same 6-month period and, as can be seen in the figure, is associated with a kinetic trace that has not evolved as far as those traces from the two samples that yield the 225 cm$^{-1}$ frequency. The 222 cm$^{-1}$ frequency is characteristic of the 7-ns photoproduct of low-affinity liganded T-state species (Friedman 1985; Rousseau and Friedman 1988; Friedman 1994; Samuni et al. 2003, 2006). As noted above, the 225 cm$^{-1}$ frequency is associated with the photoproduct of Hbs accessing the liganded form of the high-affinity T state. The photoproduct Raman frequency for the equilibrium population associated with the corresponding R state [COHbA] sample is 230 cm$^{-1}$, as seen in solution or similar samples with comparable or lower amounts of added glycerol. The final traces shown for all three [deoxyHbA]+CO samples show minimal additional change of the ensuing two-year period, during which the spectra and kinetics were monitored.

Figure 7 shows traces for the CO recombination for encapsulated samples that have the [cyanometHb]→[deoxyHb]→[COHb] history. These are the same samples whose Raman spectra are shown in Figs. 4 and 5. Thus in both cases the CO is added after the samples evolved as a deoxy Hb over a 41-day period. Traces e and f from the sample bathed in 75% glycerol show that the traces generated within 1 day and 4 months of exposure to CO (trace e) are essentially identical. Both of these traces are almost identical to that derived from fully equilibrated R state [COHbA] bathed in 75% glycerol, as seen in trace a of Fig. 6. In contrast, traces a–d show clear evolution of the kinetic trace for a similarly treated sample but differing in that it was bathed in 25% glycerol instead of 75% glycerol. In this case, the progression from day 1 to 3 months following CO saturation starts at a point where the trace resembles a [deoxyHbA]25%

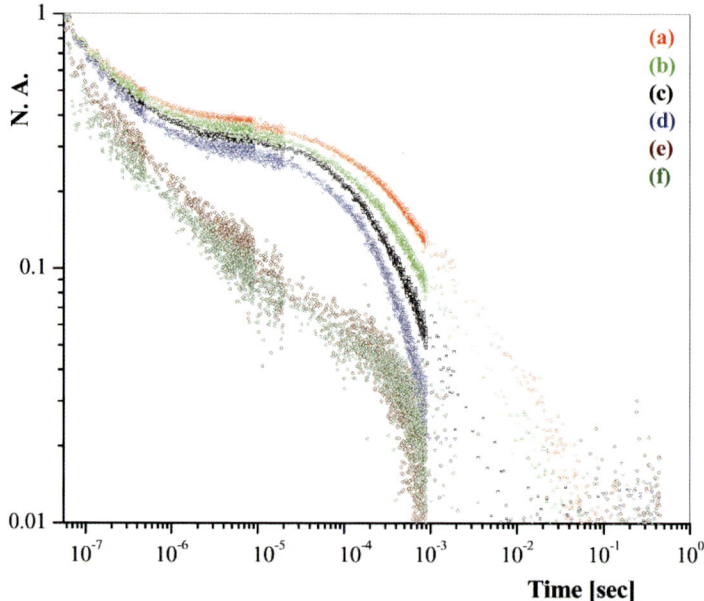

FIG. 7. Kinetic traces for the CO recombination from two encapsulated COHbA samples that were initially prepared as encapsulated cyanometHbA then converted to a deoxy derivative which was allowed to relax for 41 days after which the samples were converted to CO-saturated derivatives. One sample was bathed in 25% glycerol (traces a–d) and the other was bathed in 75% glycerol (traces e and f). Traces a, b, c and d are from the 25% glycerol bathed sample at 1, 2, 7 and 90 days, respectively after being converted to a CO derivative. Traces e and f are from the 75% glycerol bathed sample 1 and 120 days respectively after conversion to the CO derivative

glycerol+CO sample that has already undergone considerable relaxation towards the high T trace. This point is made more clear in Fig. 8 which combines a series of traces (red/orange) derived from a sample that starts as [deoxyHbA] and then converted to CO with a series of traces (blue/purple) from a sample that started as cyanomet and then converted to a deoxy sample and then after 41 days converted to a CO saturated sample. It is clear from the figure that the last of the [deoxy]+CO traces (day 52) are very close to the trace from the starting (day 0) sample that has undergone the [cyanometHbA]+dithionite→41days+CO history. Comparing this series of traces to those reported in earlier works, it is apparent that the series shown in Fig. 8 covers the full range of observed kinetic traces seen for CO derivatives of encapsulated HbA under conditions where the sample is bathed in low to moderate viscosity buffers.

The samples used for Fig. 7 were allowed to age for 41 days as a

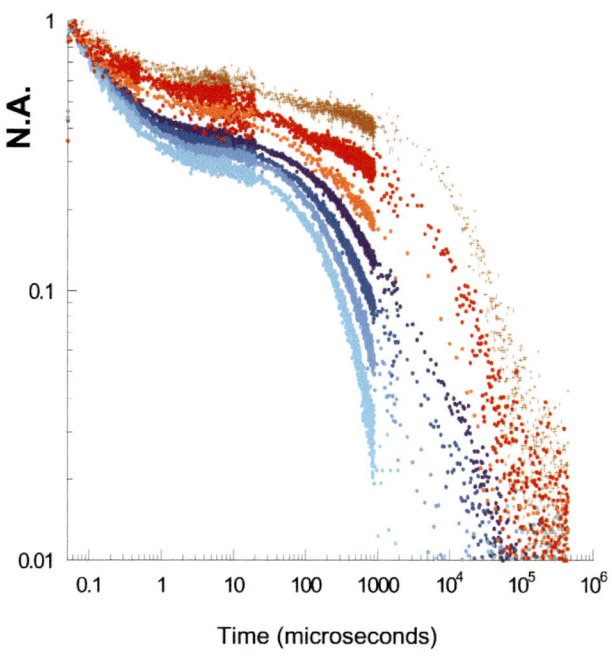

Time (microseconds)

FIG. 8. The time evolution of kinetic traces for the CO recombination from samples of [deoxyHbA]+CO derived from two different histories. The top three traces (orange/red) show the progression from day 1, day 3 and day 52 subsequent to the addition of CO to a [deoxyHbA] sample bathed in 25% glycerol buffer. The blue traces show a progression for an encapsulated sample bathed in 25% glycerol initially prepared as cyanometHbA and after ageing for several days converted to a deoxy derivative and after 41 days finally converted to the CO derivative. The progression from top to bottom is sample on day 1, day 2, day 7 and day 90 with respect to when it was converted to a CO derivative. The bottom trace is very close to traces from [COHbA] under similar conditions

cyanomet→deoxy sample prior to exposure to CO. If a sample bathed in 25% glycerol is converted to a deoxy sample (from either an oxy or cyanomet sample) and then converted to a CO sample within a few hours to days rather than weeks, the kinetic traces now resemble the full endpoint trace associated with R state [COHbA] prepared under similar conditions (not shown). For such samples the frequencies of $v$(Fe-His) for the cyanomet→deoxy sample and for the photoproduct of cyanomet→deoxy→CO are ~222 cm$^{-1}$ and 230 cm$^{-1}$ respectively. These frequencies are characteristic of deoxy R state and liganded R (specifically R$_2$) state respectively.

# Discussion

## *Tertiary Structure of T State Deoxy Hbs Under Equilibrium Conditions*

The majority of human Hbs that bind oxygen cooperatively adopt a solution phase conformation when deoxygenated that yields a $v$(Fe-His) frequency at ~214–215 cm$^{-1}$. This frequency is seen for deoxy HbA both in the presence and absence of allosteric effectors whether in solution or in sol-gel matrices. This frequency can be assigned to the classic deoxy T-state structure observed by X-ray crystallographic studies. This structure appears to be the endpoint T structure in the continuum that stretches from this lowest-affinity T structure all the way to the highest-affinity R structures. The addition of effectors has no major impact on the tertiary conformation of the deoxy T state species although they can be expected to cause alterations in local hydration at their respective binding sites. The major exceptions to this pattern occur for mutations that alter the hinge region of $\alpha_1\beta_2$ interface as in the case of the $\beta$W37E and $\alpha$Y140F mutations. For the deoxy derivatives of these mutants the frequency for $v$(Fe-His) is now higher at ~218 cm$^{-1}$ (Peterson and Friedman 1998; Juszczak, Samuni and Friedman 2005; Samuni et al. 2006). Crystallographic studies (Kavanaugh et al. 1998; Kavanaugh, Rogers and Arnone 2005) indicate that these mutations cause a specific conformational change distinct from that associated with the T to R transition. This change involves a weakening of the interactions that stabilise what has been referred to as the $\beta$37 cluster of residues that form a hydrogen-bonding network that is responsible for maintaining the endpoint deoxy T-state structure. Loosening of this hydrogen-bonding network results in a decrease in proximal strain as reflected in both the frequency increase in $v$(Fe-His) (Peterson and Friedman 1998a) and actual bond lengths seen in high-resolution crystallographic data (Kavanaugh et al. 1998). The conformational changes associated with these mutations increase ligand affinity but do not noticeably perturb the R-T defining switch region of the $\alpha_1\beta_2$ interface and as a consequence have led to the designation of these conformations as high-T or high-affinity T-state species (Kavanaugh et al. 2005).

## Tertiary Structure of T-state Liganded Hbs under Equilibrium Conditions

The frequency for $v$(Fe-His) from the 7-ns photoproduct of CO derivatives of Hbs that remain in the quaternary T state when liganded range from ~221 to 225 cm$^{-1}$. The low end of this range is associated with T-state Hbs that have the lowest ligand affinities (e.g., symmetric Fe–Zn hybrids of HbA in the presence of effectors (Samuni et al. 2003), HbA($\beta$N102A) in the presence of effectors) (Samuni et al. 2006), whereas the high-end values are derived from Hbs that are considered high-affinity T-state species (e.g. Hb($\beta$W37E) with and without $\alpha$99-$\alpha$99 crosslinking (indicating that it is not a dimer formation effect) in the presence or absence of allosteric effectors) (Peterson and Friedman 1998; Samuni et al. 2006).

## Tertiary Structure of T-state Liganded Hbs under Non-equilibrium Conditions

It has been shown that sol-gel encapsulated deoxy HbA binds oxygen non-cooperatively with either low or moderately low affinity depending on the nature of the gel and the presence or absence of effectors respectively (Bruno et al. 2001; Shibayama 2001). These and other functional and spectroscopic studies are consistent with the sol-gel inhibiting the standard ligand binding induced T to R transition observed in solution. The bottom line is that there is a variable time window extending from many hours to days depending on the gel and solution conditions where the encapsulated population of [deoxyHbA]+ligand remains in the T state. Two $v$(Fe-His) frequencies are observed for the 7-ns photoproduct of these quasi-stable liganded T-state populations (with CO as the ligand). For those populations that are associated with the low-affinity T-state conditions or samples, the frequency is ~222 cm$^{-1}$; whereas, for the high-affinity T-state population the frequency is ~225 cm$^{-1}$ (Samuni et al. 2004).

## Relaxation Properties of T-state Populations: The Influence of Mutations and Allosteric Effectors

If we combine the equilibrium T-state deoxy Hb data with that from the 7-ns photoproducts from T-state liganded Hbs, a progression in the $v$(Fe-His) frequency is apparent. The progression is as follows: low-affinity deoxy T Hb (214 cm$^{-1}$)→high-affinity deoxy T (218 cm$^{-1}$)→low-affinity liganded T (222 cm$^{-1}$)→high-affinity liganded T (225 cm$^{-1}$). Based on the X-ray crystallography (Kavanaugh et al. 1998) and UV resonance Raman data (Samuni et al. 2003, 2004; Juszczak, Samuni and Friedman 2005; Samuni et al. 2006), the progression in the decrease in proximal strain reflected in the frequency of the

$v$(Fe-His) band follows a progressive disruption of the hydrogen-bonding network of the T state $\beta37$ cluster in the hinge region of the $\alpha_1\beta_2$ interface and the progressive bending of the $\alpha_1\beta_1$ ($\alpha_2\beta_2$) dimer (Kavanaugh et al. 2005). For sol-gel-encapsulated deoxy HbA bathed in buffer containing 50% or less glycerol by volume, saturation with CO induces a rapid relaxation to the low-affinity liganded T-state conformation (with a $v$(Fe-His) for the photoproduct at ~222 $cm^{-1}$). In the absence of added effectors, the sample rapidly progresses to the high-affinity liganded T-state population followed by a much slower relaxation to the liganded R state. The addition of either L35 or bezafibrate slows this second relaxation phase considerably.

Under these conditions (glycerol content 50% or less), intermediate frequencies for $v$(Fe-His) between the initial low-affinity deoxy T-state value and the low-affinity liganded T-state value are not observed; however, as shown in the present study, higher levels of glycerol in the bathing buffer slow this initial relaxation sufficiently to observe intermediate frequencies corresponding to the high-affinity deoxy T species (~218 $cm^{-1}$). In contrast, mutations such as the $\beta$W37E that destabilise the hydrogen-bonding network of the T-state $\beta37$ cluster in the hinge region of the $\alpha_1\beta_2$ interface allow for a relatively rapid ligation-induced transition from the high-affinity deoxy T state to the high-affinity liganded T state.

## Functional Properties of T-state Hbs: Ligand Recombination Traces

In parallel with the progressive increase in the frequency of the $v$(Fe-His) band for CO-exposed encapsulated deoxy T-state Hbs are the progressive changes in the CO recombination traces. Except when the bathing buffer is extremely viscous (e.g., 99–100% glycerol) (Samuni et al. 2007b), the kinetic traces for the various intermediates are very similar. The major difference among samples is the rate at which the traces (and spectra) evolve. The traces associated with the lowest photoproduct frequencies have a very low geminate yield and a solvent phase recombination that is characteristic of T-state populations in solution. With relaxation towards higher-affinity T-state forms, the samples begin to manifest a progressive increase in the geminate yield and a distribution of solvent phase rates (as revealed using maximum entropy analysis (Samuni et al. 2006)) that progressively coalesce to a rate attributed to the high-affinity liganded T-state population. This endpoint T-state kinetic trace is similar to that of the liganded R state but with a lower geminate yield and a slightly slower solvent phase. The intermediate traces that manifest distribution of rates cannot be fit with a sum of traces from T and R state species(Samuni et al. 2006).

## Tertiary Structure of R-state Hbs under Equilibrium Conditions

Deoxy Hbs that are stabilised in the R state (e.g. NEMdesArgHbA, HbKempsey, Root effect Hbs at high pH) manifest frequencies for $\nu$(Fe-His) that are in the range of 220–223 cm$^{-1}$ (Ondrias et al. 1982; Friedman 1985; Kitagawa 1988). The corresponding frequency from the 7–10-ns photoproduct of liganded R-state Hbs in the absence of allosteric effectors including phosphate is 230 cm$^{-1}$. The same R-state photoproduct frequency is seen for CO, $O_2$ and NO derivatives (Friedman et al. 1982, 1983; Friedman 1985, 1994). This frequency is also fully apparent on the picosecond timescale (Findsen et al. 1985; Scott, Friedman and Macdonald 1985; Nagatomo et al. 2005).

## Tertiary Structure of R-state Deoxy Hbs under Non-equilibrium Conditions

Nanosecond time-resolved Raman studies revealed that for COHbA in solution, the photodissociation-initiated relaxation from the initial 230$^{-1}$ photoproduct species to the ~220–222 cm$^{-1}$ deoxy R-state intermediate (prior to relaxation to the T state), occurs within ~1 μs (Scott and Friedman 1984). This relaxation is accelerated in the presence of allosteric effectors and can be dramatically slowed for certain mutants (e.g. HbYpsilanti) (Huang et al. 1999). This same relaxation was observed for encapsulated $O_2$HbA subjected to deoxygenation through the addition of dithionite (Das et al. 1999). In that case the relaxation exhibited a dramatic temperature dependence ranging from many days at near 0°C to minutes at elevated temperatures. That study as well as the present and other studies (Findsen, Friedman and Ondrias 1988; Schiro and Cupane 2007) imply that it is temperature-dependent change in the viscosity or hydration dynamics that is the likely cause of this large temperature dependence of the relaxation within the sol-gel. The present study shows that the 220–222 cm$^{-1}$ deoxy R state intermediate can be trapped and probed over an extended time period through the use of bathing buffers containing high concentrations of glycerol. Under these high viscosity conditions, less stable deoxy intermediates are also observed as reflected in the 225 cm$^{-1}$ deoxy species seen when encapsulated $O_2$HbA bathed with 75% glycerol buffer is initially exposed to dithionite.

## Ligand Binding-induced Relaxation Properties of R-state Populations

In the present study, it is observed that when encapsulated R-state deoxy intermediates are re-saturated with ligand (CO), the photoproduct spectrum almost immediately reverts back to the endpoint 230 cm$^{-1}$ R-state liganded species. This rapid relaxation occurs even under high glycerol (75%) conditions. This behaviour is in marked contrast to the very much slower relaxation associated with T-state HbA when encapsulated deoxy HbA is saturated with CO.

## Functional Properties of R-state Hbs: Ligand Recombination Traces

Unlike the different liganded T-state populations, the liganded R-state populations show only modest variation in the kinetic traces associated with CO recombination. The major variation appears to be in the geminate yield for different equilibrium liganded R-state species. In these cases the geminate yield largely scales both with the frequency of $\nu$(Fe-His) from the 7-ns photoproduct and the effector and protein dependent rate/extent of relaxation of tertiary structure over the timescale of the geminate phase (Scott and Friedman 1984; Friedman, 1985, 1994; Huang et al. 1999; Khan et al. 2001). The solvent phase is relatively invariant except for the more than ten-fold enhancement in rate when the bathing buffer inhibits the entry of water into the vacated distal heme-pocket, as is seen when the sample is bathed in neat glycerol (Samuni et al. 2007b).

## The Impact of Sol-gel Plus Bathing Buffer on Hb Properties

This and numerous previous studies (see for example Abbruzzetti et al. 2001; Schiro and Cupane 2007) show no indication that sol-gel encapsulation alters the equilibrium population of either deoxy or liganded Hb. The major observable impact of encapsulation is on those processes that are reflective of protein dynamics. The increase in the geminate yield for encapsulated COHbs and COMbs (Abbruzetti 2001; Abbruzzetti et al. 2001; Samuni et al. 2002, 2007a, 2007b) is readily explained in terms of damping of the large amplitude solvent motions (e.g., $\alpha$ relaxations) to which large amplitude protein fluctuations are slaved (Samuni et al. 2007a, 2007b). A recent study directly demonstrates that the sol-gel damps $\alpha$ relaxations (Schirò et al. 2008). The pattern of sol-gel-induced slowdown of both T- and R-state relaxations as well as the quaternary state transitions is consistent with two interconnected factors contributing to the slow-down mechanisms. Solvent-slaving models (Fenimore et al. 2004; Frauenfelder et al. 2006; Samuni et al. 2007a, 2007b) that account for how solvent modulates protein dynamics provide a useful framework for exploring the origin of sol-gel-induced slow downs of conformational relaxations. By limiting the amplitude of large volume-changing fluctuations ($\alpha$ relaxations) within the surrounding solvent, the constrained environment of the sol-gel encapsulated protein may substantially increase the number of elemental steps required for a protein to achieve a given conformational transition. Instead of a limited number of large amplitude elemental conformational "steps/fluctuations", there are now a much larger number of smaller amplitude steps that are required to complete the full relaxation process. Concomitantly, the decreased mobility of the hydration waters within this confined space will likely increase the activation energy associated with solvent-slaved motions, with the effect being greater for those elemental motions that require the larger solvent motions of the surface waters. From this assessment, it is proposed that encapsulation will slow relaxation both by increasing the number of steps in the entropic search over a given energy landscape and by increasing the activation energy barrier for transitions among the minima on the energy landscape. The increase in activation energy will depend on both the solvent motion to which the transition is slaved (see the Protein Dynamic States model in Samuni 2007b) and the composition of the sol-gel plus bathing buffer which modulates the dynamics of the hydration shell waters.

## The Hb Energy Landscape

Each quaternary state has a minimum of two energy surfaces: one associated with the five-coordinate ferrous heme derivative (deoxy) and the other with the fully liganded derivative. Ligand binding and photodissociation occur on much faster timescales than the tertiary relaxations being discussed in this work. As a consequence, the ligand-binding or dissociation-induced relaxations are initiated by an adiabatic (Born-Oppenheimer) type transition from one surface to the next without a major change in the initial distribution of tertiary structure conformational coordinates.

## The Proposed T-state Energy Landscape

The T-state deoxy surface is dominated by a narrow distribution of very deep minima. Ligand binding induces a vertical transition to the T-state liganded surface. This "instantly" accessed narrow region of the surface is now associated with very shallow minima separated by relatively low barriers. Each protein now begins a rapid entropic search over this landscape resulting in a progressive drift towards the deeper minima associated first with the low-affinity liganded T-state distribution followed by those of the high-affinity liganded T-state distribution. There is progressive slow down both in the relaxation times and in the separation of the relaxation timescales with increasing glycerol content of the bathing buffer, for the transitions going from the initial low-affinity deoxy T state→low-affinity liganded T state→high-affinity liganded T state. This behaviour is consistent with a hierarchy of solvent-slaved activation energy barriers separating the minima from these different conformational distributions. This picture implies that the dynamics of each successive relaxation is associated with either larger amplitude solvent motions or greater numbers of elemental solvent slaved steps. Consistent with this description is the observation using high-resolution X-ray crystallography (Kavanaugh et al. 1995, 1998) showing that progressive loosening of the hydrogen-bonding network of the $\beta37$ cluster in the hinge region of the $\alpha_1\beta_2$ interface (as occurs upon T-state ligand binding) results in a corresponding increase in the amplitude of the conformational fluctuations that are likely to lead up to and facilitate the T→R transition. It has been noted (Samuni et al. 2006) that the liganded high-affinity T (HT) population is likely to be very close to the transition state species for the ligand binding induced T to R transitions.

## The R-state Energy Landscape

The energy landscape of liganded R-state HbA in the absence of allosteric effectors is dominated by the deeper minima associated with those populations giving rise to the photoproduct Raman spectrum with a 230 cm$^{-1}$ $\nu$(Fe-His) band. X-ray crystallography shows that there are multiple liganded R-state structures that can be accessed under different solution conditions (Silva, Rogers and Arnone 1992; Mueser, Rogers and Arnone 2000; Safo and Abraham 2005). NMR results (Lukin et al. 2003) indicate that multiple conformationally distinct populations (e.g., R, R$_2$) can be observed in solution in the presence of phosphate. The conclusions of these structural studies are also supported in changes observed both in the photoproduct frequency of $\nu$(Fe-His) and in the geminate yield for liganded R-state populations as a function of added effectors. Thus, unlike the very narrow distribution of T-state deoxy HbA populations, there are a number of liganded R-state structures that are sufficiently close in stability that they can be easily accessed as a function of solution conditions.

In the present study, the solution conditions for the oxy and cyano derivatives appear to overwhelmingly favour the R$_2$ population. Consequently, it can be assumed that the adiabatic transition to the five-coordinate deoxy R-state surface results in a relatively narrow distribution of shallow minima. The glycerol-dependent rate at which this initial distribution relaxes to the quasi-stable deoxy R population (characterised by a 220–222 cm$^{-1}$ $\nu$(Fe-His)) is consistent with an entropic search over a large distribution of relatively shallow minima separated by relatively low solvent-dependent barriers. Similarly, the rapidity upon ligand binding with which the quasi-stable deoxy R-state populations revert to the endpoint liganded R-state population argues for an energy landscape for liganded R-state HbA where the minima are separated by very low activation energy barriers consistent with a requirement for only small amplitude solvent motions.

## *Quaternary State Transitions for Sol-gel Encapsulated Hbs*

The T to R and R to T transitions for encapsulated Hbs occur on timescales that greatly exceed those associated with the tertiary structure transitions within either the T or R state. The sol-gel-induced slow downs are not due to a steric restriction preventing conformational change based on a tight templating of the polymer around the protein. This conclusion follows from two observations. One observation is that certain encapsulation protocols can limit even the small-amplitude tertiary conformational changes in myoglobin (Samuni et al. 2002; Shibayama 2003). The other is that different deoxy derivatives of mutant and chemically modified human Hbs that start off with the same overall T quaternary state structure undergo the ligand-binding-induced transition to the R-state structure on dramatically different timescales within similar sol-gels (Samuni et al. 2007b). Thus it is not likely that it is the magnitude of the shape change *per se* that is the determinant of the efficacy with which the sol-gel limits conformational change. Recently, it was argued that the same three relaxation processes for the R to T relaxation observed in solution are also apparent for encapsulated HbA where the distinct rates of the three processes are slowed to a process-specific degree that is dependent on the glycerol content of the bathing buffer (Schiro and Cupane 2007).

There are several non-steric associated factors that are likely to contribute to the encapsulation-induced slowdown of quaternary transitions. We first consider the nature of the hydration waters in the confined space surrounding the encapsulated protein. The hydration waters can be broken down into three domains. There are the surface waters that have a relatively strong interaction with the surface residues and accessible cavities of the protein. Adjacent to the wall of the encapsulation cavity is a layer of waters that are highly immobilised due to their interaction with the fixed oxygens that are part of the Si-O polymeric network of the sol-gel. Finally there are the relatively mobile waters that occupy the volume between the protein surface water layer and the immobile water layer adjacent to the wall of the encapsulation cavity. The quaternary transitions in HbA involve a large rearrangement of protein-associated waters. In the limit where the intermediate layer of hydration waters is sufficiently thick, the sol-gel should have little or no impact on the solvent-slaved dynamics, including those associated with quaternary transitions. In the other limit where the preparative protocol (added glycerol, extensive drying, added external PEG) greatly reduces the intermediate hydration layer of mobile waters, one would anticipate a substantial increase in the enthalpic cost both of reorganising the initial distribution of waters and of large volume-changing fluctuations. Additionally, if, as a result of either a loss of mobile waters or an addition of osmolytes (e.g., small polyols), the hydrogen-bonding interaction increases among the waters comprising the hydration layers surrounding the protein within the confined volume of the sol-gel, then the activation energy for all protein dynamics slaved to the motions of these waters will be increased. The combination of forcing these sol-gel-mediated transitions to occur via a very large number of reduced amplitude conformational fluctuations (relative to what occurs in solution) and with a higher enthalpic penalty for all solvent-slaved motions (especially large amplitude motions) could account for the greatly extended timescales for both T→R and R→T transitions within the sol-gel. The starting protein conformational distribution upon encapsulation is likely to play a significant role as well. For example, when deoxy HbA in the presence of allosteric effectors such as IHP and L35 is encapsulated, the initial "trapped" distribution of waters is in part dictated by the reduced number of waters associated with the low-affinity deoxyT state+effectors. This T-state-determined distribution makes this encapsulated population more vulnerable to added osmotic effects (through added glycerol) than the situation with liganded R state, which is associated with a much higher number of loosely associated hydration waters. This effect probably accounts for why the tertiary relaxations within the R state are less responsive to added glycerol than those within the T state, as reflected in the higher amounts of added glycerol needed to substantially slow the R-state transitions compared to those in the T state.

## Implications for Allosteric Models

The present results strongly support the earlier (Scott and Friedman 1984; Friedman 1985; Friedman 1994) and more recent (Henry et al. 2002; Viappiani et al. 2004) modifications to the two-state model of Hb reactivity that now include functionally distinct tertiary conformations within both the R and T states. The clear evidence both for functionally distinct tertiary state conformations whose stability is influenced by both ligand binding and solution conditions and for solution-dependent rates at which these tertiary state conformations relax also support the claim (Scott and Friedman 1984; Friedman 1985) that the Agmon-Hopfield type mechanism (Agmon and Hopfield 1983) is operative in the modulation of Hb reactivity. In this mechanism, the rate of conformational relaxation results in a time-dependent increase in the barrier height controlling ligand binding. As a consequence the geminate yield progressively decreases as the structure relaxes. Thus the presence of allosteric effectors that enhance the submicrosecond rate at which high affinity R (HR) relaxes to the low affinity R (LR) upon photodissociation will cause a decrease in the geminate yield which will also be reflected in ligand off-rates. Under conditions where the liganded T state is stable, the functional properties will also be determined by the effector-dependent rate at which the liganded high- and low-affinity T (HT and LT, respectively) states relax to the deoxy LT population upon ligand dissociation. These relaxation effects within the R and T states are likely to be a major contribution to the effector-dependent oxygen affinity observed for both T and R states of HbA (Imai, Tsuneshige and Yonetani 2002; Tsuneshige, Park and Yonetani 2002).

# *References*

Abbruzetti, S., Viappiani, C., Bruno, S., and Mozzarelli, A. (2001) Enhanced geminate ligand rebinding upon photo-dissociation of silica gel-embedded myoglobin-CO. Chem. Phys. Lett. 346:430–436.

Abbruzzetti, S., Viappiani, C., Bruno, S., Bettati, S., Bonaccio, M., and Mozzarelli, A. (2001) Functional characterization of heme proteins encapsulated in wet nanoporous silica gels. J. Nanosci. Nanotechnol. 1:407–415.

Agmon, N., and Hopfield, J. J (1983) CO binding to heme proteins: A model for barrier height distributions and slow conformational changes. J. Chem. Phys. 79:2042–2053.

Alpert, B., El Mohsni, S., Lindqvist, L., and Tfibel, F. (1979) Transient effects in the nanosecond laser photolysis of carbonmonoxyhemoglobin: Cage recombination and spectral evolution of the protein. Chem. Phys. Lett. 64:11–16.

Balakrishnan, G., Case, M. A., Pevsner, A., Zhao, X., Tengroth, C., McLendon, G. L., and Spiro, T. G. (2004) Time-resolved absorption and UV resonance Raman spectra reveal stepwise formation of T quaternary contacts in the allosteric pathway of hemoglobin. J. Mol. Biol. 340:843–856.

Bettati, S., and Mozzarelli, A. (1997) T state hemoglobin binds oxygen noncooperatively with allosteric effects of protons, inositol hexaphosphate, and chloride. J. Biol. Chem. 272:32050–32055.

Bjorling, S. C., Goldbeck, R. A., Paquette, S. J., Milder, S. J., and Kliger, D. S. (1996) Allosteric intermediates in hemoglobin. 1. Nanosecond time-resolved circular dichroism spectroscopy. Biochemistry 35:8619–8627.

Bruno, S., Bonaccio, M., Bettati, S., Rivetti, C., Viappiani, C., Abbruzzetti, S., and Mozzarelli, A. (2001) High and low oxygen affinity conformations of T state hemoglobin. Protein Sci. 10:2401–2407.

Dantsker, D., Roche, C., Samuni, U., Blouin, G., Olson, J. S., and Friedman, J. M. (2005a) The position 68(E11) side chain in myoglobin regulates ligand capture, bond formation with heme iron, and internal movement into the xenon cavities. J. Biol. Chem. 280:38740–38755.

Dantsker, D., Samuni, U., Friedman, J. M., and Agmon, N. (2005b) A hierarchy of functionally important relaxations within myoglobin based on solvent effects, mutations and kinetic model. Biochim. Biophys. Acta 1749:234–251.

Das, T., Khan, I., Rousseau, D., and Friedman, J. (1999) Temperature dependent quaternary state relaxation in sol-gel encapsulated hemoglobin. Biospectroscopy 5:S64–S70.

Doyle, M. L, Lew, G., De Young, A., Kwiatkowski, L., Wierzba, A., Noble, R. W., and Ackers, G. K. (1992) Functional properties of human hemoglobins synthesized from recombinant mutant beta-globins. Biochemistry 31:8629–8639.

Duddell, D., Morris, R., and Richards, J. (1979) Ultrafast recombination in nanosecond laser photolysis of carbonylhemoglobin. J. Chem. Soc. Chem. Commun. 2:75–76.

Esquerra, R. M., Goldbeck, R. A., Reaney, S. H., Batchelder, A. M., Wen, Y., Lewis, J. W., and Kliger, D. S. (2000) Multiple geminate ligand recombinations in human hemoglobin. Biophys. J. 78:3227–3239.

Fenimore, P. W., Frauenfelder, H., McMahon, B. H., and Young, R. D. (2004) Bulk-solvent and hydration-shell fluctuations, similar to alpha- and beta-fluctuations in glasses, control protein motions and functions. Proc. Natl. Acad. Sci. U.S.A. 101:14408–14413.

Findsen, E., Friedman, J., Ondrias, M., and Simon, S. (1985) Picosecond time-resolved resonance Raman studies of hemoglobin: implications for reactivity. Science 229:661–665.

Findsen, E. W., Friedman, J. M., and Ondrias, M. R. (1988) Effect of solvent viscosity on the heme-pocket dynamics of photolyzed (carbonmonoxy)hemoglobin. Biochemistry 27:8719–8724.

Frauenfelder, H., Fenimore, P. W., Chen, G., and McMahon, B. H. (2006) Protein folding is slaved to solvent motions. Proc. Natl. Acad. Sci. U.S.A. 103:15469–15472.

Friedman, J. M. (1985) Structure, dynamics, and reactivity in hemoglobin. Science 228:1273–1280.

Friedman, J. M. (1994) Time-resolved resonance Raman spectroscopy as probe of structure, dynamics, and reactivity in hemoglobin. Methods Enzymol. 232:205–231.

Friedman, J.M., and Lyons, K.B. (1980) Transient Raman study of CO-haemoprotein photolysis: origin of the quantum yield. Nature 284:570–572.

Friedman, J. M., Rousseau, D. L., Ondrias, M. R., and Stepnoski, R. A. (1982) Transient Raman study of hemoglobin: structural dependence of the iron-histidine linkage. Science 218:1244–1246.

Friedman, J. M., Scott, T. W., Stepnoski, R. A., Ikeda-Saito, M., and Yonetani, T. (1983) The iron-proximal histidine linkage and protein control of oxygen binding in hemoglobin. A transient Raman study. J. Biol. Chem. 258:10564–10572.

Ghelichkhani, E., Goldbeck, R. A., Lewis, J. W., and Kliger, D. S. (1996) Nanosecond time-resolved absorption studies of human oxyhemoglobin photolysis intermediates. Biophys. J. 71:1596–1604.

Goldbeck, R. A., Paquette, S. J., and Kliger, D. S. (2001) The effect of water on the rate of conformational change in protein allostery. Biophys. J. 81:2919–2934.

Goldbeck, R. A., Esquerra, R. M., Holt, J. M., Ackers, G. K., and Kliger, D. S. (2004) The molecular code for hemoglobin allostery revealed by linking the thermodynamics and kinetics of quaternary structural change. 1. Microstate linear free energy relations. Biochemistry 43:12048–12064.

Henry, E. R., Bettati, S., Hofrichter, J., and Eaton, W. A. (2002) A tertiary two-state allosteric model for hemoglobin. Biophys. Chem. 98:149–164.

Hofrichter, J., Sommer, J. H., Henry, E. R., and Eaton, W. A. (1983) Nanosecond absorption spectroscopy of hemoglobin: elementary processes in kinetic cooperativity. Proc. Natl. Acad. Sci. U.S.A. 80:2235–2239.

Huang, J., Juszczak, L. J., Peterson, E. S., Shannon, C. F., Yang, M., Huang, S., Vidugiris, G. V. A., and Friedman, J. M. (1999) The conformational and dynamic basis for ligand binding reactivity in hemoglobin Ypsilanti (beta 99 aspTTyr): origin of the quaternary enhancement effect. Biochemistry 38:4514–4525.

Imai, K., Tsuneshige, A., and Yonetani, T. (2002) Description of hemoglobin oxygenation under universal solution conditions by a global allostery model with a single adjustable parameter. Biophys Chem 98:79–91.

Jayaraman, V., Rodgers, K. R., Mukerji, I., and Spiro, T. G. (1995) Hemoglobin allostery: resonance Raman spectroscopy of kinetic intermediates. Science 269:1843–1848.

Jayaraman, V., and Spiro, T. G. (1995) Structure of a third cooperativity state of hemoglobin: ultraviolet resonance Raman spectroscopy of cyanomethemoglobin ligation microstates. Biochemistry 34:4511–4515.

Juszczak, L., and Friedman, J. (1999) UV resonance Raman spectra of ligand binding intermediates of sol-gel encapsulated hemoglobin. J. Biol. Chem. 274:30357–30360.

Juszczak, L., Samuni, U., and Friedman, J. M. (2005) Conformational and functional significance of the alpha140 side-chain in HbA: a UV and visible resonance Raman study of three alpha140 mutants. J. Raman Spectroscopy 36:350–358.

Kavanaugh, J. S., Chafin, D. R., Arnone, A., Mozzarelli, A., Rivetti, C., Rossi, G. L., Kwiatkowski, L. D., and Noble, R. W. (1995) Structure and oxygen affinity of crystalline desArg141 alpha human hemoglobin A in the T state. J. Mol. Biol. 248:136–150.

Kavanaugh, J. S., Weydert, J. A., Rogers, P. H., and Arnone, A. (1998) High-resolution crystal structures of human hemoglobin with mutations at tryptophan 37beta: structural basis for a high-affinity T-state. Biochemistry 37:4358–4373.

Kavanaugh, J. S., Rogers, P. H., and Arnone, A. (2005) Crystallographic evidence for a new ensemble of ligand-induced allosteric transitions in hemoglobin: the T-to-T(high) quaternary transitions. Biochemistry 44:6101–6121.

Khan, I., Shannon, C. F., Dantsker, D., Friedman, A. J., Perez-Gonzalez-de-Apodaca, J., and Friedman, J. M. (2000) Sol-gel trapping of functional intermediates of hemoglobin: geminate and bimolecular recombination studies. Biochemistry 39:16099–16109.

Khan, I., Dantsker, D., Samuni, U., Friedman, A. J., Bonaventura, C., Manjula, B., Acharya, S. A., and Friedman, J. M. (2001) Beta 93 modified hemoglobin: kinetic and conformational consequences. Biochemistry 40:7581–7592.

Kitagawa, T. (1988) The heme protein structure and the iron histidine stretching mode. In Biological Application of Raman Spectroscopy, ed. T. G. Spiro, pp. 97–131. New York: John Wiley & Sons.

Kwiatkowski, L. D, Hui, H. L, Wierzba, A., Noble, R. W, Walder, R. Y, Peterson, E. S, Sligar, S. G, and Sanders, K. E. (1998) Preparation and kinetic characterization of a series of betaW37 variants of human hemoglobin A: Evidence for high-affinity T quaternary structures. Biochemistry 37:4325–4335.

Kwiatkowski, L. D., Hui, H. L., Karasik, E., Colby, J. E., and Noble, R. W. (2007) Mutations of the betaN102 residue of HbA not only inhibit the ligand-linked T to Re state transition, but also profoundly affect the properties of the T state itself. Biochemistry 46:2037–2049.

Lalezari, I., Lalezari, P., Poyart, C., Marden, M., Kister, J., Bohn, B., Fermi, G., and Perutz, M. F. (1990) New effectors of human hemoglobin: Structure and function. Biochemistry 29:1515–1523.

Lukin, J. A., Kontaxis, G., Simplaceanu, V., Yuan, Y., Bax, A., and Ho, C. (2003) Quaternary structure of hemoglobin in solution. Proc. Natl. Acad. Sci. U.S.A. 100:517–520.

Mueser, T. C., Rogers, P. H., and Arnone, A. (2000) Interface sliding as illustrated by the multiple quaternary structures of liganded hemoglobin. Biochemistry 39:15353–15364.

Nagatomo, S., Nagai, M., Mizutani, Y., Yonetani, T., and Kitagawa, T. (2005) Quaternary structures of intermediately ligated human hemoglobin a and influences from strong allosteric effectors: Resonance Raman investigation. Biophys. J. 89:1203–1213.

Noble, R. W, Hui, H. L., Kwiatkowski, L. D., Paily, P., DeYoung, A., Wierzba, A., and Colby, J. E. (2001) Mutational effects at the subunit interfaces of human hemoglobin:

evidence for a unique sensitivity of the T quaternary state to changes in the hinge region of the alpha 1 beta 2 interface. Biochemistry 40:12357-12368.

Ondrias, M. R., Rousseau, D. L., Shelnutt, J. A., and Simon, S. R. (1982) Quaternary-transformation-induced changes at the heme in deoxyhemoglobins. Biochemistry 21:3428–3437.

Perutz, M. F., Fermi, G., Luisi, B., Shaanan, B., and Liddington, R. C. (1987) Stereochemistry of cooperative mechanisms in hemoglobin. Cold Spring Harb. Symp. Quant. Biol. 52:555–565.

Perutz, M. F., Wilkinson, A. J., Paoli, M., and Dodson, G. G. (1998) The stereochemical mechanism of the cooperative effects in hemoglobin revisited. Annu. Rev. Biophys. Biomol. Struct. 27:1–34.

Peterson, E. S., and Friedman, J. M. (1998) A possible allosteric communication pathway identified through a resonance Raman study of four beta37 mutants of human hemoglobin A. Biochemistry 37:4346–4357.

Rousseau, D. L., and Friedman, J. M. (1988) Transient and cryogenic studies of photodissociated hemoglobin and myoglobin. In Biological Applications of Raman Spectroscopy, ed. T. G. Spiro, pp. 133–215. New York: John Wiley & Sons.

Safo, M. K., and Abraham, D. J. (2005) The enigma of the liganded hemoglobin end state: a novel quaternary structure of human carbonmonoxy hemoglobin. Biochemistry 44:8347–8359.

Samuni, U., Dantsker, D., Khan, I., Friedman, A. J., Peterson, E., and Friedman, J. M. (2002) Spectroscopically and kinetically distinct conformational populations of sol-gel-encapsulated carbonmonoxy myoglobin. A comparison with hemoglobin. J. Biol. Chem. 277:25783–25790.

Samuni, U., Juszczak, L., Dantsker, D., Khan, I., Friedman, A. J., Pérez-González-de-Apodaca, J., Bruno, S., Hui, H. L., Colby, J. E., Karasik, E., Kwiatkowski, L. D., Mozzarelli, A., Noble, R., and Friedman, J. M. (2003) Functional and spectroscopic characterization of half-liganded iron-zinc hybrid hemoglobin: evidence for conformational plasticity within the T state. Biochemistry 42:8272–8288.

Samuni, U., Dantsker, D., Juszczak, L. J., Bettati, S., Ronda, L., Mozzarelli, A., and Friedman, J. M. (2004) Spectroscopic and functional characterization of T state hemoglobin conformations encapsulated in silica gels. Biochemistry 43:13674–13682.

Samuni, U., Roche, C. J., Dantsker, D., Juszczak, L. J., and Friedman, J. M. (2006) Modulation of reactivity and conformation within the T-quaternary state of human hemoglobin: the combined use of mutagenesis and sol-gel encapsulation. Biochemistry 45:2820–2835.

Samuni, U., Dantsker, D., Roche, C. J., Friedman, J. M. (2007a) Ligand recombination and a hierarchy of solvent slaved dynamics: the origin of kinetic phases in hemeproteins. Gene 398:234–248.

Samuni, U., Roche, C. J., Dantsker, D., and Friedman, J. M. (2007b) Conformational dependence of hemoglobin reactivity under high viscosity conditions: the role of solvent slaved dynamics. J. Am. Chem. Soc. 129:12756–12764.

Schiro, G., and Cupane, A. (2007) Quaternary relaxations in sol-gel encapsulated hemoglobin studied via NIR and UV spectroscopy. Biochemistry 46:11568–11576.

Schirò, G., Sclafania, M., Caronnaa, C., Natalib, F., Plazanetc, M., and Cupane, A. (2008)

Dynamics of myoglobin in confinement: an elastic and quasi-elastic neutron scattering study. Chem. Phys. (in press, corrected proof).

Scott, T. W., and Friedman, J. M. (1984) Tertiary-structure relaxation in hemoglobin: a transient Raman study. J. Am. Chem. Soc. 106:5677–5687.

Scott, T. W., Friedman, J. M., and Macdonald, V. W. (1985) Distal and proximal control of ligand reactivity: a transient Raman comparison of COHbA and COHb (Zurich). J. Am. Chem. Soc. 107:3702–3705.

Shibayama, N., and Saigo, S. (1995) Fixation of the quaternary structures of human adult haemoglobin by encapsulation in transparent porous silica gels. J. Mol. Biol. 251:203–209.

Shibayama, N., and Saigo, S. (2001) Direct observation of two distinct affinity conformations in the T state human deoxyhemoglobin. FEBS Lett. 492:50–53.

Silva, M. M., Rogers, P. H., and Arnone A (1992) A third quaternary structure of human hemoglobin A at 1.7-A resolution. J. Biol. Chem. 267:17248–17256.

Tsuneshige, A., Park, S., and Yonetani, T. (2002) Heterotropic effectors control the hemoglobin function by interacting with its T and R states – a new view on the principle of allostery. Biophys. Chem. 98:49–63.

Viappiani, C., Bettati, S., Bruno, S., Ronda, L., Abbruzzetti, S., Mozzarelli, A., and Eaton, W. A. (2004) New insights into allosteric mechanisms from trapping unstable protein conformations in silica gels. Proc. Natl. Acad. Sci. U.S.A. 101:14414–14419.

Yonetani, T., Park, S., Tsuneshige, A., Imai, K., and Kanaori, K. (2002) Global allostery model of hemoglobin: modulation of $O_2$-affinity, cooperativity, and Bohr effect by heterotropic allosteric effectors. J. Biol. Chem. 277:34508–34520.

# 14

# From O$_2$ Diffusion into Red Blood Cells to Ligand Pathways in Globins

John S. Olson

## Abstract

Jonathan and Beatrice Wittenberg have had a strong influence on much of the work done in my laboratory over the past 35 years, both indirectly through their papers and more directly through conversations and collaborations. I began reading their papers on facilitated diffusion of O$_2$ by myoglobin (Wittemberg 1965; 1970; Riveros-Moreno and Wittenberg 1972), gas exchange in fish swim bladders (Wittenberg 1958; 1961; Wittenberg and Wittenberg 1961; Wittenberg et al. 1964) and ligand binding to leghaemoglobin (Wittenberg et al. 1972) when I was a graduate student in Quentin Gibson's laboratory between 1968 and 1972. Then, as an independent investigator, I became involved in a number of studies, which built on their discoveries about the role of globins in O$_2$ storage, transport, sensing and scavenging. Four of these projects are summarised in this chapter and were chosen because of the strong influence of the Wittenbergs' work and are presented in a tribute to them and to their ideas. These studies include: (1) a demonstration that O$_2$ uptake and release by intact red blood cells is limited by diffusion through unstirred surface layers; (2) an evaluation of the factors governing sulphide binding to *Lucina pectinata* HbI; (3) a comparison of symbiotic and non-symbiotic plant haemoglobins; and (4) an experimental verification that O$_2$ enters and exits *Cerebratulus lacteus* Hb through an internal apolar channel.

## O$_2$ Uptake and Release by Intact Red Blood Cells: Role of Diffusion Through Unstirred Layers

Oxygen uptake, transport and storage by vertebrate haemoglobins (Hbs) and myoglobins (Mbs) require: (a) O$_2$ diffusion through alveolar capillaries into red cells and chemical combination with intracellular Hb; (b) circulation of the red cells to other locations; and (c) dissociation of O$_2$ from the Hb, diffusion out of the cells, through the capillary lumen and blood vessel wall, and into respiring tissues. In the case of myocytes, the incoming O$_2$ both binds to Mb and is consumed by mitochondrial cytochrome c oxidase (CcO) during blood flow. During muscle contraction, the $O_2$ stored in Mb is released to CcO to maintain oxidative phosphorylation for ATP production. For the past 40 years, the Wittenbergs have been examining the role of Mb in the last two processes and its relationship to oxygen transport and CO and NO metabolism (Wittenberg 1965; 1970; Wittenberg, Wittenberg and Caldwell 1975; Cole et al. 1982; Wittenberg and Wittenberg 1987; Wittenberg and Wittenberg 2003; Wittenberg 2007; Wittenberg and Wittenberg 2007). Their initial results suggested that Mb acts to facilitate the diffusion of $O_2$ from the surface of myocytes to mitochondria (Wittenberg 1970). In a set of technically clever experiments, they demonstrated that Mb and Hb can facilitate $O_2$ diffusion across ~50–300 μm layers of concentrated protein solutions using Millipore filters as a solid support (Wittenberg, Brown and Wittenberg 1965; Wittenberg 1966). This enhancement of $O_2$ transport could be as great as 2-fold at low $P_{O_2}$ and depended in a "bell-shaped" manner on Mb and Hb concentration, with an optimum of ~5 mM in heme, which is close to the estimated concentration of Mb in cardiac myocytes but lower than that of Hb in red cells. The reason for facilitation at low oxygen tension (1–10 μM) is that the flux of $MbO_2$ across the layer ($-D_{MbO_2} \partial[MbO_2]/\partial x$) is greater than that for free $O_2$ ($-D_{O_2} \partial[O_2]/\partial x$) due to the ~1000-fold higher concentration of protein (1–10 mM) (Wittenberg 1965; 1966; 1970; Wittenberg, Wittenberg and Caldwell 1975).

The Wittenberg work helped to motivate my group at Rice University to investigate $O_2$ uptake by human red cells to determine if facilitated diffusion is a factor in the efficiency of oxygen transport (Coin and Olson 1979; Vandegriff and Olson 1984a,b,c). By 1975, it was well established that the half-time for $O_2$ uptake by dilute suspensions of deoxygenated human red blood cells is roughly 0.080 s when mixed with air-equilibrated buffer. This half time is ~40 times larger than that for $O_2$ binding to an equivalent solution (on a per heme basis) of extracellular deoxyhaemoglobin ($t_{1/2} \approx 0.002$ s). The first red cell experiments were performed by Hartridge and Roughton in 1927, using a continuous flow rapid mixing device and a reversion spectroscope to monitor spectral shifts (Hartridge and Roughton 1927). These results were repeated over the next 50 years with more modern, rapid mixing equipment (Gibson et al. 1955; Sirs and Roughton 1963; Holland and Forster 1966; Coin and Olson 1979). A major part of the slowing is due to $O_2$ diffusion through the concentrated, highly viscous solution of Hb inside the red cell. However, calculations for $O_2$ uptake by a simple membraneless packet of Hb can only account for about a factor of 8 in slowing of the uptake process ($t_{1/2} \approx 0.016$ s for packet of concentrated Hb (Nicolson and Roughton 1951; Moll 1969; Kutchai 1975; Coin and Olson 1979)). Roughton and others suggested that the additional factor of 5 in slowing is due to either membrane resistance to $O_2$ diffusion, which he favoured as the most likely explanation (Roughton 1963), or unstirred layers of solvent adjacent to the erythrocytes due to incomplete mixing, which would increase the effective diffusion path from bulk solvent to the cell surface (Hartridge and Roughton 1927; Nicolson and Roughton 1951; Kreuzer and Yahr 1960; Moll 1969; Kutchai 1970).

We began our work wondering if we could enhance $O_2$ uptake and release by changing the membrane lipid composition. However, a number of workers, carrying out experiments similar to those of Wittenberg for facilitated diffusion, showed that $O_2$ uptake and transport by 10–100 μM layers of concentrated Hb and packed red cells were very similar (Kreuzer and Yahr 1960; Kreuzer 1970; Kutchai 1970; 1975). These results suggested that the membrane offers little extra resistance in these "static", non-mixing experiments. Building on their results, but using stopped-flow rapid mixing experiments, we were able to show unambiguously that the extra slowing of $O_2$ uptake by red cells is due to the presence of unstirred layers adjacent to the red cell surface, which increase rapidly after flow stops and approach a final steady radial value of ~3–4 μm, which is ~1000-fold greater than the thickness of the red cell membrane (Coin and Olson 1979). A one-dimensional description of this situation is shown in Fig. 1a, which approximates the profile of $[O_2]_{free}$ near the red cell surface and inside it. A complete three-dimensional description is given in Vandegriff and Olson (1984a) where cylindrically shaped erythrocytes were modelled with an initial unstirred layer of ~1 μm around the cells that grows to a 4-fold larger, steady value as the kinetic energy of cell tumbling dissipates. In these calculations, we allowed diffusion of internal Hb as well as free $O_2$ to take into account any facilitated diffusion, building on the earlier work of the Wittenbergs.

The flux of gas across the external, unstirred layer and into the red cell can be approximated by:

$$J_i = -D_{layer} \left( \frac{\partial C}{\partial x} \right)_{x=\text{cell surface}} \approx -D_{layer} \frac{(C_{cell} - C_{out})}{d_{layer}} \qquad \text{Equation 1}$$

where $C_{cell}$ is the concentration of free $O_2$ at or just inside the cell surface and is close to zero; $C_{out}$ is the concentration of $O_2$ in the bulk solvent; $D_{layer}$ is the diffusion constant of $O_2$ in the buffer layer (~$2\times10^{-5}$ cm²/s); and $\partial x$ is approximated by the thickness of the unstirred layer, which for human red cells is roughly $4\times10^{-4}$ cm (Fig. 1a). A similar expression was used by Nernst to describe the diffusion of ions up to electrodes in unstirred solutions (Atkins 1990). The rate of increase in cellular oxygen concentration by diffusion across this layer can be estimated by multiplying this flux by surface area to volume ratio of the cell, SA/V, and noting that $C_{cell}$ is given in molecules per cubic centimetre:

$$J_i \cdot \frac{SA}{V} = \frac{dC_{cell}}{dt} \approx -D_{layer} \frac{SA}{V} \frac{(C_{cell} - C_{out})}{d_{layer}} \qquad \text{Equation 2}$$

For a sphere of radius $r$, the expression for SA/V is $3/r$. The thickness of an unstirred solvent layer around a particle in suspension is proportional to its radius so $d_{Layer} \approx r$ (see Vandegriff and Olson 1984b and references therein). An approximation for the rate of change of $O_2$ concentration inside the cell is:

$$\frac{dC_{cell}}{dt}(\text{sphere}) \cong -\frac{3}{r_{sphere}^2} D_W (C_{cell} - C_{out}) \qquad \text{Equation 3}$$

Thus, the rate of $O_2$ uptake by a spherical cell will depend parabolically on its radius or more generally on the square of its surface area to volume ratio if the limiting step is diffusion through unstirred surface layers. The exact theoretical dependence of the rate of $O_2$ uptake as function of size for cylindrically shaped red cells is presented in Vandegriff and Olson (1984b).

This type of dependence on $1/r^2$ or $(SA/V)^2$ is observed experimentally for vertebrate red cells, and, as shown in Fig. 1b, the results can be dramatic. In our work (Vandegriff and Olson 1984b), we examined the apparent bimolecular rate of $O_2$ uptake, $k'_c$, by: (1) extremely small artificial red cells, prepared by encapsulating Hb in 0.2-$\mu$m diameter liposomes; (2) normal human erythrocytes, which are ~1.6-$\mu$m thick disks with a surface diameter of ~6 $\mu$m; and (3) the giant red cells of the salamander *Amphiuma means*, which are ~6-$\mu$m thick with a surface diameter of roughly 60 $\mu$m (Fig. 1c). The apparent $k'_c$ values are $1.4 \times 10^6$ M$^{-1}$ s$^{-1}$, $0.07 \times 10^6$ M$^{-1}$ s$^{-1}$ and $0.009 \times 10^6$ M$^{-1}$ s$^{-1}$, respectively, for the artificial, human and amphibian red cells. Clearly the smaller cells are more efficient at rapidly taking up $O_2$ and the $k'_c$ value for the 0.2-$\mu$m diameter liposomes is close to that for extracellular Hb in isotropic solution where k'$_{O_2}$ is ~$5 \times 10^6$ M$^{-1}$ s$^{-1}$ for the first step in ligand binding to Hb tetramers. Holland and Forster (1966) and others have argued that ruminants have small red cells to enhance the efficiency of $O_2$ uptake and release for extended running. In contrast, $O_2$ uptake and release by the salamander (and most other amphibian) red cells is too slow to support active aerobic metabolism, accounting for why amphibian muscle tissue is "white" and only supports anaerobic glycolysis to lactate for ATP production and muscle contraction (Vandegriff and Olson 1984b).

The resistance to $O_2$ uptake caused by unstirred layers around the cell surface also occurs during $O_2$ transport in capillaries and is compounded by the Fahreaus effect, in which the cells tend to accumulate in the centre of the lumen where the velocity of fluid flow is greatest (Fahraeus 1929). As a result, cell-free layers appear near the vessel walls creating another resistance to $O_2$ movement into or out of surrounding tissues. To examine these processes in detail, J. David Hellums and his students, often in collaboration with my group, examined these processes both theoretically and with an artificial capillary system, which is shown in Fig. 1d (Boland et al. 1987; Lemon et al. 1987; Page, Light and Hellums 1998; Page et al. 1998). Colour-enhanced images of a 30-$\mu$m capillary show the appearance of a cell-free layer adjacent to the vessel wall at high flow rates (Page 1997). This work has shown that unstirred plasma layers occur around the moving red cells and along the vessel walls and offer a significant resistance to both $O_2$ uptake and release. This result explains the need for capillary beds with 7–8 $\mu$m diameter vessels, which strip away both layers. However, even in the micro-capillaries, the resultant "slug" flow with gaps between the cells offers significant resistance to gas transport (Baxley and Hellums 1983; Page, Light and Hellums 1999).

Thus, in all cases, from simple mixing to capillary situations, encapsulation of Hb in cells decreases the efficiency of $O_2$ transport, and the inhibitory effect can be several-fold. This decrease in physiological efficiency is compensated

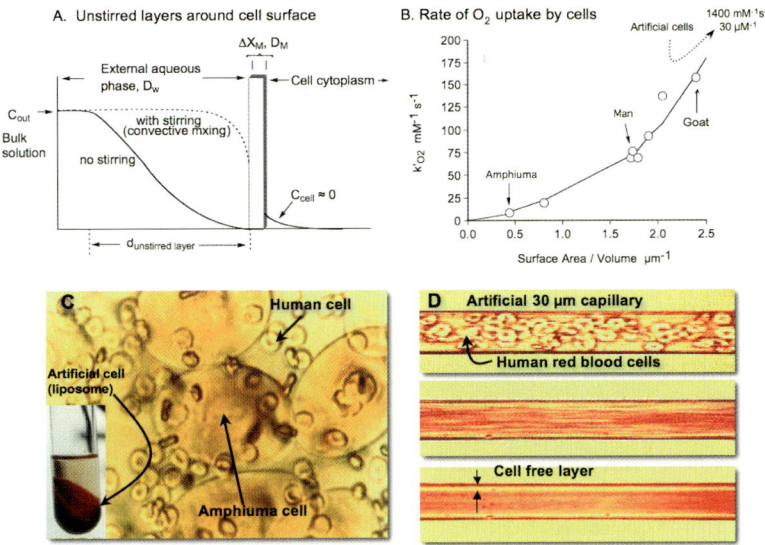

FIG. 1. $O_2$ diffusion into red blood cells and transport in capillaries. **A** Cartoon of the $O_2$ concentration distribution near the surface of red cells without convective mixing (Coin and Olson 1979). **B** Apparent bimolecular rate constant for $O_2$ uptake by vertebrate red cells and a preparation of spherical artificial red cells with an estimated diameter for 0.2 µM (Vandegriff and Olson 1984b). **C** Comparison of the sizes of *Amphiuma* and human red cells and the pellet for the artificial cells. **D** Enhanced images of red cells flowing in an artificial 30-µm diameter capillary showing the appearance of a cell-free layer near the vessel walls at high flow rates. The rates of flow for ~4 mM Hb suspensions of human red cells were 0, ~5 µl/h and ~50 µl/h (adapted from Page, 1997)

by the inhibition of oxidative reactions and degradation of Hb when the protein is encapsulated in metabolically active cells. The red cell reductase systems rapidly re-reduce any transiently oxidised iron atoms, and, if any globin and heme precipitation does occur, the resultant material is sequestered in the red cell, inhibiting oxidative stress in blood vessel walls. Perhaps even more significantly, encapsulation of the Hb prevents interference with NO signalling between endothelial cells and surrounding smooth muscle tissue in arterial blood vessels.

The reaction of NO with $HbO_2$ to produce nitrate is faster than $O_2$ binding and on the of order $70 \times 10^6$ $M^{-1}$ $s^{-1}$(Eich et al. 1996; Doherty et al. 1998; Olson et al. 2004). As discussed by the Wittenbergs in recent reviews of Mb function (Wittenberg and Wittenberg 2003; Wittenberg 2007; Wittenberg and Wittenberg 2007), this NO dioxygenation reaction appears to have been retained in Mb to detoxify NO, which is a highly potent inhibitor of mitochondrial respiration. Hb performs the same function in red cells and helps to protect all vertebrates from inhaled NO and septic shock due to high levels of NO produced by macrophages (Gladwin et al. 2000; Olson et al. 2004). Encapsulation, however, slows this reaction markedly due to unstirred layers around the red cell and cell-free layers along the vessel walls during rapid flow in arteries and arterioles. The "extra" resistance to gas transport prevents interference with NO signalling in blood vessel walls (Liu et al. 1998; Liao et al. 1999; Liu et al. 2002). Hypertensive events due to Hb scavenging of NO only occur after haemolysis (Liao et al. 1999; Reiter et al. 2002) or administration of extracellular Hb as a blood substitute (Doherty et al. 1998; Olson et al. 2004).

## Ligand discrimination in animal globins and sulphide transport

The Wittenbergs have shown that CO inhibits respiration in cardiac muscle tissue primarily by preventing $O_2$ storage and diffusion by Mb; direct inhibition of cytochrome oxidase occurs, but to a lesser extent (Wittenberg and Wittenberg 1987, 1993). Because CO is produced naturally by the first step in heme degradation, the protein portion of Mb had to evolve to discriminate strongly in favour of $O_2$ and against CO compared to simple heme compounds, which normally have a several thousand-fold higher affinity for CO than for $O_2$. The mechanism for ligand discrimination in Mbs and Hbs is now well established both experimentally and theoretically and is electrostatic in origin (Olson and Phillips 1997; Phillips et al. 1999; Spiro and Kozlowski 2001; Park and Boxer 2002). The $Fe(II)O_2$ complex is highly polar, with the bound $O_2$ having significant superoxide character, $Fe(III)^+-O_2{}^{\bullet-}$. This polar complex can be stabilised several thousand-fold by surrounding amino acid side chains with positive fields or hydrogen-bond donating capacity (Olson and Phillips 1997; Phillips et al. 1999). In contrast, the $Fe(II)CO$ complex is relatively neutral and, although its bond order is affected by surrounding electrostatic fields (Phillips et al. 1999; Spiro and Kozlowski 2001; Franzen 2002; Park and Boxer 2002), hydrogen bonding produces minimal effects on CO affinity ($\leq$5-fold).

In the case of Mb, the distal histidine (His64 at the E7 helical position, Fig. 2a) stabilises bound $O_2$ ~1000-fold by forming a strong hydrogen bond between the second ligand O atom and Nε-H of the imidazole side chain. The introduction of a polar amino acid at position E7 to enhance $O_2$ affinity selectively does have an inhibitory effect on the binding of all ligands, because a water molecule is drawn into the active site and hydrogen bonds to His64 in deoxyMb. This non-covalently bound water must be displaced before any ligand can bind.

The following equation was derived to interpret ligand affinity in terms of water displacement, hydrogen bonding, size of the distal pocket, and iron reactivity and accessibility (Olson and Phillips 1997).

$$K_{overall} = \left( \frac{1}{1 + K_{H_2O}[H_2O]} \right) K_{entry} K_{bond} (1 + K_{stabilisation})$$    Equation 4

The $1/(1+[H_2O])$ term describes the fraction of empty active sites that do not contain water, where $K_{H_2O}$ is the equilibrium association constant for water binding to the polar E7 amino acid side chain. $K_{entry}K_{bond}$ is the effective equilibrium association constant for ligand capture in the pocket and binding to the iron atom without any electrostatic interactions. The $(1+K_{stabilisaton})$ term represents the extent of stabilisation of the bound ligand by favourable electrostatic interactions. In the case of Mb, the value of $1/(1+K_{H_2O}[H_2O])$ is ~0.10, and the requirement to displace non-covalently bound water inhibits the binding of all ligands roughly 10-fold. In the case of O$_2$ binding, this inhibition is compensated by formation of a strong hydrogen bond, which results in a value of $(1+K_{stabilisation}) \approx 1000$. The net result is that the distal histidine enhances $K_{O_2}$ ~100-fold, whereas it decreases $K_{CO}$ up to 10-fold. Both distal steric hindrance, which limits the capture volume and access to the iron atom, and proximal constraints, which limit the ease of in-plane movement of the Fe(II)–His93(F8) complex, cause decreases in the product, $K_{entry}K_{bond}$, inhibiting the binding of all three diatomic ligands (O$_2$, CO and NO) uniformly.

Equation 4 serves as a framework for understanding ligand binding to a wide variety of Hbs with markedly different physiological functions. For example, the bivalve mollusc, *Lucina pectinata*, expresses several globin genes, one of which, HbI, has "normal" Mb-like O$_2$ binding properties, but a dramatically high affinity for sulphide (Kraus and Wittenberg 1990; Kraus et al. 1990). This clam is found in coastal regions of the Caribbean, which are rich in sulphur-containing sediments, and accumulates chemoautotrophic bacteria, which use sulphide as an electron source. *Lucina* HbI serves as a sulphide transporter to the colonies of these symbiotic bacteria and has evolved a sulphide affinity that is ~6000-fold greater than that of mammalian Mbs (Table 1). In the mid-1990s, Bolognesi, the Wittenbergs and coworkers discovered that *Lucina* HbI has a normal globin fold, but a remarkably different active site with three significant distal pocket substitutions compared to Mb: Leu(B10) to Phe, His(E7) to Gln, and Val(E11) to Phe (Rizzi et al. 1994, 1996; Bolognesi et al. 1999).

During this same time period, we were constructing a library of Mb mutants with aromatic amino acid substitutions to see if we could reduce the rate of NO scavenging in blood substitute prototypes (Eich et al. 1996; Doherty et al. 1998; Dou et al. 2002). One of these Mb models contained the same substitutions found in *Lucina* HbI (Fig. 2b, Table 1). When we recognised that we had inadvertently made a mimic for the clam protein, we set up a collaboration with Jonathan Wittenberg to measure the O$_2$ and H$_2$S affinities of a complete set of single, double and triple mutants to examine how the high sulphide affinity is achieved (Nguyen et al. 1998). A set of these data is given in Table 1 and serves to validate strongly the interpretations made by the Bolognesi group based on the structure of native *Lucina* HbI (Rizzi et al. 1996).

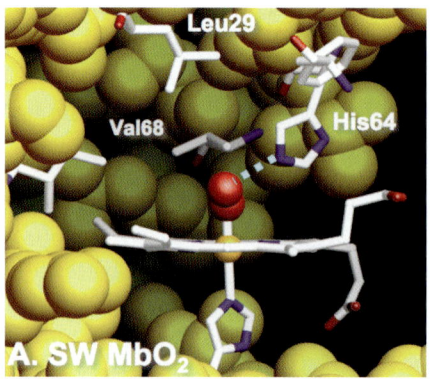 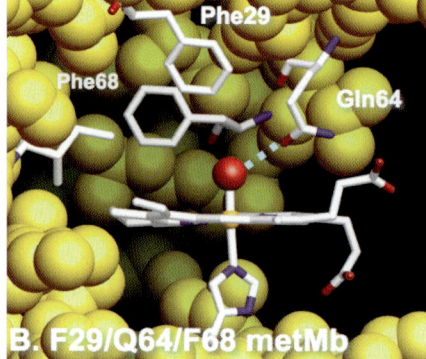

FIG. 2. Comparison of the active sites of sperm whale wild-type (2mgm) MbO$_2$ and L29F/H64Q/V68F (1obm) metMb. Dashed cyan lines indicate hydrogen bonds between NεH of His64 and bound O$_2$ in **A** and between the water OH and Oε of Gln64 in **B**

TABLE 1. Equilibrium association constants for O$_2$ binding to *Lucina pectinata* HbI and distal pocket mutants of sperm whale Mb at neutral pH, 20°C (data taken from Nguyen et al. 1998)

| Mb mutant or Hb | $K_{O_2}$ ($\mu M^{-1}$) | $K_{H_2S}$ ($\mu M^{-1}$) | $K_{H_2S}/K_{O_2}$ |
|---|---|---|---|
| Lucina HbI | 2.5 | 280 | 116 |
| Phe(B10)/Gln(E7)/Phe(E11) | | | |
| SW Mb | | | |
| Leu(B10)/His(E7)/Val(E11) | | | |
| Wild-type | 1.1 | 0.05 | 0.045 |
| L29F | 15 | 0.32 | 0.021 |
| H64Q | 0.18 | 1.2 | 6.7 |
| V68F | 0.48 | 0.33 | 0.69 |
| L29F/V68F | 74 | 5.3 | 0.072 |
| L29F/H64Q | 0.46 | 3.7 | 8.0 |
| L29F/H64Q/V68F | 3.8 | 37 | 9.7 |

First, it is clear that the distal pocket substitutions are the major cause of the dramatically increased sulphide affinity of *Lucina* HbI. There is a monotonic increase in $K_{H_2S}$ as the L29F, H64Q and V68F mutations are combined in Mb resulting in a triple mutant, which has sulphide affinity that is 700-fold greater than that of wild-type Mb and only 7-fold less than that of the clam protein. The H64Q mutation causes the largest selective increase in sulphide affinity, which we interpret as due to an expanded active site, substantial weakening of water coordination and the flexibility of the amide side chain, which can be both a hydrogen bond donor, -NεH$_2$, and acceptor, -Oε. The L29F mutation places the positive edge of the phenyl multipole directly above bound ligands, and if the ligand has a partial or complete negative change, the phenyl ring will stabilise it by favourable electrostatic interactions. The V68F substitution relieves steric hindrance due to the loss of the Cγ$_2$ atom of the Val side chain, which is in van der Waals contact with bound ligands in native Mb (Fig. 2a). The positive edge of the Phe68 benzene ring can also stabilise anion binding. In the structure of *Lucina* HbI, the size of the distal pocket is somewhat larger than the triple Mb mutant, and the edges of PheCD1, PheB10 and PheE11 are all oriented more directly toward the active site, providing an explanation of the even higher affinity for sulphide of the clam Hb (Rizzi et al. 1996). Regardless of the exact interpretation, it is clear that recombinant Mb does serve as a good model system for understanding the evolution of the sulphide binding in Lucina HbI and that the framework of Equation 4 is useful.

# Regulation of O$_2$ affinity in plant haemoglobins

Another area of overlap between our work and that of the Wittenbergs involves the discovery of non-symbiotic hexacoordinate Hbs in plants and microorganisms. It had been recognised for over 45 years that the root nodules of legumes contain a Hb that evolved both to reduce the O$_2$ tension around the *Rhizobium* infection and to facilitate O$_2$ diffusion for bacterial respiration and ATP production, which is needed for nitrogen fixation (Appleby 1962). Shortly after its discovery, Cyril Appleby and the Wittenbergs began a long and productive collaboration to characterise the ligand-binding properties and physiological function of this plant Hb (Wittenberg, Appleby and Wittenberg 1972; Appleby, Wittenberg and Wittenberg 1973a, 1973b; Wittenberg, Wittenberg and Appleby 1973; Wittenberg et al. 1975; Appleby et al. 1976; Appleby et al. 1983; Wittenberg et al. 1986; Gibson et al. 1989).

In the mid-1990s, studies of Lba structure and function were slowed by the lack of a simple recombinant system for large-scale expression and the lack of a high-resolution crystal structure. At about the same time, the appearance of plant genomic libraries led to the discovery of non-symbiotic globin genes in all plants, some of which are clearly the precursors of the leghaemoglobins. My group became involved in studies with leghaemoglobin to try to solve the first two problems and then with non-symbiotic plant haemoglobins (nsHbs) to try to study their comparative properties. As postdoctoral fellow at Rice University, Mark Hargrove carried out all of the initial work on these globins and is now a leader in the plant haemoglobin field at Iowa State University. He succeeded in cloning, expressing, crystallising and determining the structures of both recombinant soybean Lba and non-symbiotic rice Hb1 (Fig. 3, Table 2). We also worked on several other nsHbs from rice, soybeans and *Arabidopsis thaliana* in collaboration with Robert Klucas' group at Nebraska and James Peacock's and Elizabeth Dennis' groups in Australia (Arrendondo-Peter et al. 1997; Trevaskis et al. 1997). In parallel with our work, R.D. Hill's group succeeded in cloning and expressing barley nsHb and worked with the

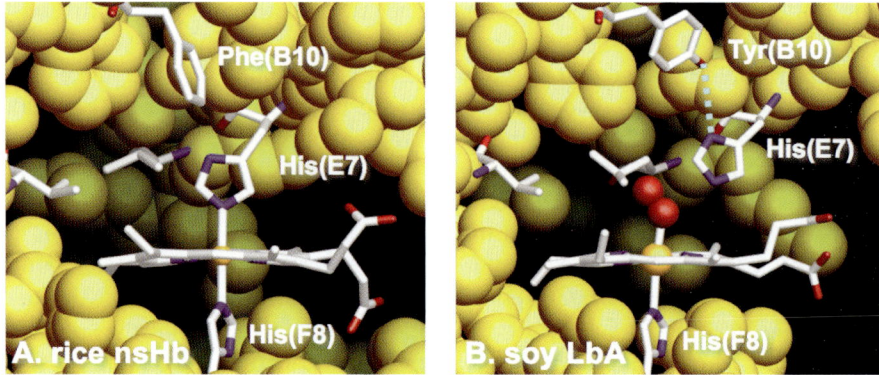

FIG. 3. Comparison of the structure of rice nsHb (1bin) and a model of soybean LbaO$_2$ based on the crystal structure of soybean metLba (1d8u), NMR data, FTIR spectra and model building (Kundu et al. 2004)

TABLE 2. Rate and equilibrium parameters for O$_2$ binding to sperm whale Mb, soybean Lba and rice nsHb. The data for SW Mb were taken from Scott, Gibson and Olson (2001). The data for soybean Lba were taken from Hargrove et al. (1997) and Kundu et al. (2002). The data for rice nsHba were taken from Hargrove et al. (2000)

| Globin | $k'_{O_2}$ ($\mu M^{-1} s^{-1}$) | $k_{O_2}$ ($s^{-1}$) | $k_{O_2}$ ($\mu M^{-1}$) | $1/(1+K_{H_2O}$ [H$_2$O]) or $1/(1+K_{hex})$ | $K_{bond}K_{entry}$ | $K_{stabilisation}$ |
|---|---|---|---|---|---|---|
| **1. SW Mb** | | | | | | |
| Wild-type | 17 | 15 | 1.1 | 0.1 | 0.011 | 1000 |
| HisE7 to Leu | 74 | 10,000 | 0.0074 | 1 | 0.0074 | 1 |
| **2. Soybean Lba** | | | | | | |
| Wild-type | 130 | 5.6 | 23 | ~0.3 | 15 | ~5 |
| HisE7 to Leu | 400 | 24 | 17 | 1 | 17 | 1 |
| **3. Rice nsHb** | | | | | | |
| Wild-type | 68 | 0.038 | 1800 | ~0.1 (0.3–0.5)[a] | 14 | 1300 |
| HisE7 to Leu | 620 | 51 | 12 | 1 | 12 | 1 |

[a]The $1/(1+K_H)$ values in parentheses for rice nsHb were measured directly by kinetic analysis of laser photolysis experiments with CO complexes at wavelengths where internal His(E7) binding and dissociation from the heme iron can be measured directly (Trent, Hvitved and Hargrove 2001)

Wittenbergs to characterise its ligand-binding properties (Duff, Wittenberg and Hill 1997).

The key structural differences between Lba and Rice nsHb are shown in Fig. 3. The most unexpected result is that the deoxygenated and oxidised forms of most plant nsHbs are hexacoordinate. The His(E7) side chain coordinates directly to the iron atom, generating hemi- and haemochrome structures. Thus, ligand binding to these proteins requires the displacement of an endogenously bound ligand instead of non-covalently bound water. The $1/(1+[H_2O])$ term in Equation 4 is replaced by a $1/(1+K_{hex})$ term, where $K_{hex}$ is the isomerisation constant for hexacoordination by the distal histidine. There are a variety of ways to obtain values for the constants in Equation 4 and the corresponding kinetic parameters for ligand binding and hexacoordination. These methods are summarised in Trent, Hvitved and Hargrove (2001) and the references therein.

The second striking difference between rice nsHb and soybean Lba is the position of the distal histidine in the symbiotic haemoglobin. In $LbaO_2$, His(E7) is held up and away from the iron atom by hydrogen bonding to the Tyr(B10) hydroxyl group. The structure shown represents a model based on our crystal structure of metLba-acetate, known features of the NMR spectrum of native $LbaO_2$, the FTIR spectra of wt and mutant LbaCO complexes, and model building by both Hargrove and Olson's (Kundu et al. 2004) and Estrin's (Marti et al. 2007) groups. This model explains the high rate of $O_2$ binding to Lba and the small effect of replacing HisE7 with apolar amino acids (Table 2). In the orientation shown in Fig. 3b, the Nε atom of the imidazole side chain is pointing away from the bound ligand, reducing its ability to hydrogen bond to and stabilise bound ligands. The effect of the His(E7) to Leu mutation on $K_{O_2}$ for Lba is very small compared to the effects of the same mutation in both sperm whale Mb and rice nsHb (Table 2).

Despite being hexacoordinate, rice nsHb has a remarkably high affinity for oxygen and a dramatically small rate of $O_2$ dissociation compared to Lba ($k_{O_2}$=0.038 s$^{-1}$ for nsHb versus 5.6 s$^{-1}$ for Lba). Both Lba and nsHb have an intrinsically reactive iron atom due to upward bowing of the heme group and a staggered His(F8)-pyrrole nitrogen geometry (Fig. 3 and Kundu et al. 2002). When the distal histidine is replaced by Leu, the two proteins have almost identical ligand-binding properties (Table 2). In the case of rice nsHb, the distal histidine forms a very strong hydrogen bond to bound $O_2$, stabilising it to the same extent as is seen in mammalian Mbs (i.e., in both cases $(1+K_{stabilisaton})\approx1000$). The net result is a protein with a nanomolar $K_d$ for $O_2$ binding and a $k_{O_2}$ value that is too small to allow transport or rapid release after storage. In Lba, Tyr(B10) hydrogen bonds to His(E7), keeping the imidazole side chain away from bound $O_2$ to maintain a high dissociation rate constant, rapid rates of ligand entry and only a moderately high $O_2$ affinity. A more detailed and elegant description of the evolution of a transport function from hexacoordinate plant haemoglobins is given in Hoy et al. (2007). However, the question still remains as to the function of the various nsHbs, most of which have affinities for $O_2$ that are extremely high and concentrations *in vivo* that are too low for a role in storage or transport (Kundu, Trent and Hargrove 2003; Hoy et al. 2007).

# Pathways for ligand entry into globins

A large number of workers, including the Wittenbergs, have been involved in the discovery of another set of globins, which have a truncated fold consisting of 2 on 2 instead of the 3 on 3 helical "sandwich" structures found in most globins (Pesce et al. 2000; Milani et al. 2001; Ouellet et al. 2002; Visca et al. 2002; Wittenberg et al. 2002; de Sanctis et al. 2004; Milani et al. 2005; Nardini et al. 2006). Although there is some controversy about the definition, types and origins of these globins, they do have highly unique structural features, one of which suggests that some of these globins have evolved an alternative pathway for binding diatomic ligands. In general, the truncated Hbs are missing the first N-terminal, A helix, which can potentially leave a gap between the E and H helices. If the amino acids lining this gap are small and apolar, a large channel occurs from the ends of the these helices to the heme group, creating what could be an alternative route for ligand entry and exit to the iron atom (Milani et al. 2004).

All of our and most other workers' kinetic, structural, computational and time-resolved crystallographic studies indicate that ligands enter and exit mammalian Mbs and Hbs by a gate mechanism, which involves the opening and closing of a channel directly above the heme propionates by rotation of the distal histidine side chain. Then the in-coming ligand is captured in the distal portion of the heme pocket in much the same way that a baseball is caught in the webbing of a player's glove (Scott, Gibson and Olson 2001; Schotte et al. 2004; Schmidt et al. 2005; Olson, Soman and Phillips 2007). In Mb, a long A helix completely fills the gap between the E and H helices, creating a solid barrier to any movement from the distal pocket to solvent through the protein interior (Fig. 4). There are internal cavities in Mb, which were originally discovered by their ability to take up electron-dense Xe atoms, and two of them, Xe4 (distal near Ile28(B9)) and Xe1 (proximal near Leu89(F4)), do become occupied by photodissociated ligands on short 100-ns time scales (Srajer et al. 2001; Schotte et al. 2004; Schmidt et al. 2005). However, experimentally, ligands never appear to escape from these locations. Instead, they return to the distal pocket either to rebind to the iron atom or to escape when the His(E7) gate is open (Schmidt et al. 2005; Olson, Soman and Phillips 2007). A simple set of mutagenesis data in support of the E7 gate/baseball glove model for Mb is given in Table 3.

Mutation of His64(E7) to Ala in Mb causes a >3-fold increase in the association rate constant for $O_2$ binding as would be expected if His64 were part of the barrier to ligand entry, and His64 to Gly and Val replacements cause even greater increases in $k'_{O_2}$ (Scott, Gibson and Olson 2001). This interpretation is supported by the 3-fold decrease in $k'_{O_2}$ when His64 is mutated to Trp (Table 3). In contrast, neither Ile28 to Ala nor to Trp mutations in the Xe4 pocket have a significant ($\geq$2-fold) effect on the rate of $O_2$ association. Most remarkable of all is that neither Leu89 to Gly nor to Trp replacements have much effect on $k'_{O_2}$, even though, in the former case, a direct hole was created, allowing water to fill the Xe1 cavity, and in the latter case, the entire volume of the Xe1 cavity is filled

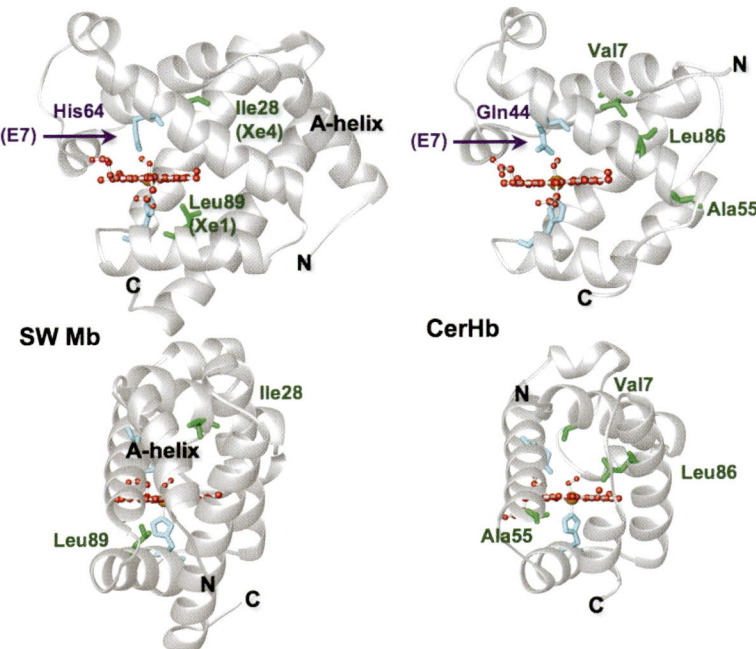

FIG. 4. Comparison of structures of sperm whale (SW) Mb (2mgm) and *Cerebratulus lacteus* (CerHb) (1kr7)

with the indole side chain (Liong et al. 2001). In our view, these results argue strongly against the Xe cavities being part of the route for ligand entry and exit.

The situation for globins missing an A-helix may be quite different. In cases where small amino acids line the gap between the E and H helices, the resulting tunnel does appear to be the route of ligand entry and exit. We have been collaborating with Luc Moen's, Martino Bolognesi's, Austen Riggs', and Karin and Uli Nienhaus' groups on studies with the mini-globin from the Nemertean worm *Cerebratulus lacteus*, CerHb (Pesce et al. 2004; Deng et al. 2007). As shown in Fig. 4, this Hb is very small, containing only 109 amino acids, and is completely missing an A-helix, creating a large gap between the E and H helices that is lined with small amino acids (Pesce et al. 2002). A small set of mutants from a library that is being constructed to map this pathway is shown in Table 3.

In contrast to Mb, the Gln44(E7) to Ala mutation in CerHb does not cause an increase in k'$_{O_2}$, suggesting that the native glutamine side chain is not part of the barrier to ligand entry. However, Trp mutations in the interior of the protein at Val7(B6) and Leu86(G12) and at the solvent entrance of the apolar channel at Ala55(E18) cause 3–10-fold decreases in both k'$_{O_2}$ and k$_{O_2}$, with little change in the overall equilibrium constant. Thus, it is clear that these mutations are increasing the barrier to ligand entry and exit, with little effect on binding in the

TABLE 3. $O_2$ binding parameters for mutants of sperm whale Mb and of CerHb at the E7 gate and internal locations in the Xe cavities and apolar tunnel, respectively. The data for SW Mb were taken from Scott, Gibson and Olson (2001) and crystal structures of the Trp mutants are given in Liong et al. (2001) and Olson, Soman and Phillips (2007). The data for CerHb were taken from Salter et al. (unpublished) and Pesce et al. (2004). Conditions are 0.1 M phosphate pH 7, 20°C

| Globin | $k'_{O_2}$ ($\mu M^{-1} s^{-1}$) | $k_{O_2}$ ($s^{-1}$) | $k'_{O_2}$ ($\mu M^{-1}$) | $k'_{O_2}$(mutant)/$k'_{O_2}$(WT) |
|---|---|---|---|---|
| 1. SW Mb | | | | |
| Wild-type | 17 | 15 | 1.1 | 1.0 |
| His64 to Ala | 53 | 2300 | 0.020 | **3.3** |
| His64 to Trp | 6.3 | 87 | 0.07 | **0.37** |
| Ile28 to Trp | 11 | 5.0 | 2.3 | 0.69 |
| Ile28 to Ala | 19 | 20 | 0.98 | 1.1 |
| Leu89 to Trp | 20 | 71 | 0.28 | 1.2 |
| Leu(89 to Gly | 12 | 9.0 | 1.3 | 0.8 |
| 3. CerHb | | | | |
| Wild-type | 240 | 180 | 1.3 | 1.0 |
| Gln44 to Ala | 180 | 30 | 6.4 | 0.75 |
| Val7 to Trp | 33 | 14 | 2.4 | **0.14** |
| Leu86 to Trp | 97 | 60 | 1.6 | **0.40** |
| Ala55 to Trp | 63 | 66 | 0.95 | **0.26** |

active site. Although more mechanistic studies and structural work are required, these initial mutagenesis results argue strongly that the *Cerebratulus* mini-globin and probably some of the truncated globins have evolved a different pathway for ligand movement into their active sites. In the case of CerHb, this new pathway evolved to allow $O_2$ storage and rapid release in the neurons and brain of the worm under hypoxic conditions (Vandergon et al. 1998; Pesce et al. 2002; Pesce et al. 2004). In contrast, the selective pressure for tunnels in the microbial and plant truncated Hbs remains controversial but, in certain cases, seems to involve NO dioxygenase functions (Ascenzi et al. 2006).

# Summary

These four examples of globin function serve to highlight the wide range of interests of Jonathan and Beatrice Wittenberg, from measurements of $O_2$ movements on millimetre to micron distances in tissues and membranes to those occurring over 5–20 Å in several hundreds of nanoseconds inside globin molecules. In each case, my interactions with them, either directly through collaborations or indirectly through the literature and meetings, were stimulating, resulted in better experiments, and led to a greater understanding for me and hopefully for them as well.

*Acknowledgements*
This work was supported by NIH Grants GM 35649 and HL 47020, and Grant C-612 from the Robert A. Welch Foundation (J.S. Olson, PI).

# References

Appleby, C. A. 1962. The oxygen equilibrium of leghemoglobin. Biochim. Biophys. Acta 60:226–235.

Appleby, C. A., Blumberg, W. E., Peisach, J., Wittenberg, B. A., and Wittenberg, J. B. 1976. Leghemoglobin. An electron paramagnetic resonance and optical spectral study of the free protein and its complexes with nicotinate and acetate. J. Biol. Chem. 251:6090–6096.

Appleby, C. A., Bradbury, J. H., Morris, R. J., Wittenberg, B. A., Wittenberg, J. B., and Wright, P. E. 1983. Leghemoglobin. Kinetic, nuclear magnetic resonance, and optical studies of pH dependence of oxygen and carbon monoxide binding. J. Biol. Chem. 258:2254–2259.

Appleby, C. A., Wittenberg, B. A., and Wittenberg, J. B. 1973a. Leghemoglobin. II. Changes in conformation and chemical reactivity linked to reaction with a dissociable low molecular weight ligand, X. J. Biol. Chem. 248:3183–3187.

Appleby, C. A., Wittenberg, B. A., and Wittenberg, J. B. 1973b. Nicotinic acid as a ligand affecting leghemoglobin structure and oxygen reactivity. Proc. Natl. Acad. Sci. USA 70:564–568.

Arrendondo-Peter, R., Hargrove, M. S., Sarath, G., Moran, J. F., Lohrman, J., Olson, J. S., and Klucas, R. V. 1997. Rice hemoglobins. Gene cloning, analysis, and O₂-binding kinetics of a recombinant protein synthesized in *Escherichia coli*. Plant Physiol. 115:1259–1266.

Ascenzi, P., Bocedi, A., Bolognesi, M., Fabozzi, G., Milani, M., and Visca, P. 2006. Nitric oxide scavenging by *Mycobacterium leprae* GlbO involves the formation of the ferric heme-bound peroxynitrite intermediate. Biochem. Biophys. Res. Commun. 339:450–456.

Atkins, P.W. 1990. Dynamic electrochemistry. In Physical chemistry (4th Edition), Chapter 30, p. 918. New York: W. H. Freeman and Company.

Baxley, P. T., and Hellums, J. D. 1983. A simple model for simulation of oxygen transport in the microcirculation. Ann. Biomed. Eng. 11:401–416.

Boland, E. J., Nair, P. K., Lemon, D. D., Olson, J. S., and Hellums, J. D. 1987. An in vitro capillary system for studies on microcirculatory O₂ transport. J. Appl. Physiol. 62:791–797.

Bolognesi, M., Rosano, C., Losso, R., Borassi, A., Rizzi, M., Wittenberg, J. B., Boffi, A., and Ascenzi, P. 1999. Cyanide binding to *Lucina pectinata* hemoglobin I and to sperm whale myoglobin: an x-ray crystallographic study. Biophys. J. 77:1093–1099.

Coin, J. T., and Olson, J. S. 1979. The rate of oxygen uptake by human red blood cells. J. Biol. Chem. 254:1178–1190.

Cole, R. P., Sukanek, P. C., Wittenberg, J. B., and Wittenberg, B. A. 1982. Mitochondrial function in the presence of myoglobin. J. Appl. Physiol. 53:1116–1124.

de Sanctis, D., Dewilde, S., Pesce, A., Moens, L., Ascenzi, P., Hankeln, T., Burmester, T., and Bolognesi, M. 2004. Mapping protein matrix cavities in human cytoglobin through Xe atom binding. Biochem. Biophys. Res. Commun. 316:1217–1221.

Deng, P., Nienhaus, K., Palladino, P., Olson, J. S., Blouin, G., Moens, L., Dewilde, S., Geuens, E., and Nienhaus, G. U. 2007. Transient ligand docking sites in *Cerebratulus lacteus* mini-hemoglobin. Gene 398:208–223.

Doherty, D. H., Doyle, M. P., Curry, S. R., Vali, R. J., Fattor, T. J., Olson, J. S., and Lemon, D. D. 1998. Rate of reaction with nitric oxide determines the hypertensive effect of cell-free hemoglobin. Nat. Biotechnol. 16:672–676.

Dou, Y., Maillett, D. H., Eich, R. F., and Olson, J. S. 2002. Myoglobin as a model system for designing heme protein based blood substitutes. Biophys. Chem. 98:127–148.

Duff, S. M., Wittenberg, J. B., and Hill, R. D. 1997. Expression, purification, and properties of recombinant barley (*Hordeum* sp.) hemoglobin. Optical spectra and reactions with gaseous ligands. J. Biol. Chem. 272:16746–16752.

Eich, R. F., Li, T., Lemon, D. D., Doherty, D. H., Curry, S. R., Aitken, J. F., Mathews, A. J., Johnson, K. A., Smith, R. D., Phillips, G. N., Jr., et al. 1996. Mechanism of NO-induced oxidation of myoglobin and hemoglobin. Biochemistry 35:6976–6983.

Fahraeus, R. 1929. The suspension stability of the blood. Physiol. Rev. 9:353–373.

Franzen, S. 2002. An electrostatic model for the frequency shifts in the carbonmonoxy stretching band of myoglobin: Correlation of hydrogen bonding and the stark tuning rate. J. Am. Chem. Soc. 124:13271–13281.

Gibson, Q. H., Kreuzer, F., Meda, E., and Roughton, F. J. 1955. The kinetics of human haemoglobin in solution and in the red cell at 37 degrees C. J. Physiol. 129:65–89.

Gibson, Q. H., Wittenberg, J. B., Wittenberg, B. A., Bogusz, D., and Appleby, C. A. 1989. The kinetics of ligand binding to plant hemoglobins. Structural implications. J. Biol. Chem. 264:100–107.

Gladwin, M. T., Ognibene, F. P., Pannell, L. K., Nichols, J. S., Pease-Fye, M. E., Shelhamer, J. H., and Schechter, A. N. 2000. Relative role of heme nitrosylation and beta-cysteine 93 nitrosation in the transport and metabolism of nitric oxide by hemoglobin in the human circulation. Proc. Natl. Acad. Sci. USA 97:9943–9948.

Hargrove, M. S., Barry, J. K., Brucker, E. A., Berry, M. B., Phillips, G. N., Jr., Olson, J. S., Arredondo-Peter, R., Dean, J. M., Klucas, R. V., and Sarath, G. 1997. Characterization of recombinant soybean leghemoglobin a and apolar distal histidine mutants. J. Mol. Biol. 266:1032–1042.

Hargrove, M. S., Brucker, E. A., Stec, B., Sarath, G., Arredondo-Peter, R., Klucas, R. V., Olson, J. S., and Phillips, G. N., Jr. 2000. Crystal structure of a nonsymbiotic plant hemoglobin. Structure (Camb) 8:1005–1014.

Hartridge, H., and Roughton, F. J. 1927. The rate of distribution of dissolved gases between the red blood corpuscle and its fluid environment: Part I. Preliminary experiments on the rate of uptake of oxygen and carbon monoxide by sheep's corpuscles. J. Physiol. 62:232–242.

Holland, R.A., and Forster, R.E. 1966. The effect of size of red cells on the kinetics of their oxygen uptake. J. Gen. Physiol. 49:727–742.

Hoy, J. A., Robinson, H., Trent, J. T., 3rd, Kakar, S., Smagghe, B. J., and Hargrove, M. S. 2007. Plant hemoglobins: A molecular fossil record for the evolution of oxygen transport. J. Mol. Biol. 371:168–179.

Kraus, D. W., and Wittenberg, J. B. 1990. Hemoglobins of the *Lucina pectinata*/bacteria symbiosis. I. Molecular properties, kinetics and equilibria of reactions with ligands. J. Biol. Chem. 265:16043–16053.

Kraus, D. W., Wittenberg, J. B., Lu, J. F., and Peisach, J. 1990. Hemoglobins of the *Lucina pectinata*/bacteria symbiosis. II. An electron paramagnetic resonance and optical spectral study of the ferric proteins. J. Biol. Chem. 265:16054–16059.

Kreuzer, F. 1970. Facilitated diffusion of oxygen and its possible significance; a review. Respir. Physiol. 9:1–30.

Kreuzer, F., and Yahr, W. Z. 1960. Influence of red cell membrane on diffusion of oxygen. J. Appl. Physiol. 15:1117–1122.

Kundu, S., Blouin, G. C., Premer, S. A., Sarath, G., Olson, J. S., and Hargrove, M. S. 2004. Tyrosine B10 inhibits stabilization of bound carbon monoxide and oxygen in soybean leghemoglobin. Biochemistry 43:6241–6252.

Kundu, S., Snyder, B., Das, K., Chowdhury, P., Park, J., Petrich, J. W., and Hargrove, M. S. 2002. The leghemoglobin proximal heme pocket directs oxygen dissociation and stabilizes bound heme. Proteins 46:268–277.

Kundu, S., Trent, J. T., 3rd, and Hargrove, M. S. 2003. Plants, humans and hemoglobins. Trends Plant Sci. 8:387–393.

Kutchai, H. 1970. Numerical study of oxygen uptake by layers of hemoglobin solution. Respir. Physiol. 10:273–284.

Kutchai, H. 1975. Role of the red cell membrane in oxygen uptake. Respir. Physiol. 23:121–132.

Lemon, D. D., Nair, P. K., Boland, E. J., Olson, J. S., and Hellums, J. D. 1987. Physiological factors affecting O₂ transport by hemoglobin in an in vitro capillary system. J. Appl. Physiol. 62:798–806.

Liao, J. C., Hein, T. W., Vaughn, M. W., Huang, K. T., and Kuo, L. 1999. Intravascular flow decreases erythrocyte consumption of nitric oxide. Proc. Natl. Acad. Sci. USA 96:8757–8761.

Liong, E. C., Dou, Y., Scott, E. E., Olson, J. S., and Phillips, G. N., Jr. 2001. Waterproofing the heme pocket. Role of proximal amino acid side chains in preventing hemin loss from myoglobin. J. Biol. Chem. 276:9093–9100.

Liu, X., Miller, M. J. S., Joshi, M. S., Krowicka, H. S., Clark, D. A., and Lancaster, J. R., Jr. 1998. Diffusion-limited reaction of free nitric oxide with erythrocytes. J. Biol. Chem. 273:18709–18713.

Liu, X., Samouilov, A., Lancaster, J. R., Jr., and Zweier, J. L. 2002. Nitric oxide uptake by erythrocytes is primarily limited by extracellular diffusion not membrane resistance. J. Biol. Chem. 277:26194–26199.

Marti, M. A., Capece, L., Bikiel, D. E., Falcone, B., and Estrin, D. A. 2007. Oxygen affinity controlled by dynamical distal conformations: the soybean leghemoglobin and the *Paramecium caudatum* hemoglobin cases. Proteins 68:480–487.

Milani, M., Pesce, A., Nardini, M., Ouellet, H., Ouellet, Y., Dewilde, S., Bocedi, A., Ascenzi, P., Guertin, M., Moens, L., et al. 2005. Structural bases for heme binding and diatomic ligand recognition in truncated hemoglobins. J. Inorg. Biochem. 99:97–109.

Milani, M., Pesce, A., Ouellet, Y., Ascenzi, P., Guertin, M., and Bolognesi, M. 2001. *Mycobacterium tuberculosis* hemoglobin N displays a protein tunnel suited for O2 diffusion to the heme. EMBO J. 20:3902–3909.

Milani, M., Pesce, A., Ouellet, Y., Dewilde, S., Friedman, J., Ascenzi, P., Guertin, M., and Bolognesi, M. 2004. Heme-ligand tunneling in group I truncated hemoglobins. J. Biol. Chem. 279:21520–21525.

Moll, W. 1969. Measurements of facilitated diffusion of oxygen in red blood cells at 37 degrees centigrade. Pflugers Arch. 305:269–278.

Nardini, M., Pesce, A., Labarre, M., Richard, C., Bolli, A., Ascenzi, P., Guertin, M., and Bolognesi, M. 2006. Structural determinants in the group III truncated hemoglobin from *Campylobacter jejuni*. J. Biol. Chem. 281:37803–37812.

Nguyen, B. D., Zhao, X., Vyas, K., La Mar, G. N., Lile, R. A., Brucker, E. A., Phillips, G. N., Jr., Olson, J. S., and Wittenberg, J. B. 1998. Solution and crystal structures of a sperm whale myoglobin triple mutant that mimics the sulfide-binding hemoglobin from *Lucina pectinata*. J. Biol. Chem. 273:9517–9526.

Nicolson, P., and Roughton, F. J. 1951. A theoretical study of the influence of diffusion and chemical reaction velocity on the rate of exchange of carbon monoxide and oxygen between the red blood corpuscle and the surrounding fluid. Proc. R. Soc. Lond. B Biol. Sci. 138:241–264.

Olson, J. S., Foley, E. W., Rogge, C., Tsai, A. L., Doyle, M. P., and Lemon, D. D. 2004. NO scavenging and the hypertensive effect of hemoglobin-based blood substitutes. Free Radic. Biol. Med. 36:685–697.

Olson, J. S., and Phillips, G. N., Jr. 1997. Myoglobin discriminates between O2, NO, and CO by electrostatic interactions with the bound ligand. J. Biol. Inorg. Chem. 2:544–552.

Olson, J. S., Soman, J., and Phillips, G. N., Jr. 2007. Ligand pathways in myoglobin: a review of Trp cavity mutations. IUBMB Life 59:552–562.

Ouellet, H., Ouellet, Y., Richard, C., Labarre, M., Wittenberg, B., Wittenberg, J., and Guertin, M. 2002. Truncated hemoglobin HbN protects *Mycobacterium bovis* from nitric oxide. Proc. Natl. Acad. Sci. USA 99:5902–5907.

Page, T. C. 1997. Oxygen transport by hemoglobin-based blood substitutes. Ph.D. Dissertation, Rice University, Houston, TX.

Page, T. C., Light, W. R., and Hellums, J. D. 1998. Prediction of microcirculatory oxygen transport by erythrocyte/hemoglobin solution mixtures. Microvasc. Res. 56:113–126.

Page, T. C., Light, W. R., and Hellums, J. D. 1999. Oxygen transport in 10 microns artificial capillaries. Adv. Exp. Med. Biol. 471:715–721.

Page, T. C., Light, W. R., McKay, C. B., and Hellums, J. D. 1998. Oxygen transport by erythrocyte/hemoglobin solution mixtures in an in vitro capillary as a model of hemoglobin-based oxygen carrier performance. Microvasc. Res. 55:54–64.

Park, E. S., and Boxer, S. G. 2002. Origins of the sensitivity of molecular vibrations to electric fields: Carbonyl and nitrosyl stretches in model compounds and proteins. J. Phys. Chem. B 106:5800–5806.

Pesce, A., Couture, M., Dewilde, S., Guertin, M., Yamauchi, K., Ascenzi, P., Moens, L., and Bolognesi, M. 2000. A novel two-over-two alpha-helical sandwich fold is characteristic of the truncated hemoglobin family. EMBO J. 19:2424–2434.

Pesce, A., Nardini, M., Ascenzi, P., Geuens, E., Dewilde, S., Moens, L., Bolognesi, M., Riggs, A. F., Hale, A., Deng, P., et al. 2004. ThrE11 regulates $O_2$ affinity in *Cerebratulus lacteus* mini-hemoglobin. J. Biol. Chem., in press.

Pesce, A., Nardini, M., Dewilde, S., Geuens, E., Yamauchi, K., Ascenzi, P., Riggs, A.F., Moens, L., and Bolognesi, M. 2002. The 109 residue nerve tissue minihemoglobin from *Cerebratulus lacteus* highlights striking structural plasticity of the alpha-helical globin fold. Structure 10:725–735.

Phillips, G. N., Jr., Teodoro, M., Li, T., Smith, B., Gilson, M. M., and Olson, J. S. 1999. Bound CO is a molecular probe of electrostatic potential in the distal pocket of myoglobin. J. Phys. Chem. B 103:8817–8829.

Reiter, C. D., Wang, X., Tanus-Santos, J. E., Hogg, N., Cannon, R. O., 3rd, Schechter, A. N., and Gladwin, M. T. 2002. Cell-free hemoglobin limits nitric oxide bioavailability in sickle-cell disease. Nat. Med. 8:1383–1389.

Riveros-Moreno, V., and Wittenberg, J. B. 1972. The self-diffusion coefficients of myoglobin and hemoglobin in concentrated solutions. J. Biol. Chem. 247:895–901.

Rizzi, M., Wittenberg, J. B., Coda, A., Ascenzi, P., and Bolognesi, M. 1996. Structural bases for sulfide recognition in *Lucina pectinata* hemoglobin I. J. Mol. Biol. 258:1–5.

Rizzi, M., Wittenberg, J. B., Coda, A., Fasano, M., Ascenzi, P., and Bolognesi, M. 1994. Structure of the sulfide-reactive hemoglobin from the clam *Lucina pectinata*. Crystallographic analysis at 1.5 A resolution. J. Mol. Biol. 244:86–99.

Roughton, F. J. 1963. Kinetics of gas transport in the blood. Br. Med. Bull. 19:80–89.

Schmidt, M., Nienhaus, K., Pahl, R., Krasselt, A., Anderson, S., Parak, F., Nienhaus, G. U., and Srajer, V. 2005. Ligand migration pathway and protein dynamics in myoglobin: A time-resolved crystallographic study on L29W MbCO. Proc. Natl. Acad. Sci. USA 102:11704–11709.

Schotte, F., Soman, J., Olson, J. S., Wulff, M., and Anfinrud, P. A. 2004. Picosecond time-resolved X-ray crystallography: Probing protein function in real time. J. Struct. Biol. 147:235–246.

Scott, E. E., Gibson, Q. H., and Olson, J. S. 2001. Mapping the pathways for $O_2$ entry into and exit from myoglobin. J. Biol. Chem. 276:5177–5188.

Sirs, J. A., and Roughton, F. J. 1963. Stopped-flow measurements of CO and $O_2$ uptake by hemoglobin in sheep erythrocytes. J. Appl. Physiol. 18:158–165.

Spiro, T. G., and Kozlowski, P. M. 2001. Is the CO adduct of myoglobin bent, and does it matter? Acc. Chem. Res. 34:137–144.

Srajer, V., Ren, Z., Teng, T. Y., Schmidt, M., Ursby, T., Bourgeois, D., Pradervand, C., Schildkamp, W., Wulff, M., and Moffat, K. 2001. Protein conformational relaxation and ligand migration in myoglobin: a nanosecond to millisecond molecular movie from time-resolved Laue X-ray diffraction. Biochemistry 40:13802–13815.

Trent, J. T., 3rd, Hvitved, A. N., and Hargrove, M. S. 2001. A model for ligand binding to hexacoordinate hemoglobins. Biochemistry 40:6155–6163.

Trevaskis, B., Watts, R. A., Andersson, C. R., Llewellyn, D. J., Hargrove, M. S., Olson, J. S., Dennis, E. S., and Peacock, W. J. 1997. Two hemoglobin genes in *Arabidopsis thaliana*: The evolutionary origins of leghemoglobins. Proc. Natl. Acad. Sci. USA 94:12230–12234.

Vandegriff, K. D., and Olson, J.S. 1984a. A quantitative description in three dimensions of oxygen uptake by human red blood cells. Biophys. J. 45:825–835.

Vandegriff, K. D., and Olson, J. S. 1984b. Morphological and physiological factors affecting oxygen uptake and release by red blood cells. J. Biol. Chem. 259:12619–12627.

Vandegriff, K. D., and Olson, J. S. 1984c. The kinetics of $O_2$ release by human red blood cells in the presence of external sodium dithionite. J. Biol. Chem. 259:12609–12618.

Vandergon, T. L., Riggs, C. K., Gorr, T. A., Colacino, J. M., and Riggs, A. F. 1998. The mini-hemoglobins in neural and body wall tissue of the nemertean worm, Cerebratulus lacteus. J. Biol. Chem. 273:16998–17011.

Visca, P., Fabozzi, G., Petrucca, A., Ciaccio, C., Coletta, M., De Sanctis, G., Bolognesi, M., Milani, M., and Ascenzi, P. 2002. The truncated hemoglobin from Mycobacterium leprae. Biochem. Biophys. Res. Commun. 294:1064–1070.

Wittenberg, B. A., and Wittenberg, J. B. 1987. Myoglobin-mediated oxygen delivery to mitochondria of isolated cardiac myocytes. Proc. Natl. Acad. Sci. USA 84:7503–7507.

Wittenberg, B. A., and Wittenberg, J. B. 1993. Effects of carbon monoxide on isolated heart muscle cells. Res. Rep. Health Eff. Inst. 1–12; discussion 13–21.

Wittenberg, B. A., Wittenberg, J. B., and Appleby, C. A. 1973. Leghemoglobin. I. Changes in conformation and chemical reactivity linked to reaction with acetic acid. J. Biol. Chem. 248:3178–3182.

Wittenberg, B. A., Wittenberg, J. B., and Caldwell, P. R. 1975. Role of myoglobin in the oxygen supply to red skeletal muscle. J. Biol. Chem. 250:9038–9043.

Wittenberg, J. B. 1958. The secretion of inert gas into the swim-bladder of fish. J. Gen. Physiol. 41:783–804.

Wittenberg, J. B. 1961. The secretion of oxygen into the swimbladder of fish. I. The transport of molecular oxygen. J. Gen. Physiol. 44:521–526.

Wittenberg, J. B. 1965. Myoglobin-facilitated diffusion of oxygen. J. Gen. Physiol. 49 (Suppl):57–74.

Wittenberg, J. B. 1966. The molecular mechanism of hemoglobin-facilitated oxygen diffusion. J. Biol. Chem. 241:104–114.

Wittenberg, J. B. 1970. Myoglobin-facilitated oxygen diffusion: role of myoglobin in oxygen entry into muscle. Physiol. Rev. 50:559–636.

Wittenberg, J. B. 2007. On optima: The case of myoglobin-facilitated oxygen diffusion. Gene 398:156–161.

Wittenberg, J. B., Appleby, C. A., Bergersen, F. J., and Turner, G. L. 1975. Leghemoglobin: the role of hemoglobin in the nitrogen-fixing legume root nodule. Ann. NY Acad. Sci. 244:28–34.

Wittenberg, J. B., Appleby, C. A., and Wittenberg, B. A. 1972. The kinetics of the reactions of leghemoglobin with oxygen and carbon monoxide. J. Biol. Chem. 247:527–531.

Wittenberg, J. B., Bolognesi, M., Wittenberg, B. A., and Guertin, M. 2002. Truncated hemoglobins: A new family of hemoglobins widely distributed in bacteria, unicellular eukaryotes, and plants. J. Biol. Chem. 277:871–874.

Wittenberg, J. B., Brown, P. K., and Wittenberg, B. A. 1965. A novel reaction of hemoglobin in invertebrate nerves. I. Observations on annelid and molluscan nerves. Biochim. Biophys. Acta 109:518–529.

Wittenberg, J. B., Schwend, M. J., and Wittenberg, B. A. 1964. The secretion of oxygen into the swim-bladder of fish. 3. The role of carbon dioxide. J. Gen. Physiol. 48:337–355.

Wittenberg, J. B., and Wittenberg, B. A. 1961. The secretion of oxygen into the swim-bladder offish. II. The simultaneous transport of carbon monoxide and oxygen. J. Gen. Physiol. 44:527–542.

Wittenberg, J. B., and Wittenberg, B. A. 2003. Myoglobin function reassessed. J. Exp. Biol. 206:2011–2020.

Wittenberg, J. B., and Wittenberg, B. A. 2007. Myoglobin-enhanced oxygen delivery to isolated cardiac mitochondria. J. Exp. Biol. 210:2082–2090.

Wittenberg, J. B., Wittenberg, B. A., Gibson, Q. H., Trinick, M. J., and Appleby, C. A. 1986. The kinetics of the reactions of *Parasponia andersonii* hemoglobin with oxygen, carbon monoxide, and nitric oxide. J. Biol. Chem. 261:13624–13631.

# 15
# Myoglobin Strikes Back

Maurizio Brunori

## Abstract

The biochemical and physiological role of myoglobin is briefly reviewed with reference to the seminal work by Wittenberg and Wittenberg. The function of myoglobin as a NO scavenger in the skeletal muscle and the heart was found to protect cellular respiration, which is known to depend on inhibition of cytochrome-c-oxidase.

In 1998 Garry et al. produced a transgenic myoglobin knockout mouse and reported that the Mb-less animals ($myo^{-/-}$) were perfectly healthy. In reading that paper I could not help but wandering about the feelings of B&J,[1] given their long-standing scientific commitment to Mb's role in the physiology of the muscle. But soon afterwards Gödecke et al. (1999) reported evidence for some compensatory events in the $myo^{-/-}$ mice which would serve to steepen the $O_2$ concentration gradient and reduce the effective diffusion pathlength between the capillaries and the mitochondria. This seemed to resuscitate the role of Mb not only as a (short-term) $O_2$ reservoir but more significantly in facilitating $O_2$ diffusion to mitochondria. The latter phenomenon was discovered by Jonathan Wittenberg for Mb (Wittenberg 1959) and independently by P.F. Scholander on haemoglobin (Scholander 1960), and subsequently considerably extended by Wittenberg (1963) and also by others. When B&J came to Rome to work with Jeffries Wyman and Eraldo Antonini in 1964, the physiology and biophysics of Mb's facilitated $O_2$ transport was discussed extensively, and elicited Wyman's interest. In 1966 he published a piece of theoretical work (Wyman 1966) demonstrating that translational diffusion of the protein should be the underlying physical mechanism; that paper appeared in the *Journal of Biological Chemistry* side by side with a companion experimental paper by Jonathan Wittenberg (1966). Only much later the essential requirement for mobility and

---

[1] Throughout this paper, I shall refer to Beatrice and Jonathan Wittenberg as B&J.

free diffusion of Mb in the cell was verified; I found the work by Chen et al. (1997) using *in vivo* nuclear magnetic resonance spectroscopy particularly convincing. Subsequent work by B&J (Wittenberg and Wittenberg 1987) aimed at defining this important function of Mb directly in cardiac myocytes, by employing sophisticated cellular biochemistry involving (among others) measurements under variable pressures of CO (which would "freeze" Mb in an $O_2$ unreactive state). The adverse effects of CO on myocytes' metabolic and physiological parameters were taken as evidence for the crucial role of $MbO_2$ in enhancing the flux of $O_2$ to mitochondria (given that CO is a very poor inhibitor of cytochrome-c-oxidase and poisoning of this enzyme only occurs at very high pCO).

The significance of Mb for muscle physiology widened several years later in connection with the discovery that cytochrome-c-oxidase, the terminal enzyme of the respiratory chain, is reversibly inhibited by NO (Carr and Ferguson 1990; Brown and Cooper 1994; Cleeter et al. 1994; Brown 1995). Inhibition of cellular respiration clearly involves a competition between $O_2$ and NO, which is not surprising as both gases are known to bind to the reduced bimetallic heme $a_3$-$Cu_B$ active site of the oxidase. NO is a potent poison of respiration, given that at physiological $O_2$ concentrations (~10–20 μM), submicromolar and even nanomolar NO is sufficient to achieve measurable inhibition of respiration. As this effect of NO is sensitive to $O_2$ concentration, Brown (1995) suggested that this may explain why the apparent $K_m$ for $O_2$ measured in cells and tissues (5–10 μM) is significantly greater than that measured in mitochondria or with the purified enzyme ($K_m$<1 μM). The detailed mechanism of inhibition, which is not a simple one, has been investigated extensively under presteady-state and steady-state conditions (Torres, Darley-Usmar and Wilson 1995; Torres, Cooper and Wilson 1998; Giuffrè et al. 1999; Mason et al. 2006). A clue to our understanding of the mechanism came from the finding that not only the fully and partially reduced heme $a_3$-$Cu_B$ binuclear centre, but also the oxidised enzyme and the intermediate oxidation states react with NO, yielding an advantage in the competition with $O_2$ (Sarti et al. 2003; Brunori et al. 2006 for reviews).

When the Mb knockout mice was produced (Garry et al. 1998; Gödecke et al. 1999) I was working on inhibition of cellular respiration by NO *in vitro* and in the cell. Therefore it seemed obvious to me that an important and hitherto overlooked function of Mb in the skeletal and cardiac muscle may be scavenging NO produced in the myocyte or diffusing from endothelial cells; thus I proposed (Brunori 2001a, Brunori 2001b) that Mb would avoid inhibition of cytochrome-c-oxidase and protect the energy-producing machinery. The mechanism whereby Mb may be an efficient intracellular NO scavenger is due to the fact that the oxygenated derivative $MbO_2$ reacts rapidly and stoichiometrically with NO, yielding nitrate (Eich et al. 1996). Since at the $O_2$ pressure in the muscle $MbO_2$ prevails and its intracellular concentration is very high (≥0.2 mM), scavenging NO should be very rapid and almost unavoidable. The reactions involved, shown in Fig. 1, were individually studied by several groups

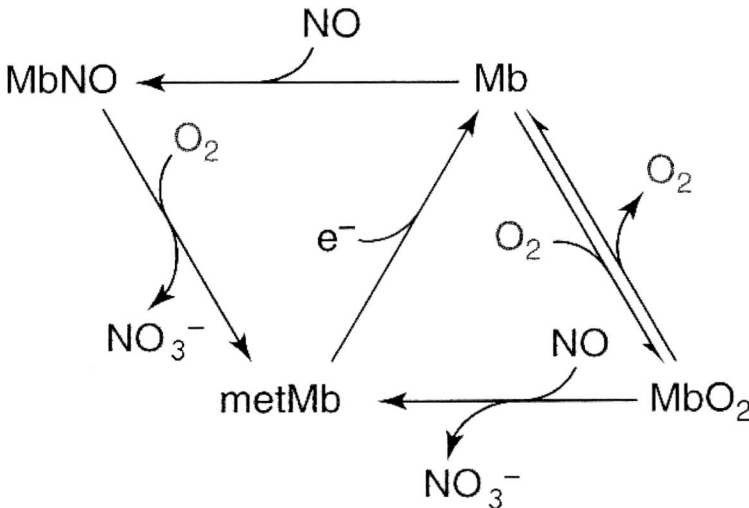

FIG. 1. The cycle of reactions of myoglobin with $O_2$, NO and electrons. The primary phys-iological role involves the reversible binding of $O_2$ to deoxymyoglobin (Mb) to yield oxymyoglobin (MbO$_2$), which acts as a (short-term) reservoir and facilitates the transport of $O_2$ from the periphery of the cell to mitochondria. MbO$_2$ reacts rapidly (and irre-versibly) with nitric oxide (NO) to yield nitrate and ferric myoglobin (metMb), thereby quenching free NO that might otherwise inhibit cytochrome-c oxidase; metMb is reduced to Mb by metMb-reductase. DeoxyMb binds rapidly NO, yielding nitrosylmyoglobin (MbNO); this pathway might be of greater significance in cellular compartments with very low $O_2$ concentrations. In the presence of $O_2$, MbNO can be converted back to metMb much more rapidly than thermal dissociation of NO (from Brunori 2001b)

round the world, including B&J. For this function to be operational under steady-state conditions *in vivo*, oxidised met-Mb should be reduced at a sus-tainable rate, as is shown to be the case.

Independently and at the same time, Flögel et al. (2001) reported direct con-vincing evidence in support of this hypothesis. Using NMR spectroscopy on isolated mouse heart, these authors found that intracoronary perfusion with increasing concentrations of NO leads to a decrease in the concentration of MbO$_2$ and a parallel increase in the formation of metMb. Even more intriguing was the observation that some cardiac physiological parameters (e.g. coronary blood flow and cardiac contractility) were more severely affected by NO in hearts from the knockout mice lacking Mb ($myo^{-/-}$) than in those from wild-type; this effect was obtained by challenging the heart with bradykinin, which increases endogenous NO flux by stimulating the NO-synthase pathway. The remarkable observation that the same NO flux (whether exogenous or endoge-nous) has greater adverse effects in the hearts from $myo^{-/-}$ mice than in those from wild-type animals may be consistent with a more substantial NO inhibi-tion of cytochrome c-oxidase in the knockouts. Although such a protective role of Mb should have been obvious in an environment experiencing a constant flux of NO, it escaped attention. It should be stressed that the NO-scavenging function of Mb is pseudo-enzymatic given that metMb produced by reaction of $MbO_2$ with NO is reduced by a specific reductase, and can therefore repeat a cycle of NO disposal as harmless nitrate (Fig. 1). It should not be neglected, however, that NO synthesised *locally* by mitochondrial NO-synthase (Bates et al. 1995; Ghafourifar and Richter 1997; Tatoyan and Giulivi 1998) may inhibit cytochrome-c-oxidase by direct regional diffusion, and thereby escape the quenching effect of $MbO_2$ (Brunori et al. 2006). This has to be taken into proper consideration also in view of the surprising finding that a PDZ-domain of mitochondrial NOS was reported to interact with subunit Va of the oxidase (Persichini et al. 2005).

In summary, if we accept that Mb protects cytochrome-c-oxidase from the adverse effects of NO, it follows that *in vivo* experiments aimed at substantiating the role of Mb in facilitating $O_2$ diffusion may have to be reassessed. In some cases (Wittenberg and Wittenberg 1987; Glabe et al. 1998) CO was used to trap Mb in the inert state MbCO, which is essentially unreactive towards $O_2$ as well as NO. If the scavenging role of $MbO_2$ is suppressed, can we exclude that the functional perturbations observed after freezing the protein as MbCO do not depend on partial inhibition of respiration by NO? This point has been recognised as significant and extensively discussed by B&J in a comprehensive and excellent review (Wittenberg and Wittenberg 2003) by the catchy title "Myoglobin function reassessed", which is fully worth reading; unfortunately its significance is still ignored by several groups working in experimental cardiology.

A final point which (as far as I know) has been cavalierly talked about but never carefully discussed is as follows. As Mb has a role in the energetic balance of cells both by enhancing $O_2$ delivery to mitochondria and by protecting cytochrome-c-oxidase from NO inhibition, why is it not expressed in other metabolically active tissues (such as the brain)? It is likely that since NO is reported to function as a messenger in the CNS (Bredt and Snyder 1994), expression of Mb at high concentrations in neurons might be deleterious by intercepting a signalling messenger. In this respect the discovery of neuroglobin (a monomeric globin present in the brain of mammals) (Burmester et al. 2000) might acquire physiological significance more for a regional control of NO fluxes than for enhancing availability of $O_2$ to mitochondria, given that it is expressed at extremely low average concentration ($<1$ μM) (see Brunori and Vallone 2006 for review). The fact that Mb is essentially absent in smooth muscle seems reasonable because an efficient scavenger would interfere with the crucial role of NO as a messenger in smooth muscle relaxation, which is the basis of vasodilation of blood vessels (Ignarro et al. 1987; Palmer, Ferrige and Moncada 1987). By contrast, in the red muscle the NO scavenging function of $MbO_2$ should be taken into consideration when describing at the molecular level the physiology of skeletal muscle and the heart; its role in protecting cytochrome-c-oxidase might be of significance in several pathophysiological states involving an increased NO flux, especially when working with $myo^{-/-}$ knockout mice.

*Acknowledgements*
I wish to express appreciation to my colleagues in the Department for their invaluable contributions over the years, and to Lucia Ugo for assistance in the preparation of this manuscript. This work was partially supported by the Ministry of University and Research of Italy.

# References

Bates, T.E., Loesch, A., Burnstock, G., and Clark, J. B. 1995. Immunocytochemical evidence for a mitochondrially located nitric oxide synthase in brain and liver. Biochem. Biophys. Res. Commun. 213:896–900.

Bredt, D.S., and Snyder, S.H. 1994. Nitric oxide: a physiologic messenger molecule. Annu. Rev. Biochem. 63:175–195.

Brown, G.C. 1995. Nitric oxide regulates mitochondrial respiration and cell functions by inhibiting cytochrome oxidase. FEBS Lett. 369:136–139.

Brown, G.C., and Cooper, C.E. 1994. Nanomolar concentrations of nitric oxide reversibly inhibit synaptosomal respiration by competing with oxygen at cytochrome oxidase. FEBS Lett. 356:295–298.

Brunori, M. 2001a. Nitric oxide, cytochrome-c oxidase and myoglobin. Trends Biochem. Sci. 26:21–23.

Brunori, M. 2001b. Nitric oxide moves myoglobin centre stage. Trends Biochem. Sci. 26:209–210.

Brunori, M., Forte, E., Arese, M., Mastronicola, D., Giuffre, A., and Sarti P. 2006. Nitric oxide and the respiratory enzyme. Biochim. Biophys. Acta 1757:1144–1154.

Brunori, M., and Vallone, B. 2006. A globin for the brain. FASEB J. 20:2192–2197.

Burmester, T., Weich, B., Reinhardt, S., and Hankein, T. 2000. A vertebrate globin expressed in the brain. Nature 407:520–523.

Carr, G.J., and Ferguson, S.J. 1990. Nitric oxide formed by nitrite reductase of *Paracoccus denitrificans* is sufficiently stable to inhibit cytochrome oxidase activity and is reduced by its reductase under aerobic conditions. Biochim. Biophys. Acta 1017:57–62.

Chen, W., Zheng, J., Eljgelshoven, M. H., Zhang, Y., Zhu, X. H., Wang, C., Cho, Y., Merkle, H., and Ugurbie, K. 1997. Determination of deoxy myoglobin changes during graded myocardial ischemia: an in vivo H-NMR spectroscopy study. Magn. Reson. Med. 38:193–197.

Cleeter, M. W., Cooper, J. M., Darley-Usmar, V. M., Moncada, S., and Schapira, A. H. 1994. Reversible inhibition of cytochrome c oxidase, the terminal enzyme of the mitochondrial respiratory chain, by nitric oxide. Implications for neurodegenerative diseases. FEBS Lett. 345:50–54.

Eich, R. F., Li, T., Lemon, D. D., Doherty, D. H., Curry, S. R., Altken, J. F., Mathews, A. J., Johnson, K. A., Smith, R. D., Phillips, G. N. Jr, Olson, J. S. 1996. Mechanism of NO-induced oxidation of myoglobin and hemoglobin. Biochemistry 35:6976–6983.

Flögel, U., Merx, M. W., Gödecke, A., Decking, U. K. M.., and Schrader, J. 2001. Myoglobin: a scavenger of bioactive NO. Proc. Natl. Acad. Sci. U.S.A. 98:735–740.

Garry, D.J., Ordway, G. A., Lorenz, J. N., Radford, N. B., Chin, E. R., Grange, R. W., Bassel-Duby, R., and Williams, R. S. 1998. Mice without myoglobin. Nature 395:905–908.

Ghafourifar, P., and Richter, C. 1997. Nitric oxide synthase activity in mitochondria. FEBS Lett. 418:291–296.

Giuffrè, A., Sarti, P., D'Itri, E., Buse, G., Soulimane, T., and Brunori, M. 1999. On the mechanism of inhibition of cytochrome c oxidase by nitric oxide. J. Biol. Chem. 271:33404–33408.

Glabe, A., Chung, Y., Xu, D., and Jue, T. 1998. Carbon monoxide inhibition of regulatory pathways in myocardium. Am. J. Physiol 234:H2143–H2151.

Gödecke, A., Flögel, U., Zanger, K., Ding, Z., Hirchenhain, J., Decking, U. K. M, and Schrader, J. 1999. Disruption of myoglobin in mice induces multiple compensatory mechanisms. Proc. Natl. Acad. Sci. U.S.A. 96:10495–10500.

Ignarro, L.J., Buga, G. M., Wood, K. S., Byrns, R. E., and Chaudhuri, G. 1987. Endothelium-derived relaxing factor produced and released from artery and vein is nitric oxide. Proc. Natl. Acad. Sci. U. S. A. 84:9265–9269.

Mason, M.G., Nicholls, P., Wilson, M. T., and Cooper, C. E. 2006. Nitric oxide inhibition of respiration involves both competitive (heme) and noncompetitive (copper) binding to cytochrome c oxidase. Proc. Natl. Acad. Sci. U. S. A. 103:708–713.

Palmer, R.M., Ferrige, A.G., and Moncada, S. 1987. Nitric oxide release accounts for the biological activity of endothelium-derived relaxing factor. Nature 327:524–526.

Persichini, T., Mazzone, V., Polticelli, F., Moreno, S., Venturini, G., Clementi, E., and Colasanti, M. 2005. Mitochondrial type I nitric oxide synthase physically interacts with cytochrome c oxidase. Neurosci. Lett. 384:254–295.

Sarti, P., Giuffre, A., Barone, M. C., Forte, E., Mastronicola, D., and Brunori, M. 2003. Nitric oxide and cytochrome oxidase: reaction mechanism from the enzyme to the cell. Free Radic. Biol. Med. 34:509–520.

Scholander, P.F. 1960. Oxygen transport through hemoglobin solutions. Science 131:585–590.

Tatoyan, A., and Giulivi, C. 1998. Purification and characterization of a nitric oxide synthase from rat liver mitochondrial. J. Biol. Chem. 273:11044–11048.

Torres, J., Cooper, C.E., and Wilson, M.T. 1998. A common mechanism for the interaction of nitric oxide with the oxidized binuclear centre and oxygen intermediates of cytochrome c oxidase. J. Biol. Chem. 273:8756–8766.

Torres, J., Darley-Usmar, V., and Wilson, M.T. 1995. Inhibition of cytochrome c oxidase in turnover by nitric oxide: mechanism and implications for control of respiration. Biochem. J. 312:169–173.

Wittenberg, B.A., and Wittenberg, J.B. 1987. Myoglobin-mediated oxygen delivery to mitochondria of isolated cardiac myocytes. Proc. Natl. Acad. Sci. U.S.A. 84:7503–7507.

Wittenberg, J.B, and Wittenberg, B.A. 2003. Myoglobin function reassessed. J. Exp. Biol. 206:2011–2020.

Wittenberg, J.B. 1959. Oxygen transport: a new function proposed for myoglobin. Biol.

Bull. 117:402–403.

Wittenberg, J.B. 1963. Facilitated diffusion of oxygen through haemerythrin solutions. Nature 199:816–817.

Wittenberg, J.B. 1966. The molecular mechanism of hemoglobin-facilitated oxygen diffusion. J. Biol. Chem. 241:104–114.

Wyman, J. 1966. Facilitated diffusion and the possible role of myoglobin as a transport mechanism. J. Biol. Chem. 241:115–121.

# 16

# The Bilatarian Sea Urchin and the Radial Starlet Sea Anemone Globins Share Strong Homologies with Vertebrate Neuroglobins

Xavier Bailly and Serge N. Vinogradov

## Abstract

A 34 834-bp gene in the genome of the sea urchin *Strongylocentrotus purpuratus* (Echinodermata) consists of 34 exons and 33 introns, and encodes a 2252aa putative chimeric Hb, comprising an unidentifiable nonglobin N-terminal domain of ~150aa and 16 globin domains (D1–D16) of ~150aa, except for D2, which lacks helices G and H. The similarity between the globin domains varies from 39 to 51% identity: they are linked tightly to one another, with interdomain segments of 10aa or less. Alignment of the globin domains with other globins shows them to have a D helix and His residues at both the E7 and F8 locations. Intron insertions within the globin domains occur at canonical positions B12.2 and G7.0. Blastp searches with each of the 16 domains showed a strong sequence similarity with vertebrate neuroglobins and globin X, the Cnidarian *Nematostella vectensis*, and with single-domain bacterial globins. A Bayesian analysis of the *S. purpuratus* globins, the 7 single domain globins from the Cnidarian *N. vectensis* and 80 globins from several metazoan groups, indicated that the *S. purpuratus* and *N. vectensis* globins share a molecular affinity with vertebrate neuroglobins and globin X, while annelid, mollusc, crustacean, lamprey, hagfish and urochordate globins form a clade with vertebrate cytoglobins. The common molecular signatures and gene structures shared between a radial metazoan, *S. purpuratus* and vertebrate globins suggest these proteins could exhibit ancestral metazoan globin properties and predate the Radiata–Bilateria split. We assume modern eumetazoan globin families could have derived from such a putative globin plesiogene.

## Introduction

The broad diversity of globins observed in extant metazoans (Vinogradov et al. 2005, 2006) reflects at least 600 million years of specialised diversification from an ancestral globin gene inherited in each emerging phylum during the Cambrian explosion, achieved via species evolution accompanied by gene duplication events (Zhang 2003). A recent model of globin evolution (Vinogradov et al., 2007) proposed that metazoan globin genes were all derived from prokaryote single-domain FHb (flavohaemoglobin)-like globin gene(s) transferred laterally prior to the emergence of multicellularity and the Cambrian explosion. Although it is known that the vertebrate globins are broadly divided into two lineages, the Ngbs (neuroglobins) (including HbX, a vertebrate globin related to neuroglobin) and the Cygbs (cytoglobins), which also comprise the derived Mbs (myoglobins) and the α- and β-globins, that emerged from the duplication of an ancestral metazoan globin (Burmester et al. 2004), the relationship of the latter to other metazoan globins remains unresolved.

The echinoderms (sea urchins) together with the hemichordates and chordates comprise the deuterostomes, and are thus the closest relatives of vertebrates. Although no haemoglobins (Hbs) have been reported in hemichordates, they are known to exist in the urochordate (ascidian) *Ciona intestinalis* (Ebner, Burmester, and Hankeln 2003) and in the cephalochordate (amphioxus) *Branchiostoma californiense* (Bishop et al. 1998). The approximately 7000 species of echinoderms occur in five classes: Asteroidea (starfish), Crinoidea (feather stars), Echinoidea (sea urchins), Holothuroidea (sea cucumbers) and Ophiuroidea (brittle stars). Intracellular Hbs occurring in coelomocytes present in the water vascular system have been reported in two of the five classes: in the holothurians *Caudina (Molpadia) arenicola* (Mauri et al. 1991; McDonald, Davidson and Kitto 1992) and *Paracaudina chilensis* (Suzuki 1989; Baker and Terwilliger 1993) and in the ophiuroid *Hemipholis elongata* (Christensen, Colacino and Bonaventura 2003).

We present below an *in silico* analysis of the gene encoding a polymeric globin, found in the recently completed genome of the echinoderm *Strongylocentrotus purpuratus* (Sodergren et al. 2006) and the results of a molecular phylogenetic analysis of the relationships of its globin domains with the globins found in the Cnidarian *Nematostella vectensis* (starlet sea anemone) and other metazoan globins.

# The Intron–Exon Structure and Coding Sequences of the *S. purpuratus* Globin Gene

The 34 834-bp putative globin gene of *S. purpuratus* (GenBank: gi|115913843| XM_001199205.1|) comprises 34 exons and 33 introns (Table 1). Apart from exons 1, 6 and 34, comprised of 312, 105 and 129 nts, respectively, the remaining exons vary between 224 and 248 nts. The introns range from 192 to 5965 nts, with the majority being 400–700 nts (Table 1). In all 16 domains of the *S. purpuratus* globin gene, the introns are inserted at the canonical B12.2 and G7.0 positions, similar to the majority of vertebrate and invertebrate Hb genes (Hardison 1996) and exhibit the consensus splicing -GT- donor and splicing -AT- acceptor motifs (Mount 1982; Keller and Noon 1984).

TABLE 1. Intron–exon structure of the 34 834-bp gene (gi|115913843|XM_001199205.1) encoding the multidomain Hb of *S. purpuratus*

| Exon no. | Location | Size, nt | Size, aa | Phase | Intron. no. | Location | Size, nt |
|---|---|---|---|---|---|---|---|
| 1 | 1–312 | 312 | 104 | 0 | 1 | 313–6277 | 5965 |
| 2 | 6278–6525 | 248 | 83 | 2 | 2 | 6526–11036 | 4511 |
| 3 | 11037–11265 | 229 | 76 | 0 | 3 | 11266–11944 | 679 |
| 4 | 11945–12168 | 224 | 75 | 2 | 4 | 12169–12616 | 448 |
| 5 | 12617–12842 | 226 | 75 | 0 | 5 | 12843–13200 | 358 |
| 6 | 13201–13304 | 104 | 35 | 2 | 6 | 13305–13712 | 408 |
| 7 | 13713–13941 | 229 | 76 | 0 | 7 | 13942–14133 | 192 |
| 8 | 14134–14360 | 227 | 76 | 2 | 8 | 14361–14710 | 350 |
| 9 | 14711–14939 | 229 | 76 | 0 | 9 | 14940–16215 | 1276 |
| 10 | 16216–16439 | 224 | 75 | 2 | 10 | 16440–18766 | 2327 |
| 11 | 18767–18995 | 229 | 76 | 0 | 11 | 18996–19335 | 340 |
| 12 | 19336–19571 | 236 | 79 | 2 | 12 | 19572–19846 | 275 |
| 13 | 19845–20075 | 229 | 77 | 0 | 13 | 20076–20404 | 329 |
| 14 | 20405–20643 | 239 | 80 | 2 | 14 | 20644–21171 | 528 |
| 15 | 21172–21400 | 229 | 76 | 0 | 15 | 21401–21664 | 264 |
| 16 | 21665–21891 | 227 | 74 | 2 | 16 | 21892–22507 | 616 |
| 17 | 22508–22736 | 229 | 76 | 0 | 17 | 22737–23344 | 608 |
| 18 | 23345–23571 | 227 | 76 | 2 | 18 | 23572–24159 | 588 |
| 19 | 24160–24388 | 229 | 76 | 0 | 19 | 24389–25012 | 624 |
| 20 | 25013–25239 | 227 | 76 | 2 | 20 | 25240–25480 | 241 |
| 21 | 25481–25709 | 229 | 76 | 0 | 21 | 25710–26333 | 624 |
| 22 | 26334–26560 | 227 | 76 | 2 | 22 | 26561–27081 | 521 |
| 23 | 27082–27310 | 229 | 76 | 0 | 23 | 27311–27755 | 445 |
| 24 | 27756–27982 | 227 | 76 | 2 | 24 | 27983–28392 | 410 |
| 25 | 28393–28621 | 229 | 76 | 0 | 25 | 28622–28979 | 358 |
| 26 | 28980–29203 | 224 | 75 | 2 | 26 | 29204–29639 | 436 |
| 27 | 29640–29868 | 229 | 76 | 0 | 27 | 29869–30294 | 426 |
| 28 | 30295–30521 | 227 | 76 | 2 | 28 | 30522–30951 | 430 |
| 29 | 30952–31180 | 229 | 76 | 0 | 29 | 31181–31615 | 435 |
| 30 | 31616–31839 | 224 | 75 | 2 | 30 | 31840–32359 | 520 |
| 31 | 32360–32588 | 229 | 76 | 0 | 31 | 32589–33215 | 627 |
| 32 | 33216–33460 | 245 | 82 | 2 | 32 | 33461–33984 | 524 |
| 33 | 33985–34213 | 229 | 76 | 0 | 33 | 34214–34705 | 492 |
| 34 | 34706–34834 | 129 | 43 | | | | |
| Total | | 7659 | 2552 | | | | 27175 |

The only other multidomain globins whose gene structure have been determined are the 9-domain chains of the extracellular Hbs from the crustaceans (brine shrimp) *Artemia* and *Parartemia* (Matthews and Trotman 1998; Coleman, Matthews and Trotman 2001). The latter has 22 introns, of which 5 occur in the interdomain regions and in N- and C-terminal extensions; 8 of the 9 chains have introns at B12.2, 6 have introns at G7.0 and 3 have introns at F3.0, F3.1 and G6.2, respectively.

The *S. purpuratus* globin gene encodes a 2252aa putative chimeric Hb, comprising a nonglobin N-terminal domain of ~150aa and 16 globin domains (D1–D16). Figure 1 shows the alignment of the amino acid sequences of the 16 putative globin domains with each other, the amino acid sequences of two *C. arenicola* globins, whose structures have been determined (Mitchell, Kitto and Hackert 1995) (Protein Data Bank (PDB): 1hlb and 1hlm), with sperm whale myoglobin (PDB: 1a6m). Note that the ~146aa N-terminal nonglobin domain consists of exon 1 and part of exon 2; its identity was not revealed by BLASTP (http://www.ncbi.nlm.nih.gov/blast/Blast), FUGUE (Shi, Blundell and Mizuguchi 2001) (http://tardis.nibio.go.jp/fugue/prfsearch.html) and SMART (Letunic et al. 2006) (http://smart.embl-heidelberg.de/smart/) searches. All 16 domains align well with each other and exhibit all the features expected of bona fide globins based on the canonical Mb fold, except for D2, which appears to lack helices G and H. Since no viable globins are known with this defect, it probably should not be considered to be a functioning globin.

# The Single-Domain Globins in the Genome of the Cnidarian Nematostella vectensis

GenBank identifies 9 putative globins in the genome of the Cnidarian (sea starlet anemone) *N. vectensis* (Sullivan et al. 2006a). Two of them are questionable: one, XP_001622964.1|GI:156353205 (415aa), has a globin domain (160–300), also identified by FUGUE ($Z=8.6$ with 1tu9), that lacks an F8His. Another, XP_001635635.1 GI:156391157 (137aa), also identified as a globin by FUGUE ($Z=12.1$ with 1yhud), lacks half of the G helix and all of the H helix. A third globin (XP_001640985.1|GI:156405932, Scaffold 5000153, 209aa) has an error in its sequence which was corrected to provide a 166aa putative globin. This and the remaining 6 globins, XP_001629477.1| GI:156373757, Scaff 141000032 (167aa), XP_001636079.1| GI:156392486 Scaff 42000019 (176aa), XP_001635310.1|GI:156390503, Scaff 50000067 (179aa), XP_001641645.1|GI:156408000, Scaff 3000224 (182aa), XP_001633127.1|GI:156384015, Scaff 76000030 (185aa) and XP_001640562.1| GI:156405084, Scaff 7000121 (195aa), were all identified as globins by FUGUE searches. Their alignment with the *S. purpuratus* domains is provided in Fig. 1. All the *N. vectensis* globin genes have two introns: one inserted at position B12.2, the other at G7.0 in five sequences and at G5.0 in the remaining globins, Scaff 42000019 and Scaff 76000030.

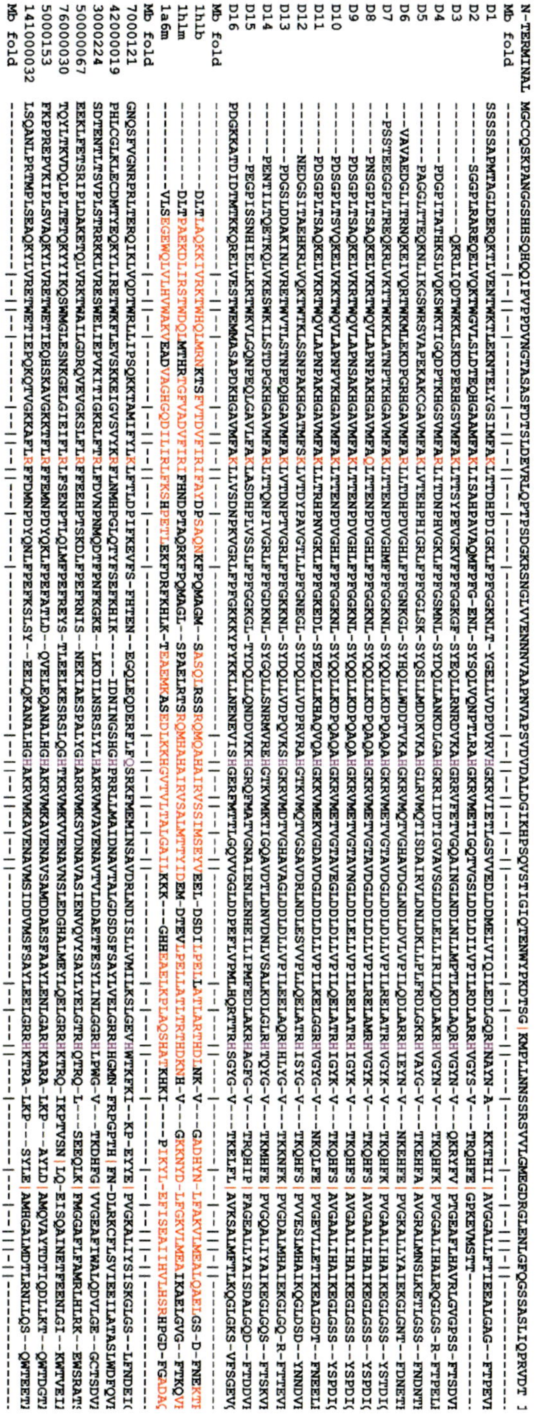

FIG. 1. The complete amino acid sequence of the multidomain Hb of *Strongylocentrotus purpuraeus* with the 16 globin domains aligned with each other and the sequences of *Caudina arenicola* chains 1hlb and 1hlm and sperm whale Mb (PDB: 1a6m) (helices A–H in red) and 7 single domain globins from *N. vectensis*. The red lines and red letters in the domains mark the intron insertion phases, 0 and 2, respectively. The Mb fold template consists of residues at 37 positions, defining helices A through H: A8, A11, A12, A15, B6, B9, B10, B13, B14, C4, CD1, CD4, E4, E7, E8, E11, E12, E15, E18, E19, F1, F4, F8, FG4, G5, G8, G11, G12, G13, G15, G16, H7, H8, H11, H12, H15, and H19. The distal and proximal His residues at E7 and F8 are in purple

# Blastp Searches with *S. purpuratus* Globins

Blastp searches (Schaffer et al. 2001) (http://www.ncbi.nlm.nih.gov/blast/) were carried out with each of the 16 *S. purpuratus* domains, except D2. They revealed that domains D1 and D3–D16 have higher bit scores for pairwise alignments with all the vertebrate Ngbs, *N. vectensis* globins and at least 24 bacterial SDgbs, ahead of any other globins, including those of other echino-derms *C. arenicola* and *P. chilensis*. The overwhelming majority of the recog-nised bacterial globins belong to the proteobacteria. Furthermore, the SDgbs from the ?-proteobacterium *Shewanella* species occur invariably among the globins with the highest bit scores. Fig. 2 shows a typical histogram of bit scores for a blastp search using D13 as query sequence. These results are in agreement with the proposal that all metazoan globins originated from one or more of prokaryotic globin genes transferred horizontally to unicellular eukaryote ancestors (Vinogradov et al. 2005, 2006, 2007).

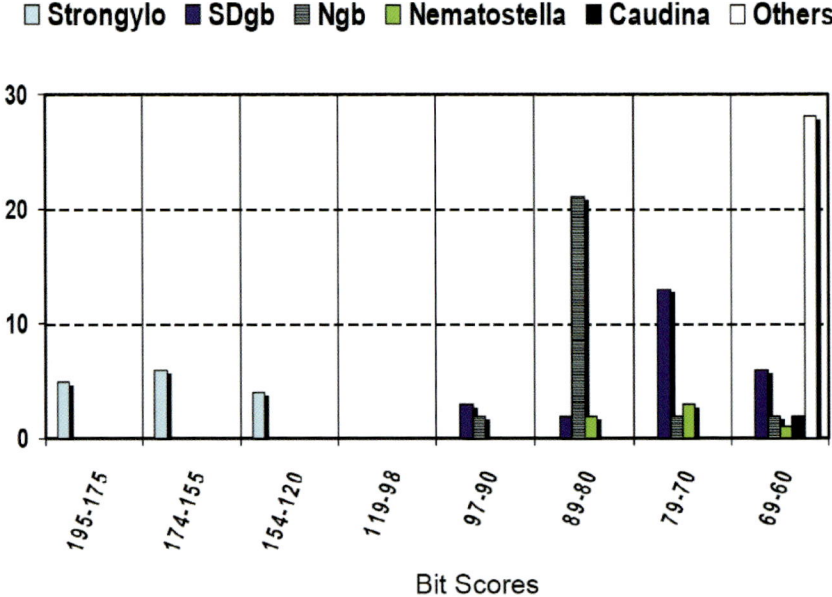

FIG. 2. Histogram of bit scores obtained in a blastp search using *S. purpuratus* globin domain D13 as the query sequence. This domain recognises the other *S. purpuratus* domains (bit score=196–120), vertebrate Ngbs (striated) (bit score=97–60), the *N. vecten-sis* globins (green) (bit score=89–60), single domain bacterial globins (dark blue) (bit score=97–60), the two *C. arenicola* globins (black) (bit score=69–60) and other metazoan globins (blank) (bit score=69–60)

# Molecular Phylogenetic Analysis of the *S. purpuratus* Globin Domains and Different Globin Families

A Bayesian phylogenetic analysis of the *S. purpuratus* globin domains, the 7 *N. vectensis* and 80 representative metazoan globins was carried out using Mr. Bayes 3.1.2 (Huelsenbeck and Ronquist 2001) (http://mrbayes.csit. fsu.edu/index.php). The Wag transition matrix was used, four chains were run simultaneously for 2 000 000 generations and trees were sampled every 100 generations. The resulting tree is shown in Fig. 3. The 80 metazoan globins comprised 4 vertebrate globinX (*Carassius auratus* CAG25724.1, *Danio rerio* CAG25723.1, *Tetraodon nigroviridis* CAG25725.1, *Xenopus tropicalis* NP_001011196.1), 5 vertebrate Ngbs (*Canis lupus familiaris* NP_001003356.2, *Cavia porcellus* CAH03122.1, *Gallus gallus* NP_001026722.1, *Homo sapiens* AJ245946.1, *Sus scrofa* NP_001001647.1), 5 vertebrates Cygbs (*G. gallus* NP_001008789.1, *Homo sapiens* NP_599030.1, *Mus musculus* NP_084482.1, *Rattus norvegicus* NP_570100.1), 19 echinoderm Hbs (*Caudina (Molpadia) arenicola* P80018 and P80017, *P. chilensis* P15161

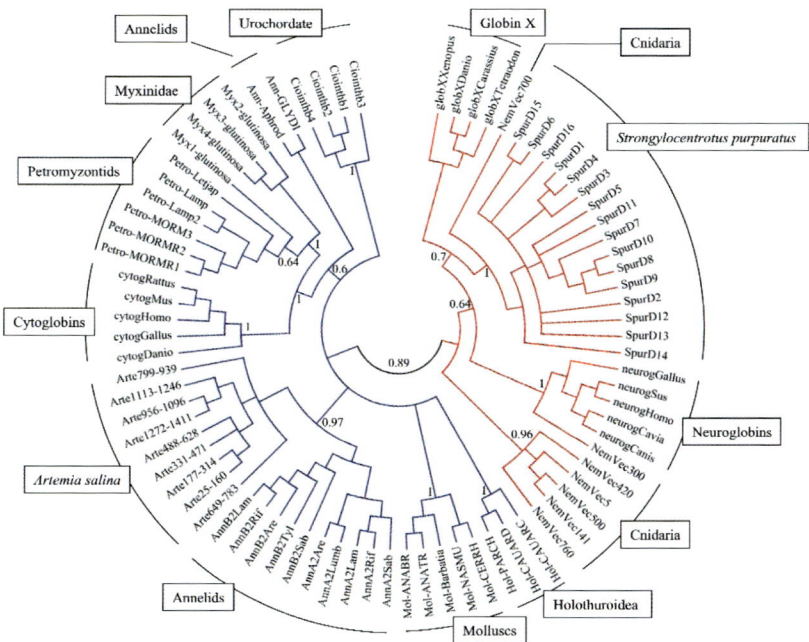

FIG. 3. Molecular phylogenetic reconstruction (Bayesian method) showing the close evolutionary relationship between the sea starlet anemone *N. vectensis*, the sea urchin *S. purpuratus* and vertebrate neuroglobins. The connection of globin X with neuroglobins (but NOT with *N. vectensis* globins) is in agreement with the postulate that globin X arose with neuroglobins by a duplication event in vertebrate lineage (Burmester et al. 2004)

and the 16 domains of *Strongylocentrotus*), 7 Cnidarian globins (*N. vectensis* in http://genome.jgi-psf.org/Nemve1/Nemve1.home.html: scaffold_3000224, scaffold_5000153, scaffold_42000019, scaffold_76000030, scaffold_141000032, scaffold_50000067, scaffold_7000121), 5 mollusc globins (*Anadara broughtonii* P02212, *Anadara trapezia* P04251, *Barbatia virescens* AAB24577.2, *Cerithidea rhizophorarum* P02215, *Nassarius mutabilis* P31331), 6 lamprey globins (*Petromyzon marinus* 0909197A and 4389013, *Lethenteron japonicum* BAF47285.1, *Mordacia mordax* P21198, P21197 and P21199), 4 hagfish globins (*Myxine glutinosa* AAD40480.1, AAD40310.1, AAD56540.1 and AAD56539.1), 4 urochordate globins (*C. intestinalis* AJ548502.1, AJ557135.1, AJ548501.1, AJ548500.1), the 9 domains of the polymeric Hb of the arthropod *Artemia salina* (4020136) and 12 annelid globins (*Aphrodite aculeata* AAC47259; *Arenicola marina* CAI56308, CAI56309; *Glycera dibranchiata* P21660; *Lamellibrachia* sp. AAP04529, P15469; *Lumbricus terrestris* P02218; *Riftia pachyptila* P80592, CAD29159; *Sabella spallanzanii* CAC37412, CAC37410; *Tylorrhynchus heterochaetus* P13578).

The Bayesian phylogenetic tree in Fig. 3 shows two separate clusters supported by a high posterior probability 0.89: one branch containing all the *S. purpuratus* globin domains, the *N. vectensis* globins and the vertebrate Ngbs and globin X, and the other branch comprising the globins from annelids, molluscs, a crustacean (*Artemia*), a urochordate (*Ciona*), other echinoderms (Holothuroidae), lamprey (Petromyzontidae), hagfish (Myxinidae) and vertebrate Cygbs. The relatedness of the sea urchin and Cnidarian globins with vertebrate Ngbs and globin X is also supported by the results of blastp searches (Fig. 2). Likewise, the location of introns is similar in the *N. vectensis* and *S. purpuratus* globins.

The genome and ESTs of *N. vectensis* have been found to have a gene repertoire and gene organisation similar to vertebrates (Miller, Ball and Technau 2005; Putnam et al. 2007). Furthermore, a comparison of the *N. vectensis*, protostomes (*Anopholes gambiae*, *Caenorhabditis elegans* and *Drosophila melanogaster*) and human genomes (Sullivan et al. 2006a) has shown a conservation of intron positions between the Radiata *N. vectensis* and the deuterostome *Homo sapiens*. These results suggest that "(1) early animal genomes were intron-rich, (2), a large fraction of introns present within the human genome likely originated early in evolution, before the Cnidarian–Bilaterian split and (3) there has been a high degree of intron loss during the evolution of the protostome lineage leading to the fruitfly, mosquito, and nematode". The current view is that introns do not represent inactive junk DNA, but rather comprise modulators allowing fine regulation of gene expression (Vasil et al. 1989; Mattick and Gagen 2001; Nott, Hir and Moore 2004). Since a recent survey of intron loss and gain in the evolution of paralogous gene families concluded that intron gain was prevalent over intron loss (Babenko et al. 2004), it is not surprising to find, in addition to the canonical positions B12.2 and G7.0, extra introns, at E11.0 in Ngbs (Burmester et al. 2000), at E10.2 and H10.0 in globin X (Roesner et al. 2005), at HC11.2 in Cygbs (Burmester et al. 2002) and at

E10.2 in *C. intestinalis* (Ebner, Burmester, and Hankeln 2003): additional introns in vertebrate neuroglobins could represent innovations for more complex and finely tuned tissue-specific expression.

Our results place the *S. purpuratus* and *N. vectensis* globins together with the Ngb of vertebrate globins and we hypothesise that a plesiomorphic function of metazoan globins could be related to the nervous system. It will therefore be of great interest to find out what globins occur in other related animal groups, such as the Myxozoa, the Ctenophora, the Placozoa, the Porifera, and also in basal Bilateria such as acoels (Ruiz-Trillo et al. 1999; Sempere et al. 2007).

## *References*

Babenko, V. N., Rogozin, I. B., Mekhedov, S. L., and Koonin, E. V. 2004. Prevalence of intron gain over intron loss in the evolution of paralogous gene families. Nucleic Acids Res. 32:3724–3733.

Baker, S., and Terwilliger, N. B. 1993. Hemoglobin structure and function in the rat-tailed sea cucumber *Paracaudina chilensis*. Biol. Bull. 185:115–122.

Bishop, J., Vandergon, T., Green, D., Doeller, J., and Kraus, D. 1998. A high-affinity hemoglobin is expressed in the notochord of amphioxus, *Branchiostoma californiense*. Biol. Bull. 195:255–259.

Burmester, T., Weich, B., Reinhardt, S., and Hankeln, T. 2000. A vertebrate globin expressed in the brain. Nature 407:520–523.

Burmester, T., Ebner, B., Weich, B., and Hankeln, T. 2002. Cytoglobin: a novel globin type ubiquitously expressed in vertebrate tissues. Mol. Biol. Evol. 19:416–421.

Burmester, T., Haberkamp, M., Mitz, S., Roesner, A., Schmidt, M., Ebner, B., Gerlach, F., Fuchs, C., and Hankeln, T. 2004. Neuroglobin and cytoglobin: genes, proteins and evolution. IUBMB Life 56:703–707.

Christensen, A., Colacino, J., and Bonaventura, C. 2003. Functional and biochemical properties of the hemoglobins of the burrowing brittle star *Hemipholis elongata* Say (Echinodermata, Ophiuroidea). Biol. Bull. 205:54–65.

Coleman, M., Matthews, C. M., and Trotman, C. N. 2001. Multimeric hemoglobin of the Australian brine shrimp *Parartemia*. Mol. Biol. Evol. 18:570–576.

Ebner, B., Burmester, T., and Hankeln, T. 2003. Globin genes are present in *Ciona intestinalis*. Mol. Biol. Evol. 20:1521–1525.

Hardison, R. C. 1996. A brief history of hemoglobins: plant, animal, protist, and bacteria. Proc. Natl. Acad. Sci. U.S.A. 93:5675–5679.

Huelsenbeck, J. P., and Ronquist, F. 2001. MRBAYES: Bayesian inference of phylogenetic trees. Bioinformatics 17:754–755.

Keller, E. B., and Noon, W. A. 1984. Intron splicing: a conserved internal signal in introns of animal pre-mRNAs. Proc. Natl. Acad. Sci. U.S.A. 81:7417–7420.

Letunic, I., Copley, R. R., Pils, B., Pinkert, S., Schultz, J., and Bork, P. 2006. SMART 5: domains in the context of genomes and networks. Nucleic Acids Res. 34:D257–260.

Matthews, C. M., and Trotman, C. N. 1998. Ancient and recent intron stability in the Artemia hemoglobin gene. J. Mol. Evol. 47:763–771.

Mattick, J. S., and Gagen, M. J. 2001. The evolution of controlled multitasked gene networks: The role of introns and other noncoding RNAs in the development of complex organisms. Mol. Biol. Evol. 18:1611–1630.

Mauri, F., Omnaas, J., Davidson, L., Whitfill, C., and Kitto, G. B. 1991. Amino acid sequence of a globin from the sea cucumber *Caudina* (Molpadia) *arenicola*. Biochim. Biophys. Acta 1078:63–67.

McDonald, G. D., Davidson, L., and Kitto, G. B. 1992. Amino acid sequence of the coelomic C globin from the sea cucumber *Caudina* (Molpadia) *arenicola*. J. Protein Chem. 11:29–37.

Miller, D. J., Ball, E. E., and Technau, U. 2005. Cnidarians and ancestral genetic complexity in the animal kingdom. Trends Genet. 21:536–539.

Mitchell, D. T., Kitto, G. B., and Hackert, M. L. 1995. Structural analysis of monomeric hemichrome and dimeric cyanomet hemoglobins from Caudina arenicola. J. Mol. Biol. 251:421–431.

Mount, S. M. 1982. A catalogue of splice junction sequences. Nucleic Acid Res. 10: 459–472.

Nott, A., Hir, H. L., and Moore, M. J. 2004. Splicing enhances translation in mammalian cells: An additional function of the exon junction complex, Genes Dev. 18:210–222.

Putnam, N. H., Srivastava, M., Hellsten, U., Dirks, B., Chapman, J., Salamov, A., Terry, A., Shapiro, H., Lindquist, E., Kapitonov, V. V., Jurka, J., Genikhovich, G., Grigoriev, I. V., Lucas, S. M., Steele, R. E., Finnerty, J. R., Technau, U., Martindale, M. Q., and Rokhsar, D. S. 2007. Sea anemone genome reveals ancestral eumetazoan gene repertoire and genomic organization. Science 317:86–94.

Roesner, A., Fuchs, C., Hankeln, T., and Burmester, T. 2005. A globin of ancient evolutionary origin in lower vertebrates: evidence for two distinct globin families in animals. Mol. Biol. Eval. 22:12–20.

Ruiz-Trillo, I., Riutort, M., Littlewood, D. T., Herniou, and E. A., Baguna, J. 1999. Acoel flatworms: earliest extant bilaterian Metazoans, not members of Platyhelminthes. Science 283:1919–1923.

Schaffer, A. A., Aravind, L., Madden, T., Shavrin, S., Spourge, J., Wolf, Y., Koonin, E.V., and Altschul, S. F. 2001. Improving the accuracy of PSI-BLAST protein database searches with composition-based statistics and other refinements. Nucleic Acids Res. 29:2994–3005.

Sempere, L. F., Martinez, P., Cole, C., Baguna, J., and Peterson, K. J. 2007. Phylogenetic distribution of microRNAs supports the basal position of acoel flatworms and the polyphyly of Platyhelminthes. Evol. Dev. 9:409–415.

Shi, J., Blundell, T., and Mizuguchi, K. 2001. FUGUE: sequence-structure homology recognition using environment-specific substitution tables and structure-dependent gap penalties. J. Mol. Biol. 310:243–257.

Sodergren, E., Weinstock, G. M., Davidson, E. H., Cameron, R. A., Gibbs, R. A., et al. 2006. The genome of the sea urchin *Strongylocentrotus purpuratus*. Science 314:941–952.

Sullivan, J. C., Ryan, J. F., Watson, J. A., Webb, J., Mullikin, J. C., Rokhsar, D., and Finnerty, J. R. 2006. StellaBase: the *Nematostella vectensis* Genomics Database. Nucleic Acids Res. 34:D495–499.

Sullivan, J. C., Reitzel, A. M., and Finnerty, J. R. 2006. A high percentage of introns in human genes were present early in animal evolution: evidence from the basal metazoan Nematostella vectensis. Genome Inform. 17:219–229.

Suzuki, T. 1989. Amino acid sequence of a major globin from the sea cucumber *Paracaudina chilensis*. Biochim. Biophys. Acta 998:292–296.

Vasil, V., Clancy, M., Ferl, R. J., Vasil, I. K., and Hannah, L.C. 1989. Increased gene expression by the first intron of maize Shrunken-1 locus in grass species. Plant Physiol. 91:1575–1579.

Vinogradov, S. N., Hoogewijs, D., Bailly, X., Arredondo-Peter, R., Gough, J., Guertin, M., Dewilde, S., Moens, L., and Vanfleteren, J. R., 2005. Three globin lineages belonging to two structural classes in genomes from the three kingdoms of life. Proc. Natl. Acad. Sci. U.S.A. 102:11385–11389.

Vinogradov, S. N., Hoogewijs, D., Bailly, X., Arredondo-Peter, R., Gough, J., Guertin, M., Dewilde, S., Moens, L., and Vanfleteren, J. R., 2006. A phylogenomic profile of globins. BMC Evol. Biol. 6:31–67.

Vinogradov, S. N., Hoogewijs, D., Bailly, X., Dewilde, S., Moens, L., and Vanfleteren, J. R. 2007. A model of globin evolution. Gene 398:132–142.

Zhang, J. 2003. Evolution by gene duplication: an update. Trends Ecol. Evol. 18:292–298.

# 17

# Whale Rider: The Co-occurrence of Haemoglobin and Haemocyanin in *Cyamus scammoni*

Nora Terwilliger

## Abstract

Beatrice and Jonathan Wittenberg have inspired generations of scientists to explore the mysteries of heme proteins. Their papers on the high-affinity haemoglobins of *Ascaris*, a parasite living in the gut of a mammal, prompted discussions among Bob Terwilliger, Bob Garlick and the author in the 1970s, and led to investigations on numerous invertebrate myoglobins and haemoglobins in the Terwilliger laboratory. In this chapter, I review the structure and function of several of the outstanding molecules and their source organisms, including the haemoglobin and haemocyanin from *Cyamus scammoni*, the crustacean amphipod that rides on a whale. The presence of a high $M_w$ extracellular haemoglobin in this arthropod is the first report of haemoglobin expression in the more advanced crustaceans. The co-occurrence of haemoglobin and haemocyanin is also extremely unusual and unprecedented in the Arthropoda.

The red patches on black are visible as the whales tango through the waters of Scammon's Lagoon, on the western shore of Baja, Mexico. An unusual haemoglobin in *Cyamus scammoni*, a crustacean amphipod that lives on the surface of the grey whale, *Eschrictius robustus*, is the source of the red colour (Terwilliger 1991). The whales arrive in late winter to give birth and mate before swimming back to the colder waters of Alaska. Occasionally, a grey whale lingers along the Oregon coast on its way north, feeding in the nutrient-rich nearshore waters. Even from the headlands (using binoculars), one can spot the haemoglobin-red patches on the whale.

Haemoglobins like this one from *Cyamus* have tweaked the enthusiasm and curiosity of Beatrice and Jonathan Wittenberg for years. This chapter is dedicated to these two scientists, who have inspired many of us to join them in investigating the structure and function of heme proteins. I recall conversations with Bob Terwilliger and Bob Garlick in the 1970s, talking in our lab about the Wittenbergs' papers on the oxygen binding of the haemoglobins of the nematode, *Ascaris lumbricoides* (Okazaki et al. 1965, 1967; Okazaki and Wittenberg 1965; Wittenberg, Okazaki and Wittenberg 1965). We speculated about living conditions in the gut of a mammal, where the parasitic *Ascaris* is found, and thought about dark, anoxic conditions and the potential usefulness of a high-affinity haemoglobin in the nematode. We went on to study annelid worm haemoglobins and chlorocruorins, molluscan haemoglobins, myoglobins and haemocyanins, and arthropod blood proteins too. One of the weirdest animals was a barnacle that parasitises the king crab and transfers oxygen from the crab to the barnacle's embryos via an extracellular haemoglobin in the barnacle's circulatory system (Fox 1953; Terwilliger, Terwilliger and Schabtach 1986). Whenever we presented our findings at a conference, we could count on interested responses from Beatrice and Jonathan.

Haemoglobin expression in an amphipod is unique. While haemoglobins have been described in a limited number of crustaceans, including Ostracoda, Copepoda, Cirrepedia, and especially Branchiopoda such as *Artemia*, *Daphnia* and *Lepidurus*, haemoglobins have not been found in the higher crustaceans, including the Malacostraca (Terwilliger and Ryan 2001; Weber and Vinogradov 2001 for review). Haemocyanin, the blue copper-containing protein, is the oxygen transport molecule that typically occurs in the more advanced crustaceans, as well as in the chelicerates. We initially set out to characterise the haemoglobin of *Cyamus*. Recently we were surprised to discover that *Cyamus* also expresses haemocyanin (Terwilliger and Ryan 2006). The co-occurrence of haemoglobin and haemocyanin is extremely unusual, and it was unprecedented in the Arthropoda. Here, I review the haemoglobin, introduce the haemocyanin and discuss its possible function in the amphipod, *C. scammoni*.

On the whale, the dorso-ventrally flattened cyamids tend to cluster in folds of the skin or around clumps of barnacles (*Cryptolepas rhachianecti*), holding on with their hooked dactyls (Figs. 1 and 2). As the haemoglobin circulates through the corkscrew shaped gills on the ventral surface of the cyamid (Fig. 2), it can bind oxygen and then transport it throughout the body of adult and juvenile cyamids. Red haemolymph in the swollen lobes or oostegites that form the female's marsupial pouch suggests that oxygen transfer to the developing embryos within the marsupium occurs as well. The ~1800-kDa multisubunit, multidomain haemoglobin (Terwilliger, 1991) has a distinctive chevron shape when examined by transmission electron microscopy (Fig. 3). Some of the molecules resemble igloos with a low opening on one side of the chevron, and the chevrons sometimes aggregate in clusters or strings of 3–6. The apparent subunit size of 175 kDa as determined by SDS PAGE in the presence of reducing agents is larger than that of the multidomain haemoglobin subunit of another crustacean, *Artemia*. Proteolysis experiments and heme analysis indicate the subunit of *Cyamus* haemoglobin is composed of a series of smaller domains; however, the low heme/protein ratio of the subunit suggests that not all domains contain a heme. Oxygen equilibrium studies show that the haemoglobin has a relatively low oxygen affinity ($P_{50}$=14–24 mmHg between pH 7.0 and 7.9; 1 mmHg=133 Pa) with a slight Bohr effect and cooperativity of about 1.5. These oxygenation properties indicate that the haemoglobin functions in oxygen transport rather than in oxygen storage.

FIG. 1. Adult and juvenile *Cyamus scammoni* amphipods crawl over coronulid barnacles *Cryptolepas rhachianecti* on the surface of a grey whale *Eschrictius robustus*. The large cyamids are about 30 cm long

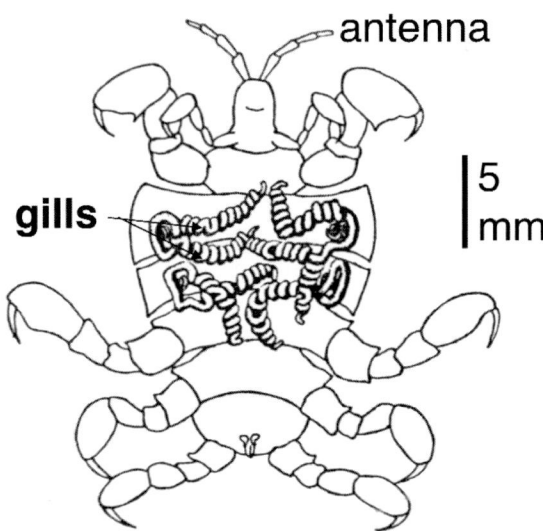

FIG. 2. Ventral view of *Cyamus scammoni* Dall, 1872, male with species-specific corkscrew gills. Medial orientation of hooked dactyls on tips of appendages helps ensure a tight grip of amphipod onto surface of whale or barnacle. Sketch adapted from Leung 1967

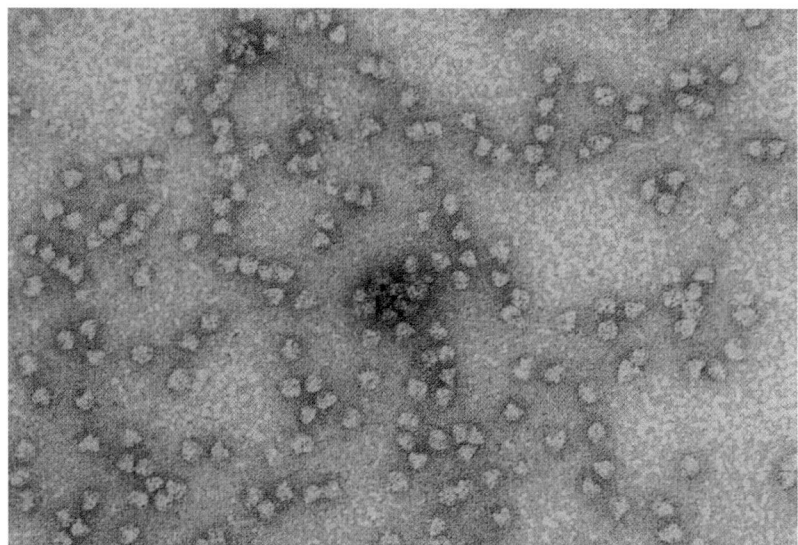

Fɪɢ. 3. TEM of negatively stained *Cyamus scammoni* haemolymph. The majority of the molecules are chevron-shaped haemoglobins. Square or hexagonal shapes that may correspond to haemocyanin hexamers are also present. Microscopy by E. Schabtach

We have identified haemocyanin expression in *Cyamus* using PCR and degenerate crustacean haemocyanin primers that were designed based on conserved regions near the Copper A site (Terwilliger and Ryan 2006). Total RNA was obtained from fresh specimens of *C. scammoni*, reverse transcribed and amplified. The 350-bp products were cloned, sequenced and used to design additional primers to further amplify the haemocyanin using the Clontech SMART RACE protocol (BD Biosciences). The resulting *C. scammoni* haemocyanin cDNA (GenBank DQ230982) was analysed by the BLAST search programs and ClustalW alignments. In a phylogenetic comparison with 72 other members of the arthropod haemocyanin gene superfamily, *Cyamus* haemocyanin clusters with the haemocyanin from another amphipod, *Gammarus roeseli* (AJ937836), in a clade of crustacean haemocyanins and cryptocyanins (Terwilliger and Ryan 2006, figure 7). Native PAGE of both whole haemolymph and a 500-kDa fraction (BioGel A-5m Fraction III, Terwilliger 1991) shows a band with mobility similar to *Cancer magister* hexameric haemocyanin, in addition to the more slowly migrating red band of haemoglobin. This putative haemocyanin band re-electrophoreses on SDS PAGE as a doublet corresponding to 81 and 79 kDa, the expected size of haemocyanin subunits (data not shown). In electron micrographs of whole haemolymph and Fraction III, occasional square and hexameric shapes, the typical appearance of crustacean haemocyanin hexamers, may represent *Cyamus* haemocyanin (Fig. 3).

The expression of both haemoglobin and haemocyanin in *Cyamus* was unexpected, as the haemolymph of organisms in most phyla studied have either the iron-based or copper-based oxygen transport protein, but not both. Cytoplasmic globins and circulating haemocyanins do co-occur in the Mollusca. Gastropods and chitons, for example, often have radular muscle myoglobin, neuroglobin and circulating haemocyanin, while primitive bivalves express cytoplasmic haemoglobins in nerve and/or gill tissues plus circulating haemocyanin (Terwilliger 1998). At least one deep-sea gastropod has a gill cytoplasmic haemoglobin (Wittenberg and Stein 1995); it probably has a circulating haemocyanin also. However, a planorbid snail, *Biomphalaria glabrata*, is the only other animal reported to date with both circulating haemoglobin and haemocyanin (Lieb et al. 2006). The examples of both red and blue oxygen transport proteins in the arthropod *Cyamus* and the mollusc *Biomphalaria* will doubtless spur further investigations into dual expression patterns and debates about the evolution of the proteins.

Does *Cyamus* also express prophenoloxidase, an enzyme phylogenetically related to haemocyanin, whose active form catalyses the melanisation and sclerotisation of the arthropod exoskeleton? Like haemocyanin, prophenoloxidase contains the dioxygen-copper binding site. We searched for prophenoloxidase mRNA using degenerate chelicerate haemocyanin primers (Parkinson et al. 2001; Kusche, Ruhberg and Burmester 2002). These primers have been successful in obtaining prophenoloxidase sequences in haemocyanin-containing crabs and in branchiopods with circulating haemoglobins (Terwilliger and Ryan 2006), but did not amplify a prophenoloxidase from *Cyamus*.

The apparent absence of a prophenoloxidase in the amphipod *Cyamus* is consistent with results from other peracaridan crustaceans, including the isopods *Bathynomus giganteus* and *Cirolana harfordi* (Pless et al. 2003; Arellano and Terwilliger 2004; Terwilliger 2007) and the gammarid amphipod *Megalorchestia californiana* (Terwilliger, unpublished), as well as several chelicerates (Decker and Jaenicke 2004). In the peracaridans and chelicerates, phenoloxidase activity of the haemocyanin appears to compensate for the lack of a phenoloxidase protein. It is likely that in *Cyamus*, also, the haemocyanin functions as a phenoloxidase, participating in moulting, wound healing and immunochallenge. Haemocyanin's typical role in oxygen transport in other peracaridans may be minimal in this unique amphipod, whose haemoglobin functions as the primary oxygen carrier.

*Acknowledgements*

This study has especially benefited from the contributions of M. R. Ryan, E. Schabtach and R. C. Terwilliger. I also thank numerous colleagues who have helped harvest live cyamids from stranded whales.

# References

Arellano, S., and Terwilliger, N. 2004. Hemocyanin, cryptocyanin and phenoloxidase in deep sea (*Bathynomus giganteus*) and intertidal (*Cirolana harfordi*) isopods. Integr. Comp. Biol. 43:961A.

Decker, H., and Jaenicke, E. 2004. Recent findings on phenoloxidase activity and antimicrobial activity of hemocyanins. Devel. Comp. Immun. 28:673–687.

Fox, H. 1953. Hemoglobin and biliverdin in parasitic cirripede Crustacea. Nature 171:162–163.

Kusche, K., Ruhberg, H., and Burmester, T. 2002. A hemocyanin from the Onychophora and the emergence of respiratory proteins. Proc. Natl. Acad. Sci. U.S.A. 98:10545–10548.

Leung, Y. 1967. An illustrated key to the species of whale-lice (Amphipoda, Cyamidae), ectoparasites of cetacea, with a guide to the literature. Crustaceana 12:279–291.

Lieb, B., Dimitrova, K., Kang, H., Braun, S., Gebauer, W., Martin, A., Hanelt, B., Saenz, S., Adema, C., and Markl, J. 2006. Red blood with blue-blood ancestry: Intriguing structure of a snail hemoglobin. Proc. Natl. Acad. Sci. U.S.A. 103:12011–12016.

Okazaki, T. and Wittenberg, J. B. 1965. The hemoglobin of *Ascaris* perienteric fluid. III. Equilibria with oxygen and carbon monoxide. Biochim. Biophy. Acta 111:503–511.

Okazaki, T., Briehl, R., Wittenberg, J., and Wittenberg, B. 1965. The hemoglobin of *Ascaris* perienteric fluid. II. Molecular weight and subunits. Biochim. Biophys. Acta 111:496–502.

Okazaki, T., Wittenberg, B., Briehl, R., and Wittenberg, J. 1967. The hemoglobin of *Ascaris* body walls. Biochim. Biophys. Acta 140:258–265.

Parkinson, N., Smith, I., Weaver, R., and Edwards, J. 2001. A new form of arthropod phenoloxidase is abundant in venom of the parasitoid wasp *Pimpla hypochondriaca*. Insect Biochem. Mol. Biol. 31:57–63.

Pless, D., Aguilar, M., Falcon, A., Lozano-Alvarez, E., and Heimer de la Cotera, E. 2003. Latent phenoloxidase activity and N-terminal amino acid sequence of hemocyanin from *Bathynomus giganteus*, a primitive crustacean. Arch. Biochem. Biophys. 409:402–410.

Terwilliger, N. B. 1991. Arthropod (*Cyamus scammoni*, Amphipoda) hemoglobin structure and function. In Structure and Function of Invertebrate Oxygen Carriers, eds. S. Vinogradov and O. Kapp, pp. 59–63. Berlin, Heidelberg, New York: Springer.

Terwilliger, N. B. 1998. Functional adaptations of oxygen-transport proteins. J. Exp. Biol. 201:1085–1098.

Terwilliger, N. B. 2007. Hemocyanins and the immune response: Defense against the dark arts. Integr. Comp. Biol. 47:662–665.

Terwilliger, N. B. and Ryan, M. 2001. Ontogeny of crustacean respiratory proteins. Am. Zool. 41:1057–1067.

Terwilliger, N. B. and Ryan, M. 2006. Functional and phylogenetic analyses of phenoloxidases from brachyuran (*Cancer magister*) and branchiopod (*Artemia franciscana*, *Triops longicaudatus*) crustaceans. Biol. Bull. 210:38–50.

Terwilliger, R. C., Terwilliger, N. B., and Schabtach, E. 1986. Hemoglobin from the parasitic barnacle, *Briarosaccus callosus*. In Invertebrate Oxygen Carriers, ed. B. Linzen, pp. 125–127. New York: Springer.

Weber, R. and Vinogradov, S. 2001. Nonvertebrate hemoglobins: Functions and molecular adaptations. Physiol. Rev. 81:569–628.

Wittenberg, B., Okazaki, T., and Wittenberg, J. 1965. The hemoglobin of *Ascaris* perienteric fluid. I. Purification and spectra. Biochim. Biophys. Acta 111:485–495.

Wittenberg, J. and Stein, J. 1995. Hemoglobin in the symbiotic-harboring gill of the marine gastropod *Alviniconcha hessleri*. Biol. Bull. 188:5–7.

# 18

# Neuroglobin and Other Nerve Haemoglobins

Thorsten Burmester and Thomas Hankeln

## Abstract

The nervous system of animals requires huge amounts of metabolic energy and thus oxygen. Intracellular haemoglobins sporadically occur in glial cells and neurons of various invertebrate taxa, including Annelida, Arthropoda, Echiura, Mollusca, Nematoda and Nemertea. At least some of these respiratory proteins sustain the aerobic metabolism and thus the excitability of the nervous system. Recently, we have identified neuroglobin as an oxygen-binding protein of vertebrate neurons. The physiological role of neuroglobin, which is apparently present in much lower amounts than many invertebrate nerve haemoglobins, is less well established. Phylogenetic analyses have shown that neuroglobin is orthologous to at least some of the invertebrate respiratory proteins, while other invertebrate nerve globins may have divergent evolutionary origins. Functional changes thus appear to be common in animal globin evolution. The occurrence of yet other types of intracellular globins in nervous tissue of vertebrates is functionally unexplained and requires further studies.

The nervous system of most animals consumes large amounts of metabolic energy which requires a continuous supply with sufficient oxygen. Certain pathological and environmental conditions can cause hypoxic phases and may therefore result in serious damage to the nervous system. Intracellular respiratory proteins may sustain the oxidative metabolism either by supporting a continuous flow of oxygen to the mitochondria or by providing a short-term store for hypoxic phases (Wittenberg and Wittenberg 1990; Wittenberg 1992). The most prominent respiratory proteins are haemoglobin (Hb), which transports $O_2$ in the body fluid, and myoglobin (Mb), which supplies $O_2$ within muscle cells (Wittenberg and Wittenberg 2003). Intracellular globins also occur in the nervous system of various animals. While known for more than 130 years among invertebrates (Lankester 1872), the discovery of globins in the nervous system of vertebrates (Burmester et al. 2000) has created much interest in unravelling the cellular functions of these respiratory proteins (Reviews: Burmester and Hankeln 2004; Hankeln et al. 2005; Hankeln and Burmester 2008).

# Invertebrate Nerve Haemoglobins

As early as 1872, Lankester noted a bright red colour in the nerve cord from the polychaete annelid *Aphrodite aculeata*, which reminded him of the appearance of human blood. In later years, intracellular haemoglobins were discovered in the nervous systems of several invertebrates, although they are by no means common (Wittenberg 1992, Weber and Vinogradov 2001). Phyla in which nerve haemoglobins (nHbs) have been observed include Annelida, Arthropoda, Echiura, Mollusca, Nematoda and Nemertea. nHbs can either reside in neurons themselves or in supporting glial cells (Table 1; Wittenberg 1992). In *A. aculeata*, the molluscs *Tellina alternata*, *Spisula solidissima*, *Lymnaea stagnalis* and *Planorbis corneus* (Schindelmeiser, Kuhlmann and Nolte 1979; Kraus, Doeller and Smith 1988), and in the nemertean *Cerebratulus lacteus* (Vandergon et al. 1998) the nHb is located in glia. Neuronal nHbs have been identified in the marine gastropod *Aplysia* sp. and the air-breathing gastropods *Helix pomatia* and *Cepaea nemoralis* (Wittenberg, Briehl and Wittenberg 1965; Schindelmeiser, Kuhlmann and Nolte 1979). The physiological implications of this cell type-specific expression of nHbs in the nervous system are uncertain. However, by rule of thumb, glia-based nHbs accumulate to concentrations of several mM (Doeller and Kraus 1988; Wittenberg 1992), whereas in neurons nHb concentrations are substantially lower. Interestingly, some nerve-specific nHbs of molluscs are bound to intracellular granules (Arvanitaki and Chalazonitis 1960; Wittenberg 1992). Initially it was assumed that these nHbs have photoreceptor function, but later studies found no supporting evidence (Wittenberg 1992).

TABLE 1. Characteristics of selected invertebrate nerve haemoglobins and vertebrate neuroglobin

| Species | Phylum | Localisation | $P_{50}$ (Torr) | $Fe^{2+}$-coordination | Reference |
|---|---|---|---|---|---|
| *Aphrodite aculeata* | Annelida | Glia | 1.1 | penta | Wittenberg et al. 1965 |
| *Cerebratulus lacteus* | Nemertea | Glia | 2.9 | penta | Vandergon et al. 1998 |
| *Spisula solidissima* | Mollusca | Glia | 2.3 | hexa | Doeller and Kraus 1988 |
| *Tellina alternata* | Mollusca | Glia | 1.3 | hexa | Doeller and Kraus 1988 |
| *Macrocalesta nimbosa* | Mollusca | | | | Doeller and Kraus 1988 |
| *Aplysia depilans* | Mollusca | Neurons | 4 | penta | Wittenberg et al. 1965 |
| *Lymnaea stagnalis* | Mollusca | Glia | | | Schindelmeiser et al. 1979 |
| *Planorbarius corneus* | Mollusca | Glia | | | Schindelmeiser et al. 1979 |
| *Helix pomatia* | Mollusca | Neurons | | | Schindelmeiser et al. 1979 |
| *Cepaea nemoralis* | Mollusca | Neurons | | | Schindelmeiser et al. 1979 |
| *Homo sapiens* | Chordata | Neurons | 1 | hexa | Burmester et al. 2000 |

In 1965, Beatrice and Jonathan Wittenberg studied the biochemical properties of nHbs from two annelids (*A. aculeata, Halosydna* sp.) and one molluscan species (*Aplysia californica*) (Wittenberg, Briehl and Wittenberg 1965). They concluded that nHbs have similar properties like myoglobin in muscle. Oxygen affinities are moderate with $P_{50}$ values ranging between 1.1 and 4 Torr (Table 1). nHbs may be Mb-like monomers or associate to quaternary structures up to tetramers with cooperative $O_2$-binding (Doeller and Kraus 1988; Wittenberg 1992; Weber and Vinogradov 2001). Spectral analyses revealed that nHbs belong to two distinct structure types. Like vertebrate haemoglobin and myoglobin, the $Fe^{2+}$ of the heme in deoxygenated nHbs of *A. aculeata, Aplysia* sp. and *C. lacteus* nHbs is pentacoordinated (Wittenberg, Briehl and Wittenberg 1965; Vandergon et al. 1998). By contrast, the nHbs of the bivalves *T. alternata* and *S. solidissima* display a cytochrome b-like absorption spectrum, indicating hexacoordination of the heme-iron ion (Doeller and Kraus 1988; Dewilde et al. 2006). Here, the distal E7 histidine occupies the sixth coordinating position and thus competes with the binding ligand. The functional consequences of this structural property are currently unknown.

Three invertebrate nHbs have been characterised at the molecular level. The pentacoordinate nHb of the sea mouse *A. aculaeata* is a homodimer with subunits of 15.6 kDa (150 amino acids). The globin chains contain the typically conserved residues PheCD1, HisE7 and HisF8 (Dewilde et al. 1996). While coloration indicates the primary presence of nHb in the nerve cord, nHb mRNA could also be detected in other tissues such as longitudinal muscle, gut and pharynx. The gene has a single intron in position G7.0. Sequence comparison showed that *A. aculaeata* nHb displays the highest similarity to the haemoglobins from the bloodworm *Glycera dibranchiata* (Annelida) and to vertebrate neuroglobin (Dewilde et al. 1996; Burmester et al. 2000); these three globin types also form a common clade in phylogenetic analyses (see below). The hexacoordinate nHb of the Atlantic surf clam *S. solidissima* is composed of 162 amino acids, and also features the typical residues in CD1, E7 and F8 and harbours the two standard introns in B12.2 and G7.0 (Dewilde et al. 2006). It is phylogenetically not related to the annelid nHb, but groups with haemoglobins and myoglobins from other molluscs (see below). The nHb from the ribbon worm *C. lacteus* is the smallest functional globin characterised so far (Vandergon et al. 1998). Although it comprises only 109 amino acids and parts of helices A, B and H are deleted, all crucial residues (such as PheCD1, HisE7 and HisF8) are present and the globin fold is conserved (Pesce et al. 2004). The gene contains the globin-typical introns in positions B12.2 and G7.0. The sequence of *C. lacteus* nHb is very similar to that of the Hb isolated from the body wall of this species, suggesting that the nerves recruited such protein for respiratory purposes only recently. The phylogenetic affiliation of the nemertean nHb is ill-defined, due to its divergent and shortened primary sequence.

There is little doubt that the principal function of invertebrate nHbs is $O_2$ supply (Fig. 1a). In the mud-dwelling seashell *T. alternata* nHb guarantees neuron function for 30 min under anoxia, while no such effect could be demonstrated in the related, but nHb-less species *Tagelus plebeius* (Kraus and Colacino 1986). Moreover, when the $O_2$-delivery function of *T. alternata* nHb was abolished by poisoning with CO, neural excitability under anoxia ceased within a few minutes. Nerves of mollusc species that contain nHbs (*T. alternata*, *S. solidissima* and *Macrocallista nimbosa*) consume less $O_2$ during generation of action potential than nerves from mollusc species without nHb (*T. plebeius* and *Geukensia demissa*) (Kraus and Doeller 1988). In gastropods, the appearance of red-stained ganglia coincides with water-breathing and potentially hypoxic habitats, while air-breathing snails appear to have much less nHb (Schindelmeiser, Kuhlmann and Nolte 1979). Taken together, these data provide strong evidence that many invertebrate nHbs have a similar role like myoglobin in muscles, i.e., enhancing $O_2$ delivery to the mitochondria or providing a short-term $O_2$ store that enables the animals to better survive hypoxic periods.

## Neuroglobin: The Vertebrate Nerve Haemoglobin with Elusive Function

For a long time, Hb and Mb had been considered as the only types of globins in man and other vertebrates. However, in 2000 we identified neuroglobin (Ngb) in neuronal tissues of mice and humans (Burmester et al. 2000). Since then, Ngb has also been discovered in fish (Awenius, Hankeln and Burmester 2001), amphibians (Fuchs, Burmester and Hankeln 2006), reptiles (Milton et al. 2006) and birds (Kugelstadt et al. 2004), suggesting its ubiquitous occurrence in vertebrates. Ngb is a monomeric globin with a typical mass in the range of 16 kDa (151 to 164 amino acids). Recombinant Ngb binds $O_2$ with an affinity of a $P_{50}$ in the range of 1 Torr (Dewilde et al. 2001). In addition to the globin-typical introns in positions B12.2 and G7.0, the Ngb genes have conserved introns in E11.0. Like some invertebrate nHbs, Ngb belongs to the class of hexacoordinated globins; in the deoxy state the distal heme provides the sixth ligand which must be replaced upon $O_2$ binding (Pesce et al. 2003). Ngb is preferentially expressed in the neurons of the central and peripheral nervous systems, as well as some endocrine tissues (Burmester et al. 2000; Reuss et al. 2002). The highest Ngb concentration has been found in the neuronal retina (Schmidt et al. 2003). While it is generally accepted that invertebrate nHbs have a similar role to Mb in the vertebrate muscle and enhance the supply of $O_2$ to mitochondria (see above), Ngb has been implied to be involved in various different aspects of $O_2$-dependent metabolism (for review, see Burmester and Hankeln 2004; Hankeln et al. 2005; Hankeln and Burmester 2008).

First, Ngb may be in fact a typical intracellular $O_2$ supply protein that either acts as short-term $O_2$ buffer or that enhances $O_2$ diffusion to the mitochondria of metabolically highly active neurons (Fig. 1a; Burmester et al. 2000; Wittenberg and Wittenberg 2003). For certain highly active regions and the

strongly Ngb-expressing retina, local Ngb concentrations may in fact be sufficient for an $O_2$ supply function (Schmidt et al. 2003). There is also a strong
correlation between Ngb expression levels and metabolic activity, most notably
in the retina, where Ngb distribution perfectly matches $O_2$ consumption rates
and the protein is strictly associated with mitochondria (Schmidt et al. 2003;
Bentmann et al. 2005). In line with a function in $O_2$ supply, overexpression of
Ngb in cultured neurons (Sun et al. 2001) or in total rat brain (Sun et al. 2003)
enhances viability of the nerve cells under hypoxia and ischaemia. Ngb expression is upregulated at low oxygen conditions in hypoxia-tolerant zebrafish
(Roesner, Hankeln and Burmester 2006) but not in mammals, which never
experience hypoxia in their adult life (for review, see Burmester, Gerlach and
Hankeln 2007). However, Ngb rapidly autoxidises to metNgb, and in order to
function in $O_2$ supply, an Ngb-reducing enzymatic activity has to be postulated (Dewilde et al. 2001).

Second, Ngb may have a role in the decomposition of reactive oxygen or
nitrogen species (ROS, RNS) (Fig. 1b; Herold et al. 2004; Brunori et al. 2005;
Fordel et al. 2007). Such a function of Ngb would also be in line with its neuroprotective effect after ischaemia and reperfusion of brain tissue (Sun et al.
2003), when ROS/RNS are known to form, as well as with the presence of Ngb
adjacent to mitochondria. However, our own data do not reveal any tight association of Ngb tissue distribution and expression levels with some known sites
of preferential ROS/RNS production (unpublished). Ngb synthesis is not
induced by exposure of neurons to ROS (unpublished) or RNS (Sun et al.
2001). It remains to be shown whether the RNS-decomposing reactions of Ngb
found *in vitro* are of importance *in vivo*.

Third, Fago et al. (2006) proposed that the neuroprotective effect of Ngb
under hypoxia is due to the reduction of ferric ($Fe^{3+}$) cytochrome c (Cyt c) by
ferrous ($Fe^{2+}$) Ngb (Fig. 1c). This would prevent the induction of apoptosis by
Cyt c. This hypothesis, which is based on *in vitro* experiments with purified
proteins, is in line with many expression data, including the co-localisation of
Ngb and mitochondria, which actually release Cyt c upon cell death. Future
studies must show whether an electron transfer from Ngb to Cyt c occurs *in
vivo* and has any functional significance.

Fourth, Ngb has been suggested to be part of a signalling chain. Wakasugi
and co-workers proposed that oxidised Ngb ($Fe^{3+}$) is involved in intracellular
signalling by inhibiting dissociation of GDP from G$\alpha$ proteins and triggering
release of the G$\beta\gamma$ complex, thereby enhancing cell survival under oxidative
stress (Fig. 1d; Wakasugi, Nakano and Morishima 2003). The interaction of
Ngb with G proteins was postulated on the basis of an alleged sequence similarity between Ngb and regulators of G protein signalling (RGS) and RGS
domains of G protein-coupled receptor kinases, but we could not confirm this
(Burmester and Hankeln 2004; Hankeln et al. 2005). Moreover, the GDP dis-
sociation-inhibiting feature is not conserved in zebrafish Ngb (Wakasugi and
Morishima 2005), casting additional doubt on the *in vivo* relevance of these
interactions. Other studies have suggested more, functionally very diverse

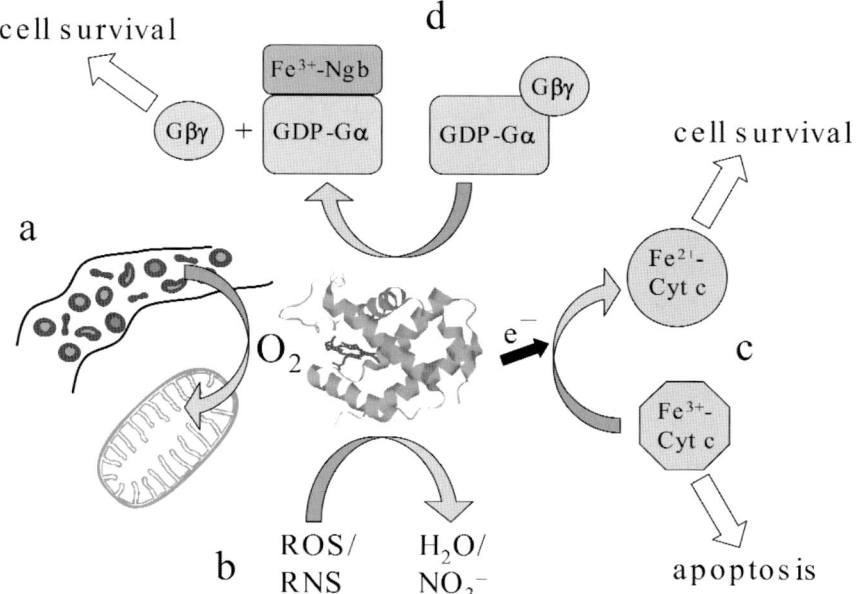

FIG. 1. Some postulated functions of neuroglobin. Ngb may (a) supply $O_2$ to the oxidative chain in mitochondria, (b) detoxify reactive oxygen and nitrogen species (ROS/RNS), (c) prevent hypoxia-induced apoptosis via reduction of Cyt c or (d) act as signal protein by inhibiting the release of GDP from $G\alpha$

interaction partners for Ngb (for review, see Hankeln et al. 2005). As these data are difficult to reconcile with the known cellular and intracellular distribution of Ngb, and have not been confirmed by independent evidence, it remains rather uncertain whether these *in vitro* interactions explain the function of Ngb *in vivo*.

In this discussion we should acknowledge the fact that, by analogy to myoglobin (Wittenberg and Wittenberg 2003), also Ngb may exert different functions in a tissue- or taxon-specific way.

## More Globins in the Vertebrate Nervous System

Besides neuroglobin, other globins may occur in vertebrate nervous systems. For example, cytoglobin (Cygb) is a cytoplasmic protein in fibroblasts and related cell types, but is located in the nuclei and cytoplasm of yet to be defined nerve cell populations in the PNS and CNS (Schmidt et al. 2004). The physiological role of Cygb in neurons is uncertain. On the one hand, Cygb shows structural and phylogenetic affinities to Mb (Burmester et al. 2002; De Sanctis et al. 2004a). On the other hand, Cygb features peculiarities like its N- and C-terminal extensions, which might be mediating special protein–protein interactions, heme hexacoordination and a redox-dependent $O_2$ affinity, and special cavities for ligand diffusion (Burmester et al. 2002; Hamdane et al. 2003; De Sanctis et al. 2004a,b). It has been postulated that in the fibroblast cell-lineage Cygb may be involved in cell proliferation, probably related to collagen synthesis (Schmidt et al. 2004; Hankeln et al. 2005; Hankeln and Burmester 2008). The expression pattern in only some neurons suggests a distinct function of Cygb in the nervous system, a hypothesis that is supported by the occurrence of two paralogous Cygb genes in fish, only one of which is predominantly expressed in the brain (Fuchs, Burmester and Hankeln 2006). Interestingly, the neuronal cells that harbour Cygb are also positive for the expression of neuronal NO synthase (nNOS) and thus produce NO (Stefan Reuss, Sylvia Wystub, Thorsten Burmester and Thomas Hankeln, unpublished). We therefore hypothesise that Cygb in these cells either provides $O_2$ to nNOS for making NO, or detoxifies NO as a dioxygenase.

Still other globins are located in nerve tissues, although it remains to be demonstrated that they occur in neurons or glia. Globin E (GbE), which is a distant relative of Mb and Cygb, has been exclusively found to be expressed in the eye of chicken (*Gallus gallus*) (Kugelstadt et al. 2004). No orthologue has so far been identified in other vertebrates. Globin X (GbX) is restricted to "lower" vertebrates (fish and Amphibia) and is distantly related to Ngb (Roesner et al. 2005; Fuchs, Burmester and Hankeln 2006). It is ubiquitously expressed in various tissues, including *Xenopus* brain and eye (Fuchs, Burmester and Hankeln 2006). A distinct isoform of Mb was identified in brain tissue of the hypoxia-tolerant carp, *Cyprinus carpio* (Fraser et al. 2006). No information on the physiological functions of these globin types in neurons or other cell types is available at the moment.

Surprisingly, several studies demonstrated the expression of Hb in mammalian brain neurons. mRNA of Hb α and β chains was detected in the neurons (Humphries, Windass and Williamson 1976; Ohyagi, Yamada and Goto 1994). One possible explanation for the presence of Hb in brain offers the observation by Moeller et al. (1997), who isolated a decapeptide ($NH_2$-LVVYPWTQRF-COOH) from the sheep's brain that actually derived from the Hb β polypeptide. This Hb fragment has been named LVV-haemorphin-7 and binds with high affinity to the angiotensin IV receptor. This finding may point to a divergent function of Hb in the brain that is not related to gas binding.

# Independent Evolution of Nerve Haemoglobins

Heme hexacoordination has been observed for Ngb (Pesce et al. 2003), Cygb (De Sanctis et al. 2004a), GbX (unpublished) and *S. solidissima* nHb (Dewilde et al. 2006). Phylogenetic analyses show that hexacoordination is not restricted to any particular globin type and has arisen independently in the evolution of distinct globin lineages (Fig. 2). It had been proposed that invertebrate nHbs are phylogenetically orthologous to vertebrate Ngb (Burmester et al. 2000). In

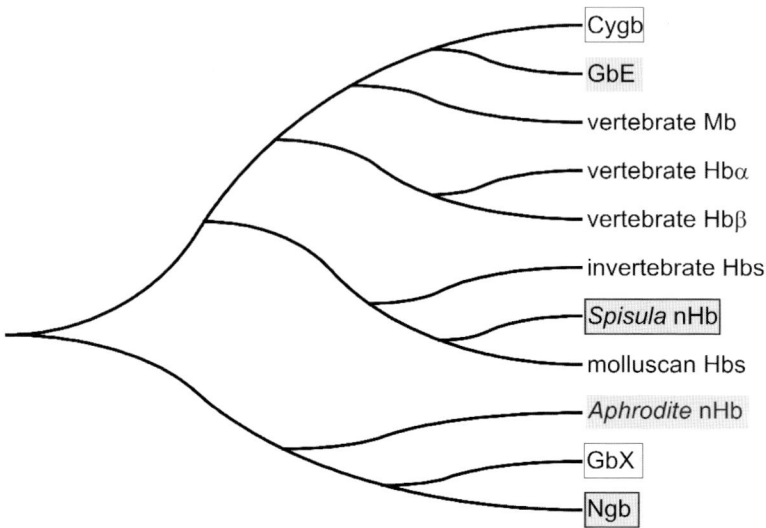

FIG. 2. Simplified phylogenetic tree of selected globins. Globins residing in the nervous system are shaded in grey, globins with a hexacoordinated deoxy form are framed. Branch lengths are arbitrary, i.e., no time scale was assumed

fact, the phylogenetic tree shows a close relationship of Ngb with the nHb of the annelid *A. aculeata* (Fig. 2). This clade also includes the Hbs of the annelid *G. dibranchiata*, as well as GbX and the Hbs of the tunicate *C. intestinalis* (Roesner et al. 2005). These globin types are distinct from the other invertebrate and vertebrate Hbs and Mbs, as well as from Cygb and GbE, suggesting that there are two separate globin families in Bilateria. However, recently it has been demonstrated that *S. solidissima* nHb does not belong to the Ngb+*A. aculeata* nHb clade, but groups with the other mollusc Hbs and Mbs (note that the nHb from the nemertean worm *C. lacteus* could not be placed within any clade with sufficient confidence). Therefore, nerve globins are actually of polyphyletic origin: we propose that, in addition to the existence of an ancient nerve globin lineage, several times during evolution an Hb or Mb has been secondarily recruited to function in the nervous system. This finding demonstrates the enormous evolutionary flexibility of globin-type proteins. It may also predict that nerve globins fulfil different physiological roles in response to taxon-specific adaptive needs.

*Acknowledgements*

We are very grateful to Beatrice and Jonathan Wittenberg for their inspiration and numerous opportunities for scientific discussion. We wish to thank Sylvia Dewilde and Luc Moens (Antwerp, Belgium), Angela Fago and Roy Weber (Aarhus), Martino Bolognesi (Milano), Michael C. Marden (Paris), Aaron Avivi and Eviatar Nevo (Haifa), Bettina Ebner, Christine Fuchs, Anja Roesner, Sigrid Saaler-Reinhardt, Stefan Reuss,

Marc Schmidt, Bettina Weich, Sylvia Wystub (Mainz), Frank Gerlach and Stephanie Mitz (Hamburg), who have contributed to our studies.
Our research has been supported by grants from the DFG (Bu956/11 and Ha2103/3), the European Union (QLG3-CT2002-01548), the Stiftung für Innovation Rheinland-Pfalz (695) and the Fonds der Chemischen Industrie.

# References

Arvanitaki, A., and Chalazonitis, N. 1960. Photopotentiels d'excitation et d'inhibition de différents somata identifiables (Aplysia). Activations monochromatiques. Bull. Inst. Oceanog. 57:1–83.

Awenius, C., Hankeln, T., and Burmester, T. 2001. Neuroglobins from the zebrafish *Danio rerio* and the pufferfish *Tetraodon nigroviridis*. Biochem. Biophys. Res. Commun. 287:418–421.

Bentmann, A., Schmidt, M., Reuss, S., Wolfrum, U., Hankeln, T., and Burmester, T. 2005. Divergent distribution in vascular and avascular mammalian retinae links neuroglobin to cellular respiration. J. Biol. Chem. 280:20660–20665.

Brunori, M., Giuffre, A., Nienhaus, K., Nienhaus, G. U., Scandurra, F. M., and Vallone, B. 2005. Neuroglobin, nitric oxide, and oxygen: functional pathways and conformational changes. Proc. Natl. Acad. Sci. U.S.A. 102:8483–8488.

Burmester, T., and Hankeln, T. 2004. Neuroglobin: A respiratory protein of the nervous system. News Phys. Sci. 19:110–113.

Burmester, T., Ebner, B., Weich, B., and Hankeln, T. 2002. Cytoglobin: a novel globin type ubiquitously expressed in vertebrate tissues. Mol. Biol. Evol. 19:416–421.

Burmester, T., Gerlach, F., and Hankeln, T. 2007. Regulation and role of neuroglobin and cytoglobin under hypoxia. Adv. Exp. Med. Biol. 618:169–180.

Burmester, T., Weich, B., Reinhardt, S., and Hankeln, T. 2000. A vertebrate globin expressed in the brain. Nature 407:520–523.

de Sanctis, D., Dewilde, S., Pesce, A., Moens, L., Ascenzi, P., Hankeln, T., Burmester, T., and Bolognesi, M. 2004a. Crystal structure of cytoglobin: the fourth globin type discovered in man displays heme hexa-coordination. J. Mol. Biol. 336:917–927.

de Sanctis, D., Dewilde, S., Pesce, A., Moens, L., Ascenzi, P., Hankeln, T., Burmester, T., and Bolognesi, M. 2004b. Mapping protein matrix cavities in human cytoglobin through Xe atom binding: a crystallographic investigation. Biochem. Biophys. Res. Commun. 316:1217–1221.

Dewilde, S., Blaxter, M., Van, Hauwaert, M. L., Vanfleteren, J., Esmans, E. L., Marden, M., Griffon, N., and Moens, L. 1996. Globin and globin gene structure of the nerve myoglobin of *Aphrodite aculeata*. J. Biol. Chem. 271:19865–19870.

Dewilde, S., Ebner, B., Vinck, E., Gilany, K., Hankeln, T., Burmester, T., Kreiling, J., Reinisch, C., Vanfleteren, J., Kiger, L., Marden, M.C, Hundahl, C., Fago, A., Van Doorslaer, S., and Moens, L. 2006. The nerve hemoglobin of the bivalve mollusc *Spisula solidissima*: molecular cloning, ligand binding studies and phylogenetic analysis. J. Biol. Chem. 281:5364–5372.

Dewilde, S., Kiger, L., Burmester, T., Hankeln, T., Baudin-Creuza, V., Aerts, T., Marden,

M.C., Caubergs, R., and Moens, L. 2001. Biochemical characterization and ligand-binding properties of neuroglobin, a novel member of the globin family. J. Biol. Chem. 276:38949–38955.

Doeller, J. E., and Kraus, D. W. 1988. A physiological comparison of bivalve mollusc cerebro-visceral connectives with and without neurohemoglobin. II. Neurohemoglobin characteristics. Biol. Bull. 174:67–76.

Fago, A., Mathews, A. J., Dewilde, S., Moens, L., and Brittain T. 2006. The reactions of neuroglobin with CO: evidence for two forms of the ferrous protein. J. Inorg. Biochem. 100: 1339–1343.

Fordel, E., Thijs, L., Moens, L., and Dewilde, S. 2007. Neuroglobin and cytoglobin expression in mice. FEBS J. 274:1312–1317.

Fraser, J., de Mello, L. V., Ward, D., Rees, H. H., Williams, D. R., Fang, Y., Brass, A., Gracey, A. Y., and Cossins, A. R. 2006. Hypoxia-inducible myoglobin expression in nonmuscle tissue. Proc. Natl. Acad. Sci. U.S.A. 103:2977–2981.

Fuchs, C., Burmester, T., and Hankeln, T. 2006. The amphibian globin gene repertoire as revealed by the Xenopus genome sequence. Cytogenet. Genome Res. 112:296–306.

Hamdane, D., Kiger, L., Dewilde, S., Green, B. N., Pesce, A., Uzan, J., Burmester, T., Hankeln, T., Bolognesi, M., Moens, L., and Marden, M. C. 2003. The redox state of the cell regulates the ligand binding affinity of human neuroglobin and cytoglobin. J. Biol. Chem. 278:51713–51721.

Hankeln, T., and Burmester, T. 2008. Neuroglobin and cytoglobin. In: The Smallest Biomolecules: Diatomics and Their Interactions with Heme Proteins, ed. A. Ghosh, pp. 203–218.

Hankeln, T., Ebner, B., Fuchs, C. Gerlach, F., Haberkamp, M., Laufs, T., Roesner, A., Schmidt, M., Weich, B., Wystub, S., Saaler-Reinhardt, S., Reuss, S., Bolognesi, M., De Sanctis, D., Marden, M.C., Kiger, L., Dewilde, S., Moens, L., Nevo, E., Avivi, A., Weber, R. E., Fago, A., and Burmester, T. 2005. Neuroglobin and cytoglobin in search of their role in the vertebrate globin family. J. Inorg. Biochem. 99:110–119.

Herold, S., Fago, A., Weber, R. E., Dewilde, S., and Moens, L. 2004. Reactivity studies of the Fe(III) and Fe(II)NO forms of human neuroglobin reveal a potential role against oxidative stress. J. Biol. Chem. 279:22841–22847.

Humphries, S., Windass, J., and Williamson, R. 1976. Mouse globin gene expression in erythroid and non-erythroid tissues. Cell 7:267–277.

Kraus, D. W., Doeller, J. E., and Smith, P. R. 1988. A physiological comparison of bivalve mollusc cerebro-visceral connectives with and without neurohemoglobin. I. Ultrastructural and electrophysiological characteristics. Biol. Bull. 174:54–66.

Kraus, D. W., and Doeller, J. E. 1988. A physiological comparison of bivalve mollusc cerebro-visceral connectives with and without neurohemoglobin. III. Oxygen demand. Biol. Bull. 174: 346–354.

Kraus, D. W., and Colacino, J. M. 1986. Extended oxygen delivery from the nerve hemoglobin of Tellina alternata (Bivalvia). Science 232:90–92.

Kugelstadt, D., Haberkamp, M., Hankeln, T., and Burmester, T. 2004. Neuroglobin, cytoglobin and a novel, retina-specific globin from chicken. Biochem. Biophys. Res. Commun. 325:719–725.

Lankester, E. R. 1872. A contribution to the knowledge of haemoglobin. Proc. Royal Soc. London 21:70–81.

Milton, S. L., Nayak, G., Lutz, P. L., and Prentice, H. M. 2006. Gene transcription of neuroglobin is upregulated by hypoxia and anoxia in the brain of the anoxia-tolerant turtle *Trachemys scripta*. J. Biomed. Sci. 13:509–514.

Moeller, I., Lew, R. A., Mendelsohn, F. A., Smith, A. I., Brennan, M. F., Tetaz, T. J., and Chai, S. Y., 1997. The globin fragment LVV-hemorphin-7 is an endogenous ligand for the AT4 receptor in the brain. J. Neurochem. 68:2530–2537.

Ohyagi, Y., Yamada, T., and Goto, I. 1994. Hemoglobin as a novel protein developmentally regulated in neurons. Brain Res. 635:323–327.

Pesce, A., Dewilde, S., Nardini, M., Moens, L., Ascenzi, P., Hankeln, T., Burmester, T., and Bolognesi, M. 2003. Human brain neuroglobin structure reveals a distinct mode of controlling oxygen affinity. Structure 11:1087–1095.

Pesce, A., Nardini, M., Ascenzi, P., Geuens, E., Dewilde, S., Moens, L., Bolognesi, M., Riggs, A. F., Hale, A., Deng, P., Nienhaus, G. U., Olson, J. S., and Nienhaus, K., 2004. Thr-E11 regulates $O_2$ affinity in *Cerebratulus lacteus* mini-hemoglobin. J. Biol. Chem. 279:33662–33672.

Reuss, S., Saaler-Reinhardt, S., Weich, B., Wystub, S., Reuss, M., Burmester, T., and Hankeln, T. 2002. Expression analysis of neuroglobin mRNA in rodent tissues. Neuroscience 115:645–656.

Roesner, A., Fuchs, C., Hankeln, T., and Burmester, T. 2005. A globin gene of ancient evolutionary origin in lower vertebrates: evidence for two distinct globin families in animals. Mol. Biol. Evol. 22:12–22.

Roesner, A., Hankeln, T., and Burmester, T. 2006. Hypoxia induces a complex response of globin expression in zebrafish (*Danio rerio*). J. Exp. Biol. 209:2129–2137.

Schindelmeiser, I., Kuhlmann, D., and Nolte, A. 1979. Localization and characterization of hemoproteins in the central nervous tissue of some gastropods. Comp. Biochem. Physiol. B 64:149–154.

Schmidt, M., Gerlach, F., Avivi, A., Laufs, T., Wystub, S., Simpson, J. C., Nevo, E., Saaler-Reinhardt, S., Reuss, S., Hankeln, T., and Burmester, T. 2004. Cytoglobin is a respiratory protein expressed in connective tissue and neurons that is up-regulated by hypoxia. J. Biol. Chem. 279:8063–8069.

Schmidt, M., Gießl, A., Laufs, T., Hankeln, T., Wolfrum, U., and Burmester, T. 2003. How does the eye breathe? Evidence for neuroglobin-mediated oxygen supply of the mammalian retina. J. Biol. Chem. 278:1932–1935.

Sun, Y., Jin, K., Mao, X. O., Zhu, Y., and Greenberg, D. A. 2001. Neuroglobin is up-regulated by and protects neurons from hypoxic-ischemic injury. Proc. Natl. Acad. Sci. U.S.A. 98:15306–15311.

Sun, Y., Jin, K., Peel, A., Mao, X. O., Xie, L., and Greenberg, D. A. 2003. Neuroglobin protects the brain from experimental stroke in vivo. Proc. Natl. Acad. Sci. U.S.A. 100:3497–3500.

Vandergon, T. L., Riggs, C. K., Gorr, T. A., Colacino, J. M., and Riggs, A. F. 1998. The mini-hemoglobins in neural and body wall tissue of the nemertean worm, *Cerebratulus lacteus*. J. Biol. Chem. 273:16998–17011.

Wakasugi, K., and Morishima, I. 2005. Identification of residues in human neuroglobin crucial for Guanine nucleotide dissociation inhibitor activity. Biochemistry 44:2943–2948.

Wakasugi, K., Nakano, T., and Morishima, I. 2003. Oxidized human neuroglobin acts as a heterotrimeric Ga protein guanine nucleotide dissociation inhibitor. J. Biol. Chem. 278:36505–36512.

Weber, R. E., and Vinogradov, S. N. 2001. Nonvertebrate hemoglobins: functions and molecular adaptations. Physiol. Rev. 81:569–628.

Wittenberg, B. A., Briehl, R.W., and Wittenberg, J. B. 1965. Haemoglobin of invertebrate tissues. Nerve haemoglobins of *Aphrodite*, *Aplysia*, and *Halosydna*. Biochem. J. 96:363–371.

Wittenberg, J. B. 1992. Functions of cytoplasmatic hemoglobins and myohemerythrin. Adv. Comp. Environ. Physiol. 13:60–85.

Wittenberg, J. B., and Wittenberg, B. A. 1990. Mechanisms of cytoplasmic hemoglobin and myoglobin function. Annu. Rev. Biophys. Chem. 19:217–241.

Wittenberg, J. B., and Wittenberg, B. A. 2003. Myoglobin function reassessed. J. Exp. Biol. 206:2011–2020.

# 19

# Ever Surprising Nematode Globins

David Hoogewijs, Eva Geuens, Lesley Tilleman,
Jacques R. Vanfleteren, Luc Moens and Sylvia Dewilde

## Abstract

Nematodes express pseudocoelomic, body wall and cuticle globin isoforms. All globin isoforms display the major determinants of the globin fold and a B10Tyr/E7Gln residue pair, which is a signature of high oxygen affinity. The hitherto studied pseudocoelomic globins are octamers of covalently linked didomain globin chains. Body wall globins so far are monomeric, whereas cuticle globins are tetrameric. The extremely high oxygen affinity of the pseudo-coelomic globins is caused by a network of three H-bonds between the bound ligand, B10Tyr and E7Gln resulting in a very low dissociation rate. The body wall and cuticle globins, albeit also displaying B10Tyr and E7Gln, have more moderate oxygen affinities. The structural reason for the latter observation is unknown. Although many hypotheses have been put forward, the real function of the nematode globins remains illusive.

*Caenorhabditis elegans* expresses 33 globin-like proteins. They display the major determinants of the globin fold and are expressed at very low levels. Most of them have N- and C-terminal extensions as well as interhelical insertions of variable length. Orthologues of these globins have been identified in closely related species and also in other nematode taxa.

Introns inserted at B12.2 and G7.0 are common in nematode globin genes and the E-helix is also interrupted by an intron, however at more variable positions. The globins of *C. elegans* are unique in having more introns that seem to be inserted rather randomly. Thus the intron insertion pattern of the nematode globin introns substantially deviates from the conserved intron/exon pattern seen in vertebrates.

Phylogenetic analysis of all nematode globin sequences reveals two strictly separated clades, one comprises all *C. elegans* globins except ZK637.13 and the other groups ZK637.13 and all other nematode globins. This might suggest that the globins in both clades have acquired different functions.

# Nematode Globins

Although the globin superfamily was extensively studied in the past, the last decade has revealed new vertebrate globin types e.g., neuro- and cytoglobin (Burmester et al. 2000, 2002), as well as new functions e.g., a role in the NO metabolism (Brunori 2001; Flogel et al. 2001), demonstrating the diversity and versatility of these proteins.

Non-vertebrate globins exhibit even a much greater variability in their primary and quaternary structure and functions than their vertebrate counterparts. Proven functions include $O_2$-sensing, -scavenging, -storage, -transport, oxidase or peroxidase activities, NO detoxification and even a function as shadowing pigment (for review see: Wittenberg and Wittenberg 1990; Wittenberg 1992; Weber and Vinogradov 2001; Vinogradov and Moens 2008).

It has long been known that nematodes possess globins (Davenport 1949; Smith and Lee 1964). Based on their tissue localisation, Blaxter (1993) grouped these globins in three different classes: pseudocoelomic globins, body wall and cuticle globins. However, the three different classes are not necessarily present in all nematodes.

The *pseudocoelomic or perienteric fluid haemoglobin* (Hb) of *Ascaris suum* is an octamer (Mr ~350 kDa) consisting of two layers of four subunits stacked in an eclipsed orientation (Darawshe, Tsafadyah and Daniel 1987; Darawshe and Daniel 1991). Each subunit of 43-kDa molecular mass consists of two covalently linked, highly similar globin domains (Mr ~17 kD for each domain; 62% similarity) followed by a C-terminal polar zipper tail of 23 amino acids, necessary but not sufficient alone for octamerisation (De Baere et al. 1992; Kloek et al. 1993; Minning et al. 1995). The structure of the pseudocoelomic Hb of the closely related ascarid species, *Pseudoterranova decipiens*, is similar (Dixon et al. 1991). *A. suum* Hb has an exceptionally high $O_2$ affinity due to a very slow dissociation rate and a normal association rate (Table 1) (Gibson and Smith 1965; De Baere et al. 1992) resulting in a fully oxygenated protein even in the micro-aerobic environment of the host's gut (Davenport 1949). The tertiary structure of the recombinant N-terminal domain reveals that this high $O_2$ affinity can be explained by the presence of a hydrogen bond network stabilising the bound ligand (Yang et al. 1995). Three hydrogen bonds are identified: a strong one between B10Tyr and the distal oxygen atom, a weak one between E7Gln and the proximal oxygen atom, and one connecting both amino acid residues stabilising the $O_2$ molecule in the heme pocket (De Baere et al. 1994; Yang et al. 1995; Xia et al. 1999; Das et al. 2004) (Fig. 1). Mutation of B10Tyr to Leu in *A. suum* and *P. decipiens* Hb domain1 abolished the high $O_2$ affinity supporting the hydrogen bond network hypothesis (Gibson et al. 1993; Kloek et al. 1994).

This hydrogen bond network stabilising the bound ligand was extensively studied in *A. suum* wild-type Hb and distal pocket mutants by resonance Raman, kinetic and $^1$H NMR experiments. Probing with CO results in the detection by resonance Raman spectroscopy of two active site conformers, H

TABLE 1. Ligand-binding kinetics of nematode globins and some reference globins

| Species | $O_2$ | | | | CO | | | NO | Ref. |
|---|---|---|---|---|---|---|---|---|---|
| | $k_{on}$ ($\mu M^{-1}$ $s^{-1}$) | $k_{off}$ ($s^{-1}$) | $K$ ($\mu M^{-1}$) | $P_{50}$ (mmHg) | $k_{on}$ ($\mu M^{-1}$ $s^{-1}$) | $k_{off}$ ($s^{-1}$) | $K$ ($\mu M^{-1}$) | $k_{on}$ ($\mu M^{-1}$ $s^{-1}$) | |
| **1. Vertebrates** | | | | | | | | | |
| H. sapiens Hb | 10–20 | 9–24 | ~0.9 | 21.4 | | | | | (Olson et al. 1988; Perutz 1990) |
| H. sapiens neuroglobin | 250 | 0.8 | 313 | 1 | 65 | 0.014 | 4642 | 250 | (Dewilde et al. 2001; Fago et al. 2004; Uzan et al. 2004) |
| H. sapiens cytoglobin | 27 | 0.9 | 30 | | 5 | | | | (Fago et al. 2004; Uzan et al. 2004) |
| P. catodon Mb | 17 | 15 | 1.1 | 1 | 0.51 | 0.019 | 26.84 | | (Gibson et al. 1986; Rohlfs et al. 1990) |
| **2. Plants** | | | | | | | | | |
| P. sativum HbI | 250 | 16 | 15.63 | | 57 | 0.0091 | 6263 | 250 | (Gibson et al. 1989) |
| A. thaliana GlB1 | 74 | 0.12 | 617 | | 0.55 | 0.0012 | 458 | | (Watts et al. 2001) |
| GLB2 | 86 | 0.014 | 6000 | | 22 | 0.0013 | 16923 | | (Watts et al. 2001) |
| GLB3 | 0.2 | 0.3 | 0.66 | | 0.014 | 0.001 | 14 | | (Watts et al. 2001) |
| **3. Nematodes** | | | | | | | | | |
| A. suum perienteric fluid Hb | 1.5 | 0.0041 | 375 | 0.001–0.004 | 1.71 | 0.018 | 95 | 6.5 | (Gibson and Smith 1965; De Baere et al.1992; Gibson et al. 1993) |
| A. suum perienteric fluid Hb dom1 WT | 2.8 | 0.013 | 215.3 | 0.0072 | 0.35 | | | | (De Baere et al. 1994) |
| Tyr(B10)Leu | 9.0 | 5.0 | 1.8 | 0.82 | 0.75 | | | | (De Baere et al. 1994) |
| Tyr(B10)Phe | 40.3 | 2.0 | 20.1 | 0.6 | 2.7 | | | | (De Baere et al. 1994) |
| A. suum Mb | 1.2 | 0.23 | 5.2 | 0.1 | 0.15 | | | | (Okazaki and Wittenberg 1965; Blaxter, Ingram and Tweedie 1994) |
| P. decipiens perienteric fluid Hb | 1.1 | 0.4 | 2.5 | | | | | 2.1 | (Gibson et al. 1989) |
| P. equorum Hb | 1.0 | 0.5 | | | 10 | 0.01 | 1000 | | (Coletta, Ascenzi and Brunori 1988) |
| C. elegans Hb ZK637.13 | 23 | | 46 | 0.03 | 5.8 | | | | (Geuens 2007) |
| C. elegans Hb T22C1.2 | – | – | – | | 23 | 0.05 | 460 | | (Geuens 2007) |
| M. nigrescens Hb body wall isoform | 1 | | | | 0.18 | | | | (Marden et al., personal communication) |
| eye isoform | 20 | | | | 2 | | | | (Marden et al., personal communication) |
| S. trachea Hb | | | | 9.4 | | | | | (Rose and Kaplan 1972) |

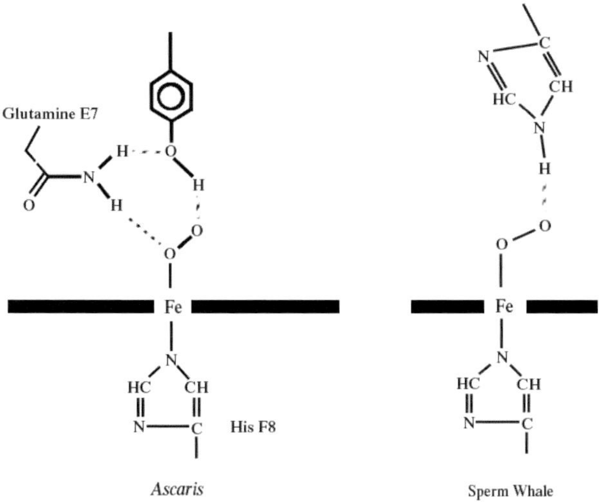

FIG. 1. Comparison of the proposed hydrogen bonding pattern within the distal pocket in
*A. suum* Hb to that of sperm whale myoglobin (Peterson et al. 1997). Reproduced from
*Biochemistry* with permission from ACS

and L, with different Fe–CO stretching modes ($\nu_{\text{Fe–CO}}$ H conformer: 543 cm$^{-1}$;
L conformer: 515 cm$^{-1}$). The relative abundance of both conformers is dependent on the ligand. Oxygen binding induces a tightly packed configuration that
favours the network of strong H-bonding interactions (H conformer), strongly
suggesting an $O_2$ scavenging function. The active site conformer distribution
upon CO binding is dependent on the sample history (Das et al. 2004).

$^1$H NMR confirms the 3D structure of *A. suum* Hb described by Yang et al.
(1995) with the exception of the rotation over 180° along the α–γ meso axis of
the heme relative to the orientation observed in the crystal structure (Zhang et
al. 1997; Xia et al. 1999).

Construction of myoglobin (Mb) double, triple and quadruple mutants at
the positions B10, E7, E10 and G8 demonstrates that the B10Tyr and E7Gln are
necessary for the high $O_2$ affinity, low $k_{\text{off}}$ and monophasic behaviour. In addition, the G8Ile to Phe mutation of Mb enhances the geminate recombination by
limiting the access to the distal cavity, Xe4, demonstrating the importance of
the cavities for ligand binding (Draghi et al. 2002).

The function of the *A. suum* pseudocoelomic Hb is still a matter of debate
(for review see: Blaxter 1993; Goldberg 1995; Weber and Vinogradov 2001).
Because of its extremely high $O_2$ affinity, it is never deoxygenated *in vivo* and
therefore cannot function as an $O_2$ carrier or store. Several alternative functions
have been proposed e.g. $O_2$ scavenger, iron and/or heme store, catalyst in the
sterol biosynthesis, a buffer and/or osmotic function, etc. An $O_2$ scavenging
function was considered assuming that the metabolism of *A. suum* requires
semi-anaerobic conditions (Davenport 1949; Smith and Lee 1964). This could

be realised by the enzymatic consumption of $O_2$ driven by NO, which means that the Hb will work as a "deoxygenase". However no nitrate accumulation in the perienteric fluid could be demonstrated (Vanfleteren, personal communication). The proposed chemistry is, as such, unusual (Minning et al. 1999) and it has been demonstrated that *A. suum* cannot survive in fully anaerobic conditions (Davenport 1949).

Genes coding for the *A. suum* and *P. decipiens* pseudocoelomic Hb have remarkable intron/exon patterns. The N-terminal globin domain has, besides the well conserved B12.2 and G7.0 introns, an additional "central intron" inserted at position E8.1, quite dissimilar from the plant central intron location (E15.0). In addition, introns separating the leader from the globin sequence or in the sequence linking both domains, are present. Other nematodes display altered patterns. Indeed the "central intron" position is variable as well as the total numbers of introns itself (Table 2). This strongly suggests that intron gain and loss occurred during the evolution of the nematode globin genes (Gilbert 1978; Cavalier-Smith 1985; Gilbert 1987; Rodriguez-Trelles, Tarrio and Ayala 2006; Roy and Gilbert 2006).

Native *body wall globins*, occurring in the hypodermis, the lateral cords, the nerve cord and body muscles, are homodimers (35–37 kD) of monomeric (~17 kD) globin chains. They lack a signal sequence and can thus be classified as Mb-like molecules (Okazaki and Wittenberg 1965; Okazaki et al. 1965; Blaxter, Ingram and Tweedie 1994). This cellular isoform has been described in many species (Blaxter 1993) and is characterised at the molecular level in the strongylids *Nippostrongylus brasiliensis* and *Trichostrongylus colubriformis* and in the ascarid *A. suum* (Frenkel et al. 1992; Wittenberg 1992; Mansell et al. 1993; Blaxter, Ingram and Tweedie 1994; Vanfleteren et al. 1994). Interestingly, the body wall globin from *A. suum* also displays B10Tyr/E7Gln, the signature for high $O_2$ affinity, yet its affinity for $O_2$ is much lower than that of the pseudocoelomic isoform (Sherman et al. 1992; De Baere et al. 1994). Structural information will be needed to clarify the difference in affinity.

*Mermis nigrescens*, a parasitic nematode of grasshoppers, expresses two intracellular monodomain (Mr ~17 000) globin isoforms, a body wall isoform and an ocellus or eye specific isoform. Alignment of the translated cDNAs of these globins with reference globin sequences reveals that both display all essential determinants of the globin fold as well as the nematode-specific B10Tyr/E7Gln. Both globins are 84% identical, suggesting that they result from a relatively recent duplication event and that the eye isoform is a specialised form of the body wall type (Burr et al. 2000). The eye globin is present in high concentration (~10 mM) in the anterior part of gravid adult females, forming a strongly pigmented cylinder which surrounds the ocellus. Electron microscopy reveals that the cells are densely packed with needle-like crystals. The patterns and dimensions are consistent with parallel rows of protein molecules forming hexagonal tubes with shared walls. Spectrophotometry, morphology and behavioural experiments strongly suggest that this globin has a

TABLE 2. Intron insertion position in nematode and some reference globin genes

| Species | Pre/inter domain intron (Leader seq +A helix) | N-terminal intron (B-helix) | Central intron (E, F-helix) | C-terminal intron (G-helix) | Ref. |
|---|---|---|---|---|---|
| **1. Vertebrates** | | | | | |
| Mb, α, β chain | – | B12.2 | – | G7.0 | (Hardison 1996) |
| Neuroglobin | – | B12.2 | E11.0 | G7.0 | (Burmester et al. 2000) |
| Cytoglobin | – | B12.2 | – | G7.0;HC11.2 | (Burmester et al. 2002) |
| **2. Plants** | | | | | |
| G. maxima Leg-Hb | – | B12.2 | E15.0 | G7.0 | (Hyldig-Nielsen et al. 1982) |
| A. thaliana Glb1 | – | B12.2 | E15.0 | G7.0 | (Trevaskis et al. 1997; Hunt et al. 2001; Watts et al. 2001) |
| Glb2 | | B12.2 | E15.0 | G7.0 | |
| Glb3 | | B12.2 | E15.0 | G7.0 | |
| **2. Nematodes** | | | | | |
| A. suum perienteric fluid Hb | | | | | |
| domain 1 | A4.1 | B12.2 | E8.1 | G7.0 | (Sherman et al. 1992) |
| domain 2 | A4.1 | B12.2 | E8.1 | G7.0 | |
| A. suum Mb | – | – | – | – | (Blaxter, Ingram and Tweedie 1994) |
| P. decipiens perienteric fluid Hb | | | | | |
| domain 1 | A4.1 | B12.2 | E8.1 | G7.0 | (Dixon et al. 1991) |
| domain 2 | A4.1 | B12.2 | – | – | |
| N. brasiliensis Hb body wall isoform | – | B12.2 | E3.2 | G7.0 | (Blaxter, Ingram and Tweedie 1994) |
| cuticular isoform | A4.1 | B12.2 | E3.2 | G7.0 | |

TABLE 2 continue

| Species | Prae/inter domain intron (Leader seq +A helix) | N-terminal intron (B-helix) | Central intron (E, F-helix) | C-terminal intron (G-helix) | Ref. |
|---|---|---|---|---|---|
| S. trachea Hb body wall isoform | – | B12.2 | E3.2; F3.0 | G7.0 | (Hunt and Blaxter, personal communication) |
| cuticular isoform | L17.1; L25.2 | B12.2 | E3.2; EF 5.0 | G7.0 | |
| C. elegans Hb ZK637.13 | – | – | E3.2 | – | (Kloek, Sherman and Goldberg 1993; Moens et al. 1992) |
| C. briggsae | – | – | E3.2 | – | (Kloek et al. 1996) |
| C. remanei 1 | – | – | E3.2 | – | (Kloek et al. 1996) |
| 2 | – | – | E3.2 | – | |
| T. canis Hb body wall isoform | – | B12.2 | E8.1 | G7.0 | (Hunt and Blaxter, personal communication) |
| perienteric fluid Hb | A4.1 | B12.2 | E8.1 | G7.0 | |
| M. nigrescens Hb body wall isoform | – | B12.2 | – | G7.0 | |
| eye isoform | – | B12.2 | – | G7.0 | (Burr et al. 2000) |

shadow-casting function during positive photo taxis (Burr, Schiefke and Bollerup 1975; Burr and Harosi 1985; Burr et al. 2000; Mohamed, Burr, and Burr 2007). These crystals are fundamentally different from the paracrystalline arrays of aggregated fibres formed by human deoxy sickle cell Hb (Dickerson and Geis 1983).

The genes coding for body wall globins differ in intron/exon pattern from those of the pseudocoelomic Hbs. In *A. suum* Mb no introns were observed, whereas in *M. nigrescens* globin genes only display the conservative intron insertion positions, B12.2 and G7.0 (Burr et al. 2000) (Table 2).

The *cuticle Hb* of *N. brasiliensis* is a homotetramer (Mr ~70 kDa) of 18-kDa subunits with a typical signal sequence at the N-terminus which is a hallmark of extracellular proteins. This isoform is found in the fluid-filled zone of the adult cuticle (Lee and Smith 1965; Sharpe and Lee 1981). Cuticle globin mRNA is expressed at high levels throughout adulthood and the globin protein is continuously exported to the cuticle after the moult. The native protein has a pI of 8.0, which is lower than that of the body wall globin. It is most likely post-translational modified, which results in the generation of multiple spots on 2DE gels (Blaxter, Ingram and Tweedie 1994). Further information on this isoform is lacking.

The $O_2$ affinity of the body wall and cuticle globin of nematodes is much lower than that of pseudocoelomic Hb, despite the presence of a B10Tyr/E7Gln, suggesting that these globin isoforms might be involved in shuttling $O_2$ from the gut to the muscles (Wittenberg 1992).

The potential use of Hbs of parasitic organisms as vaccination antigens to induce protection against parasitic infection has been suggested previously (Blaxter 1993). Single-domain (Mr ~17 kDa) Hbs have been detected in the excretory/secretory products of *Trichostrongylus colubriformis* and *Clonorchis sinensis* (Frenkel et al. 1992; Sim, Park and Yong 2003) and successful vaccination of guinea pigs against *T. colubriformis* was reported (Frenkel et al. 1992).

# The *Caenorhabditis* Globin Family

The existence of globin-like proteins in *Caenorhabditis elegans* was not expected given its small size, whereby sufficient amounts of $O_2$ were supposed to reach the sites of $O_2$ consumption by simple diffusion. The early data of the *C. elegans* genome sequencing, however, revealed the presence of a first globin-like protein encoded by the gene ZK637.13 (Sulston et al. 1992). Recent careful screening of the complete genome of *C. elegans* detected 33 globins, all having orthologues in the closely related species *C. briggsae* and *C. remanei* (Hoogewijs et al. 2007). They show wide diversity in gene structure and amino acid sequence, suggesting a long evolutionary history. At the protein level they all display the invariant proximal histidine at position F8, but other determinants of the globin fold are slightly relaxed. For example, *C. elegans* globins can have Leu (R01E6.6), Ile (T22C1.2), Tyr (T06A1.3), Met (C26C6.7, Y57G7A.9) or Val (Y75B7AL.1) instead of Phe at CD1 and four of them have Ile (T06A1.3, Y75B7AL.1, C09H10.8) or Val (C06E4.7) instead of His or Gln at E7. These amino acid residues are fully conserved in *C. briggsae* and *C. remanei*. Another feature shared by all three *Caenorhabditis* species is the occurrence of interhelical, N- and/or C-terminal extensions. Internal GH interhelical extensions of unusual length are seen in F49E2.4 (40aa), R102.9 (32aa), C52A11.2 (21aa), C06E4.7 (21aa), C09H10.8 (22aa) and Y22D7AR.5 (17aa). The external extensions are also very well conserved in *C. briggsae* and *C. remanei* and vary largely in length between the various globins. Database searches identified no discernible similarity to any known sequence. Surprisingly, several *C. elegans* globins still show significant sequence similarity with standard vertebrate globins as α- and β-chains, Mb and neuro- and cytoglobin (Hoogewijs et al. 2007).

The intron/exon patterns of these *C. elegans* globin coding genes are unique in the number of introns and in their insertion position, in contrast with the highly conserved intron/exon pattern of vertebrate globins (B12.2 and G7.0) (Hankeln et al. 1997; Dewilde et al. 1998; Hoogewijs et al. 2004, 2007). The number of introns interrupting the globin domain ranges from 1 (ZK637.13) to 5 (C06E4.7) and 1 to 5 and 1 to 3 introns are inserted in the pre-A and post-H extensions of the proteins, respectively (Hoogewijs et al. 2007). Nine introns are located in the interhelices/interhelix extensions. In all, more than 70 different intron insertion positions are scattered throughout the 33 *C. elegans* globin domains (Table 3). Only one globin gene (F21A3.6) features both conserved vertebrate insertion sites B12.2 and G7.0. Fourteen of the 33 putative globin genes display 3 intron insertions in the globin domain and 7 of them feature 2 interruptions in the globin domain. Although the globin gene repertoire is highly conserved in *C. elegans*, *C. briggsae* and *C. remanei*, a comparison of all genomic structures of these species indicates that variation can be found in intron size.

Comparative codon analysis, a measure for determining selective pressure, indicates that all globins are under purifying selection with very low ratios of non-synonymous to synonymous substitutions suggesting functional constraint (Miyata, Yasunaga and Nishida 1980). An evolutionary tree based on Bayesian inference of all 33 *C. elegans* globins, 11 previously identified nematode globins and EST sequences of several parasitic nematodes is shown in Fig. 2. Both the monophyly of the nematode globin sequences and their resolution into two separate clades are highly supported. One clade comprises all *C. elegans* globins except ZK637.13, the other clade groups ZK637.13 with all previously identified nematode globins and numerous ESTs (not shown) from parasitic nematodes (see also Vinogradov et al. 2006). The separation of nematode globins into two well resolved clades is consistent with the huge variability of intron positions in clade 1 globins, whereas the canonical intron insertion positions are generally fairly conserved in clade 2 globins. Clade 1 globins also generally have longer branch lengths relative to clade 2 globins, likely suggesting a faster rate of evolution. The *C. elegans* globin genes are widely scattered

TABLE 3: Lack of conservation in *C. elegans* globin intron positions

| Globin gene | Intron insertion positions |
| --- | --- |
| C06E4.7 | A14.2 CD3.1 E18.0 GH7.1 H26.0 |
| C06H2.5 | A2.2  G15.0 |
| C09H10.8 | A9.2 B11.2 F7.2 GH6.2 HC13.2 HC42.1 HC93.2 |
| C18C4.1 | NA20.0 NA43.1 NA67.0 B3.0 E6.2 H11.0 |
| C18C4.9 | NA44.1 NA72.1 NA106.1 NA137.2 NA185.1 FG6.1 H3.0 HC12.0 |
| C23H5.2 | C7.0 E14.0 F6.2 GH4.0 HC76.1 |
| C26C6.7 | NA4.1 NA76.2 NA106.0 NA187.0 C4.0 GH4.2 HC1.0 |
| C28F5.2 | NA23.1 NA45.1 E13.2 FG5.0 HC5.2 |
| C29F5.7 | A14.2 E14.0 H10.2 |
| C36E8.2 | NA26.2 EF14.2 H22.2 HC36.2 |
| C52A11.2 | NA14.0 E18.0 FG2.2 GH2.0 HC31.0 |
| F19H6.2 | AB2.0 E10.1 H12.0 |
| F21A3.6 | NA75.1 B12.2 E5.2 G7.0 |
| F35B12.8 | NA28.2 E17.1 F2.2 H2.0 HC1.0 |
| F46C8.7 | NA45.1  B9.0  F5.1  HC26.0  HC69.0 |
| F49E2.4 | A14.2 E8.0 GH30.0 H22.1 |
| F52A8.4 | NA20.2 NA59.0 A6.0 E14.0 G10.0 |
| F56C4.3 | C1.0 E14.0 G1.2 H21.0 |
| R01E6.6 | NA31.0 NA80.2  B9.0 E16.0 H1.0 HC27.0 HC55.0 |
| R102.9 | A14.2 G6.2 H22.1 |
| R11H6.3 | NA87.0  C4.0  D3.2 GH1.0 HC10.2 HC37.2 |
| R13A1.8 | NA37.0 NA81.0 A4.0 FG4.0 H13.2 |
| R90.5 | NA22.0 NA46.2 NA82.0 NA119.0 NA142.1 H22.0 |
| T06A1.3 | A3.2 B12.2 E14.0 FG2.0 |
| T22C1.2 | A5.0  E18.1 FG9.0 H18.1 |
| W01C9.5 | B13.0 EF4.0 HC22.2 |
| Y15E3A.2 | NA23.0 A9.2 G1.2 NA14.2 |
| Y17G7B.6 | NA12.2 A10.0 C5.2 FG1.0 H21.0 |
| Y22D7AR.5 | AB1.2 CD6.2 E10.2 G1.2 GH6.2 NA10.0 |
| Y57G7A.9 | B9.0 E15.0 G2.0 |
| Y58A7A.6 | NA28.0 NA85.1 F3.0 |
| Y75B7AL.1 | NA35.2 NA134.0 NA372.2 CD2.0 EF10.0 FG7.0 |
| ZK637.13 | E3.2 |

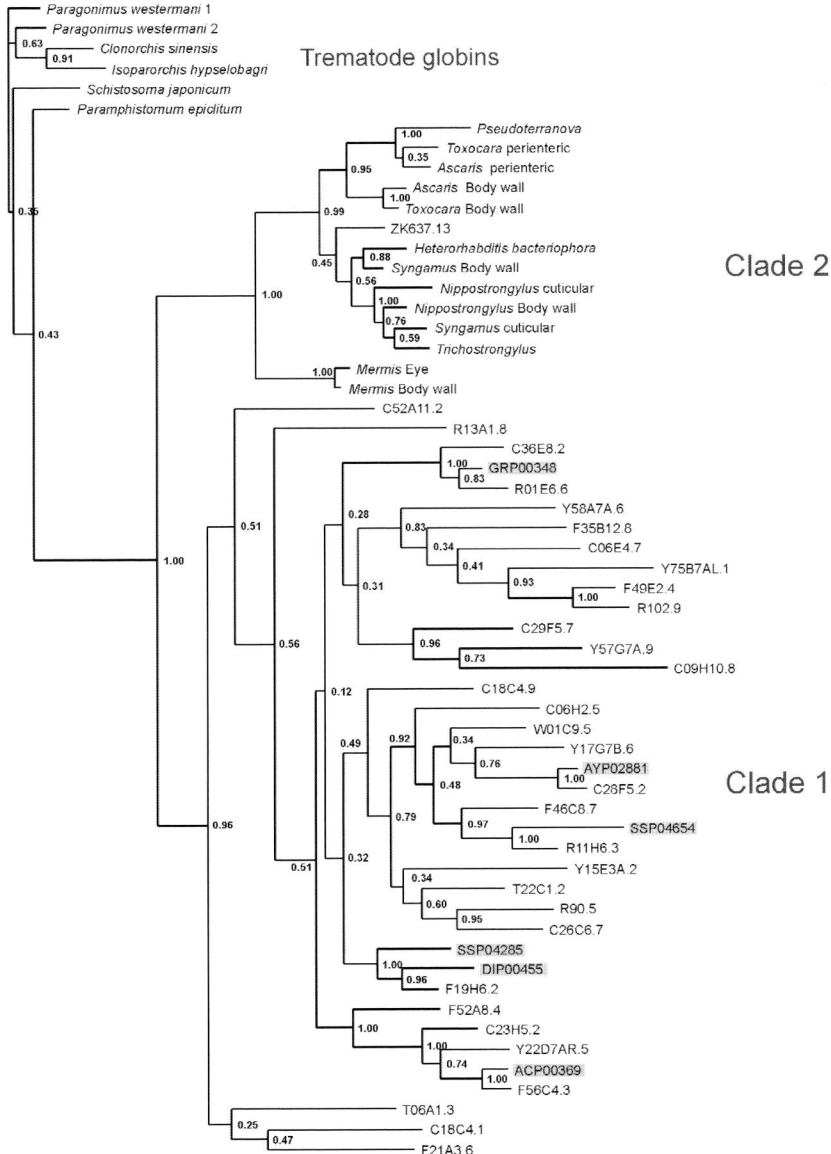

FIG. 2. Bayesian phylogenetic inference of all *C. elegans* globins, a selection of parasitic globins based on EST analysis and other known nematode globins. Six trematode globins were included as outgroup: *Schistosoma japonicum* (AAP06216), *Paramphistomum epiclitum* (AAG48877), *Paragonimus westermani* (AAX11352 and AAX11353), *Clonorchis sinensis* (AAM18464), *Isoparorchis hypselobagri* (P80722). The numbers at the nodes represent Bayesian posterior probabilities. Parasitic ESTs are shaded: SSP04654 and SSP04285: *Strongyloides stercoralis*; DIP00455: *Dirofilaria immitis*; AYP02881: *Ancylostoma ceylanicum*; ACP00369: *Ancylostoma caninum*; GRP00348: *Globodera rostochiensis*

all over the 6 chromosomes, indicating that they are not the result of recent gene duplications. Taken together these features suggest that the divergence of both globin clades was a major event of nematode globin evolution. It is likely that the globins of clade 1 acquired novel and highly specific functions. Orthologues of these globins have already been identified in other nematode taxa, as shown by presence of some globin ESTs from animal parasites among clade 1 globins (Fig. 2), and there is little doubt that many more are to follow.

All 33 putative *C. elegans* globins are expressed *in vivo*, albeit at low or very low levels, except T22C1.2 and ZK637.13, which are expressed more prominently. Given the large number of *C. elegans* globins and their variability in primary amino acid and gene structure, it is unlikely that they all serve simple $O_2$ binding. Ten *C. elegans* globins are responsive to $O_2$ deprivation in an interacting HIF-1- and DAF-16-dependent manner. Globin ZK637.13 is not responsive to $O_2$ deprivation and is regulated by the Ins/IGF-1 pathway, only suggesting that this globin may contribute to the life maintenance programme (Hoogewijs et al. 2007). Others, like Y75B7AL.1, are chimeric proteins consisting of a G-coupled receptor domain containing 7 transmembrane helices and a globin domain, which may suggest a role as $O_2$ sensor (Hoogewijs et al. 2007). In order to characterise and unravel their functions, a selection of *C. elegans* globins were cloned, expressed *in vitro* and studied by optical, resonance Raman and electron paramagnetic resonance spectroscopy, kinetics and $O_2$-equilibrium techniques. Initial results already provided evidence for the presence of 2 distinct functional classes (Geuens 2007; Kiger et al., personal communication).

## Conclusion

During the last 15 years our knowledge of nematode globins has greatly expanded and some globins (e.g. from *A. suum*) are structurally well characterised. Nevertheless the function of these globins remains essentially unknown. Recent work on *C. elegans* unveiled the presence of numerous hitherto unknown globin sequences in this species and there is little doubt that homologues will be found in the phylum Nematoda as additional nematode genome sequences become available. This definitely prompts further research.

*Acknowledgement*
This work is supported by the Fund for Scientific Research Flanders (Grand nr G.0331.04N). SD is a post-doctoral researcher of this Fund.

## References

Blaxter, M. 1993. Nemoglobins: Divergent nematode globins. Parasitol. Today 9:353–360.
Blaxter, M. L., Ingram, L., and Tweedie, S. 1994. Sequence, expression and evolution of the globins of the parasitic nematode *Nippostrongylus brasiliensis*. Mol. Biochem. Parasitol. 68:1–14.

Brunori, M. 2001. Nitric oxide moves myoglobin centre stage. Trends Biochem. Sci. 26:209–210.

Burmester, T., Ebner, B., Weich, B., and Hankeln, T. 2002. Cytoglobin: a novel globin type ubiquitously expressed in vertebrate tissues. Mol. Biol. Evol. 19:416–421.

Burmester, T., Weich, B., Reinhardt, S., and Hankeln, T. 2000. A vertebrate globin expressed in the brain. Nature 407:520–523.

Burr, A. H., and Harosi, F. I. 1985. Naturally crystalline hemoglobin of the nematode *Mermis nigrescens*. An in situ microspectrophotometric study of chemical properties and dichroism. Biophys. J. 47:527–536.

Burr, A. H., Hunt, P., Wagar, D. R., Dewilde, S., Blaxter, M. L., Vanfleteren, J. R., and Moens, L. 2000. A hemoglobin with an optical function. J. Biol. Chem. 275:4810–4815.

Burr, A. H., Schiefke, R., and Bollerup, G. 1975. Properties of a hemoglobin from the chromatrope of the nematode *Mermis nigrescens*. Biochim. Biophys. Acta 405:404–411.

Cavalier-Smith, T. 1985. Selfish DNA and the origin of introns. Nature 315:283–284.

Coletta, M., Ascenzi, P., and Brunori, M. 1988. Kinetic evidence for a role of heme geometry on the modulation of carbon monoxide reactivity in human hemoglobin. J. Biol. Chem. 263:18286–18289.

Darawshe, S., and Daniel, E. 1991. Molecular symmetry and arrangement of subunits in extracellular hemoglobin from the nematode *Ascaris suum*. Eur. J. Biochem. 201:169–173.

Darawshe, S., Tsafadyah, Y., and Daniel, E. 1987. Quaternary structure of erythrocruorin from the nematode *Ascaris suum*. Evidence for unsaturated haem-binding sites. Biochem. J. 242:689–694.

Das, T. K., Samuni, U., Lin, Y., Goldberg, D. E., Rousseau, D. L., and Friedman, J. M. 2004. Distal heme pocket conformers of carbonmonoxy derivatives of *Ascaris* hemoglobin: evidence of conformational trapping in porous sol-gel matrices. J. Biol. Chem. 279:10433–10441.

Davenport, H. E. 1949. Hemoglobins of *Ascaris lumbricoides*. Proc. R. Soc. Lond. Biol. Sci. 136:255–270.

De Baere, I., Liu, L., Moens, L., Van Beeumen, J., Gielens, C., Richelle, J., Trotman, C., Finch, J., Gerstein, M., and Perutz, M. 1992. Polar zipper sequence in the high-affinity hemoglobin of *Ascaris suum*: amino acid sequence and structural interpretation. Proc. Natl. Acad. Sci. U.S.A. 89:4638–4642.

De Baere, I., Perutz, M. F., Kiger, L., Marden, M. C., and Poyart, C. 1994. Formation of two hydrogen bonds from the globin to the heme-linked oxygen molecule in *Ascaris* hemoglobin. Proc. Natl. Acad. Sci. U.S.A. 91:1594–1597.

Dewilde, S., Blaxter, M., Van Hauwaert, M. L., Van Houte, K., Pesce, A., Griffon, N., Kiger, L., Marden, M. C., Vermeire, S., Vanfleteren, J., Esmans, E., and Moens, L. 1998. Structural, functional, and genetic characterization of *Gastrophilus* hemoglobin. J. Biol. Chem. 273:32467–32474.

Dewilde, S., Kiger, L., Burmester, T., Hankeln, T., Baudin-Creuza, V., Aerts, T., Marden, M.C., Caubergs, R., and Moens, L. 2001. Biochemical characterization and ligand binding properties of neuroglobin, a novel member of the globin family. J. Biol. Chem. 276:38949–38955.

Dickerson, R. E., and Geis, I. (1983) Hemoglobin, 1st edn. Amsterdam: Benjamin/Cummings Inc.

Dixon, B., Walker, B., Kimmins, W., and Pohajdak, B. 1991. Isolation and sequencing of a cDNA for an unusual hemoglobin from the parasitic nematode *Pseudoterranova decipiens*. Proc. Natl. Acad. Sci. U.S.A. 88:5655–5659.

Draghi, F., Miele, A. E., Travaglini-Allocatelli, C., Vallone, B., Brunori, M., Gibson, Q. H., and Olson, J. S. 2002. Controlling ligand binding in myoglobin by mutagenesis. J. Biol. Chem. 277:7509–7519.

Fago, A., Hundahl, C., Dewilde, S., Gilany, K., Moens, L., and Weber, R. E. 2004. Allosteric regulation and temperature dependence of oxygen binding in human neuroglobin and cytoglobin. Molecular mechanisms and physiological significance. J. Biol. Chem. 279:44417–44426.

Flogel, U., Merx, M. W., Godecke, A., Decking, U. K., and Schrader, J. 2001. Myoglobin: a scavenger of bioactive NO. Proc. Natl. Acad. Sci. U.S.A. 98:735–740.

Frenkel, M. J., Dopheide, T. A., Wagland, B. M., and Ward, C. W. 1992. The isolation, characterization and cloning of a globin-like, host-protective antigen from the excretory-secretory products of *Trichostrongylus colubriformis*. Mol. Biochem. Parasitol. 50:27–36.

Geuens, E. 2007. A structural and functional study of globins in vertebrates and nonvertebrates. Doctorate thesis, Universiteit Antwerpen.

Gibson, Q. H., Olson, J. S., McKinnie, R. E., and Rohlfs, R. J. 1986. A kinetic description of ligand binding to sperm whale myoglobin. J. Biol. Chem. 261:10228–10239.

Gibson, Q. H., Regan, R., Olson, J. S., Carver, T. E., Dixon, B., Pohajdak, B., Sharma, P. K., and Vinogradov, S. N. 1993. Kinetics of ligand binding to *Pseudoterranova decipiens* and *Ascaris suum* hemoglobins and to Leu-29—>Tyr sperm whale myoglobin mutant. J. Biol. Chem. 268:16993–16998.

Gibson, Q. H., and Smith, M. H. 1965. Rates of reaction of *Ascaris* haemoglobins with ligands. Proc. R. Soc. Lond. B Biol. Sci. 163:206–214.

Gibson, Q. H., Wittenberg, J. B., Wittenberg, B. A., Bogusz, D., and Appleby, C. A. 1989. The kinetics of ligand binding to plant hemoglobins. Structural implications. J. Biol. Chem. 264:100–107.

Gilbert, W. 1978. Why genes in pieces? Nature 271:501.

Gilbert, W. 1987. The exon theory of genes. Cold Spring Harb. Symp. Quant. Biol. 52:901–905.

Goldberg, D. E. 1995. The enigmatic oxygen-avid hemoglobin of *Ascaris*. Bioessays 17:177–182.

Hankeln, T., Friedl, H., Ebersberger, I., Martin, J., and Schmidt, E. R. 1997. A variable intron distribution in globin genes of *Chironomus*: evidence for recent intron gain. Gene 205:151–160.

Hardison, R. C. 1996. A brief history of hemoglobins: plant, animal, protist, and bacteria. Proc. Natl. Acad. Sci. U.S.A. 93:5675–5679.

Hoogewijs, D., Geuens, E., Dewilde, S., Moens, L., Vierstraete, A., Vinogradov, S., and Vanfleteren, J. 2004. Genome-wide analysis of the globin gene family of *C. elegans*. IUBMB Life 56:697–702.

Hoogewijs, D., Geuens, E., Dewilde, S., Vierstraete, A., Moens, L., Vinogradov, S., and Vanfleteren, J. R. 2007. Wide diversity in structure and expression profiles among members of the *Caenorhabditis elegans* globin protein family. BMC Genomics 8:356.

Hunt, P. W., Watts, R. A., Trevaskis, B., Llewelyn, D. J., Burnell, J., Dennis, E. S., and Peacock, W. J. 2001. Expression and evolution of functionally distinct haemoglobin genes in plants. Plant Mol. Biol. 47:677–692.

Hyldig-Nielsen, J. J., Jensen, E. O., Paludan, K., Wiborg, O., Garrett, R., Jorgensen, P., and Marcker, K. A. 1982. The primary structures of two leghemoglobin genes from soybean. Nucleic Acids Res. 10:689–701.

Kloek, A. P., McCarter, J. P., Setterquist, R. A., Schedl, T., and Goldberg, D. E. 1996. Caenorhabditis globin genes: rapid intronic divergence contrasts with conservation of silent exonic sites. J. Mol. Evol. 43:101–108.

Kloek, A. P., Sherman, D. R., and Goldberg, D. E. 1993. Novel gene structure and evolutionary context of *Caenorhabditis elegans* globin. Gene 129:215–221.

Kloek, A. P., Yang, J., Mathews, F. S., Frieden, C., and Goldberg, D. E. 1994. The tyrosine B10 hydroxyl is crucial for oxygen avidity of *Ascaris* hemoglobin. J. Biol. Chem. 269:2377–2379.

Kloek, A. P., Yang, J., Mathews, F. S., and Goldberg, D. E. 1993. Expression, characterization, and crystallization of oxygen-avid *Ascaris* hemoglobin domains. J. Biol. Chem. 268:17669–17671.

Lee, D. L., and Smith, M. H. 1965. Hemoglobins of parasitic animals. Exp. Parasitol. 16:392–424.

Mansell, J. B., Timms, K., Tate, W. P., Moens, L., and Trotman, C. N. 1993. Expression of a globin gene in *Caenorhabditis elegans*. Biochem. Mol. Biol. Int. 30:643–647.

Minning, D. M., Gow, A. J., Bonaventura, J., Braun, R., Dewhirst, M., Goldberg, D. E., and Stamler, J. S. 1999. *Ascaris* haemoglobin is a nitric oxide-activated 'deoxygenase'. Nature 401:497–502.

Minning, D. M., Kloek, A. P., Yang, J., Mathews, F. S., and Goldberg, D. E. 1995. Subunit interactions in *Ascaris* hemoglobin octamer formation. J. Biol. Chem. 270:22248–22253.

Miyata, T., Yasunaga, T., and Nishida, T. 1980. Nucleotide sequence divergence and functional constraint in mRNA evolution. Proc. Natl. Acad. Sci. U.S.A. 77:7328–7332.

Moens, L., Vanfleteren, J., De Baere, I., Jellie, A. M., Tate, W., and Trotman, C. N. 1992. Unexpected intron location in non-vertebrate globin genes. FEBS Lett. 312:105–109.

Mohamed, A. K., Burr, C., and Burr, A. H. 2007. Unique two-photoreceptor scanning eye of the nematode *Mermis nigrescens*. Biol. Bull. 212:206–221.

Okazaki, T., Briehl, R. W., Wittenberg, J. B., and Wittenberg, B. A. 1965. The hemoglobin of *Ascaris* perienteric fluid. II. Molecular weight and subunits. Biochim. Biophys. Acta 111:496–502.

Okazaki, T., and Wittenberg, J. B. 1965. The hemoglobin of *Ascaris* perienteric fluid. 3. Equilibria with oxygen and carbon monoxide. Biochim. Biophys. Acta 111:503–511.

Olson, J. S., Mathews, A. J., Rohlfs, R. J., Springer, B. A., Egeberg, K. D., Sligar, S. G., Tame, J., Renaud, J. P., and Nagai, K. 1988. The role of the distal histidine in myoglobin and haemoglobin. Nature 336:265–266.

Perutz, M. F. 1990. Mechanisms regulating the reactions of human hemoglobin with oxygen and carbon monoxide. Annu. Rev. Physiol. 52:1–25.

Peterson, E. S., Huang, S., Wang, J., Miller, L. M., Vidugiris, G., Kloek, A. P., Goldberg, D. E., Chance, M. R., Wittenberg, J. B., and Friedman, J. M. 1997. A comparison of functional and structural consequences of the tyrosine B10 and glutamine E7 motifs in two invertebrate hemoglobins (*Ascaris suum* and *Lucina pectinata*). Biochemistry 36:13110–13121.

Rodriguez-Trelles, F., Tarrio, R., and Ayala, F. J. 2006. Origins and evolution of spliceosomal introns. Annu. Rev. Genet. 40:47–76.

Rohlfs, R. J., Mathews, A. J., Carver, T. E., Olson, J. S., Springer, B. A., Egeberg, K. D., and Sligar, S. G. 1990. The effects of amino acid substitution at position E7 (residue 64) on the kinetics of ligand binding to sperm whale myoglobin. J. Biol. Chem. 265:3168–3176.

Rose, J. E., and Kaplan, K. L. 1972. Purification, molecular weight, and oxygen equilibrium of hemoglobin from *Syngamus trachea*, the poultry gapeworm. J. Parasitol. 58:903–906.

Roy, S. W., and Gilbert, W. 2006. The evolution of spliceosomal introns: patterns, puzzles and progress. Nat. Rev. Genet. 7:211–221.

Sharpe, M. J., and Lee, D. L. 1981. Changes in the level of acetylcholinesterase of nematospiroides dubius and *Trichostrongylus colubriformis* following paralysis by levamisole in vivo. Mol. Biochem. Parasitol. 3:57–60.

Sherman, D. R., Kloek, A. P., Krishnan, B. R., Guinn, B., and Goldberg, D. E. 1992. *Ascaris* hemoglobin gene: plant-like structure reflects the ancestral globin gene. Proc. Natl. Acad. Sci. U.S.A. 89:11696–11700.

Sim, S., Park, G. M., and Yong, T. S. 2003. Cloning and characterization of *Clonorchis sinensis* myoglobin using immune sera against excretory-secretory antigens. Parasitol. Res. 91:338–343.

Smith, M. H., and Lee, D. L. 1964. Metabolism of haemoglobin and haemitin compounds in *Ascaris lumbrivcoides*. Proc. R. Soc. Lond. B Biol. Sci. 157:234–257.

Sulston, J., Du, Z., Thomas, K., Wilson, R., Hillier, L., Staden, R., Halloran, N., Green, P., Thierry-Mieg, J., Qiu, L., Dear, S., Coulson, A., Craxton, M., Durbin, R., Berks, M., Metzstein, M., Hawkins, T., Ainscough, R., and Waterston, R. 1992. The *C. elegans* genome sequencing project: a beginning. Nature 356:37–41.

Trevaskis, B., Watts, R. A., Andersson, C. R., Llewellyn, D. J., Hargrove, M. S., Olson, J. S., Dennis, E. S., and Peacock, W. J. 1997. Two hemoglobin genes in *Arabidopsis thaliana*: the evolutionary origins of leghemoglobins. Proc. Natl. Acad. Sci. U.S.A. 94:12230–12234.

Uzan, J., Dewilde, S., Burmester, T., Hankeln, T., Moens, L., Hamdane, D., Marden, M. C., and Kiger, L. 2004. Neuroglobin and other hexacoordinated hemoglobins show a weak temperature dependence of oxygen binding. Biophys. J. 87:1196–1204.

Vanfleteren, J. R., Van de Peer, Y., Blaxter, M. L., Tweedie, S. A., Trotman, C., Lu, L., Van Hauwaert, M. L., and Moens, L. 1994. Molecular genealogy of some nematode taxa as based on cytochrome c and globin amino acid sequences. Mol. Phylogenet. Evol. 3:92–101.

Vinogradov, S., and Moens, L. 2008. Diversity of globin function: enzymatic, transport, storage and sensing. J. Biol. Chem. (in press).

Vinogradov, S. N., Hoogewijs, D., Bailly, X., Arredondo-Peter, R., Gough, J., Dewilde, S., Moens, L., and Vanfleteren, J. R. 2006. A phylogenomic profile of globins. BMC. Evol. Biol. 6:31.

Watts, R. A., Hunt, P. W., Hvitved, A. N., Hargrove, M. S., Peacock, W. J., and Dennis, E. S. 2001. A hemoglobin from plants homologous to truncated hemoglobins of microorganisms. Proc. Natl. Acad. Sci. U.S.A. 98:10119–10124.

Weber, R. E., and Vinogradov, S. N. 2001. Nonvertebrate hemoglobins: functions and molecular adaptations. Physiol. Rev. 81:569–628.

Wittenberg, J. B. (1992) Functions of Cytoplasmic Hemoglobins and Myohemerythrin. Berlin, Heidelberg: Springer-Verlag.

Wittenberg, J. B., and Wittenberg, B. A. 1990. Mechanisms of cytoplasmic hemoglobin and myoglobin function. Annu. Rev. Biophys. Biophys. Chem. 19:217–241.

Xia, Z., Zhang, W., Nguyen, B. D., Mar, G. N., Kloek, A. P., and Goldberg, D. E. 1999. 1H NMR investigation of the distal hydrogen bonding network and ligand tilt in the cyanomet complex of oxygen-avid *Ascaris suum* hemoglobin. J. Biol. Chem. 274:31819–31826.

Yang, J., Kloek, A. P., Goldberg, D. E., and Mathews, F. S. 1995. The structure of Ascaris hemoglobin domain I at 2.2 A resolution: molecular features of oxygen avidity. Proc. Natl. Acad. Sci. U.S.A. 92:4224–4228.

Zhang, W., Rashid, K. A., Haque, M., Siddiqi, A. H., Vinogradov, S. N., Moens, L., and Mar, G. N. 1997. Solution of 1H NMR structure of the heme cavity in the oxygen-avid myoglobin from the trematode *Paramphistomum epiclitum*. J. Biol. Chem. 272:3000–3006.

# 20

# Microbial Haemoglobins: Proteins at the Crossroads of Oxygen and Nitric Oxide Metabolism

Robert K. Poole

## Abstract

The globins of microorganisms were ignored for many decades after their discovery by Warburg in the 1930s and rediscovery by Keilin in the 1950s. Three classes of microbial globin are now recognised, all having features of the classical globin protein fold. The first is typified by the myoglobin-like protein, Vgb, from the bacterium *Vitreoscilla*, and by the Cgb protein of *Campylobacter jejuni*. Second, the truncated globins, widely distributed in bacteria, microbial eukaryotes and plants, are characterised by a two-over-two helical structure while retaining the essential features of the globin superfamily. The third and best understood class are the flavohaemoglobins, possessing an additional domain with binding sites for FAD and NAD(P)H. Flavohaemoglobins have no known physiological role in oxygen metabolism but undoubtedly confer protection from NO, as do some, but not all, of the myoglobin-like and truncated microbial globins. This chapter honours the contributions of Beatrice and Jonathan Wittenberg to globin research, specifically their work on bacterial truncated globins and their thoughtful consideration of the role of the bacterial myoglobin-like proteins. The focus of the chapter is on recent and current work from the Poole laboratory, with reference to earlier studies by the Wittenbergs. We have investigated all three classes of bacterial globins and used physiological and genetic methods to yield insights into globin function and reveal new roles for these old proteins in pathogenicity.

## Haemoglobins and the Wittenbergs

From my perspective, there are two notable aspects to this volume and my contribution to it. First, it may surprise some that, in 2008, the bioscience community is *still* interested in dioxygen-binding proteins in general and in haemoglobins in particular. Indeed, in some communities (such as that of microbiology – my own – and plant science), interest in haemoglobins has never been greater, for the reasons expounded in this contribution. The second is the pleasure it gives to me to record my own appreciation of the work of Beatrice and

Jonathan, and to reflect on their distinguished contributions to this field for more than 40 years. In addition to their individual contributions, over 70 publications have been co-authored by Beatrice and Jonathan, a remarkable example of cooperativity in the haemoglobin field!

The globins are arguably the most intensively studied proteins in biochemistry and the Wittenbergs have done much to advance this field. Yet, textbooks of biochemistry and molecular biology have been slow to communicate the sustained and accelerating interest in haemoglobins and the exciting discoveries made, focusing instead on well established roles in oxygen transport and buffering in higher animals. In this chapter, I present a brief, personal, and necessarily selective, outline of microbial haemoglobins, illustrated largely by examples from my laboratory and highlighting research on the enterobacteria *Escherichia coli* and *Salmonella enterica* serovar typhimurium ('*S. typhimurium*') and the food-borne pathogen *Campylobacter jejuni*. Happily, these bacteria collectively possess all three major classes of microbial haemoglobins currently recognised.

# A trio of microbial globins

The diversity evident in the globin family, especially the globins of multicellular animals, is familiar and examples of some classes are given in this volume. However, in prokaryotic and eukaryotic microorganisms, diversity is perhaps equally striking, and three distinct but broad classes can be distinguished (Wu, Wainwright and Poole 2003) (Fig. 1). These are (1) the myoglobin-like, 'single-domain' haemoglobins, (2) the truncated haemoglobins (trHbs) and (3) the flavohaemoglobins ('two-domain' haemoglobins).

It is probable that the first description of any microbial haemoglobin was of a flavohaemoglobin in yeast, during a series of groundbreaking papers by Keilin in 1953 (Keilin 1953; Keilin and Ryley 1953; Keilin and Tissieres 1953). The flavohaemoglobins have received the most attention in recent years, due to unequivocal demonstrations of the physiological role of such proteins. Such apparent certainty regarding function for this class of globins has been missing from parallel studies of the myoglobin-like and truncated globins. Interest in flavohaemoglobins in particular has been heightened by the fact that their role is detoxification of nitric oxide (NO), a molecule that is itself the subject of intense scrutiny. As Fig. 1 shows, current evidence suggests that flavohaemoglobins are unique to microorganisms and have one function – NO detoxification. I am unaware of any convincing evidence that such proteins play important roles in oxygen management. In contrast, some members of the 'single-domain' haemoglobin class appear to function in NO detoxification, whilst others are involved in cellular oxygen management.

The truncated haemoglobins probably have no single function. In an influential minireview (Wittenberg et al. 2002), the Wittenbergs, with Martino Bolognesi and Michel Guertin, reviewed the information available up to 2002

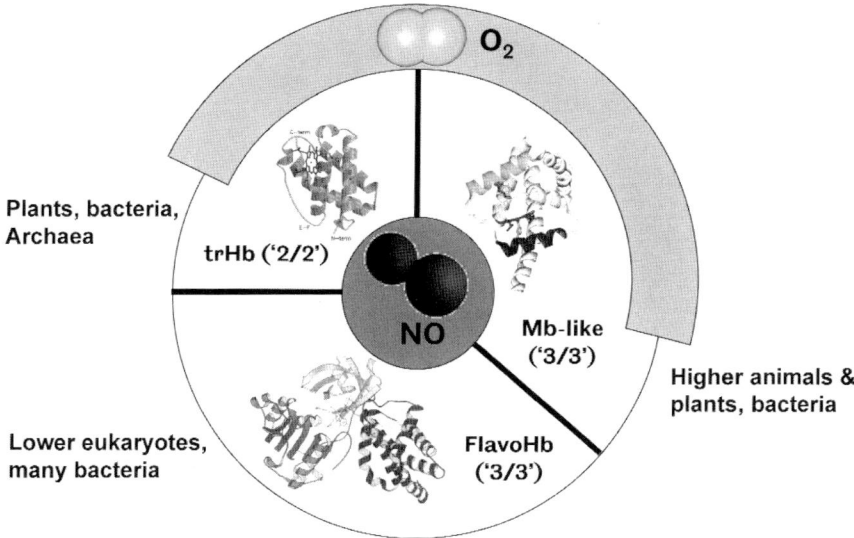

FIG. 1. The trio of microbial globins

on the truncated globins and proposed a sub-classification that has been wide-ly adopted. Phylogenetic analysis of 40 extant, putative trHb genes emphasised the distinctiveness of the trHb class and distinguished three sub-groups. In the intervening years, members of all three trHb groups have been described in detail. An example of the least well understood group (III) is the protein named Ctb in *C. jejuni*, which is described below. The Wittenbergs have made impor-tant contributions to the study of several truncated haemoglobins, notably *Mycobacterium tuberculosis* trHbO (Ouellet et al. 2003) and trHbN (Dantsker et al. 2004), *M. bovis* trHbN (Ouellett et al. 2002), *Paramecium caudatum* haemoglobin (Das et al. 2000), *Synechocystis* haemoglobin (Couture et al. 2000) and *Chlamydomonas eugametos* chloroplast haemoglobin (Couture et al. 1999; Das et al. 1999). Wittenberg et al. (2002) prudently suggested that 'the functional roles of trHbs…may be various' and more recent work has tended to confirm this. The Wittenbergs and others have focussed on the truncated haemoglobins of *M. tuberculosis*; as they point out, it is remarkable that this bacterium possesses examples of both groups I (trHbN) and II (trHbO), which presumably have distinct functions. Unfortunately, this bacterium and some other microorganisms with interesting truncated globins – *Nostoc commune, C. eugametos* and *P. caudatum* – are not readily amenable to genetic manipula-tion, impeding the direct testing of hypotheses concerning biological function. The pathogenic bacterium *C. jejuni*, however, although having only one trun-cated globin (in group III, Ctb), also possesses a myoglobin-like globin (Cgb). This bacterium *is* (to a limited extent) amenable to genetic approaches and knockout mutations of each globin have been constructed. This work is described below.

# Recent Advances in NO Biology and Enterobacterial NO Detoxification

Nitric oxide (NO) is one of the most important small molecules in biology. The 1998 Nobel Prize in Physiology recognised the seminal demonstration that NO generated by endothelial cells relaxes vascular smooth muscle through activation of guanylate cyclase. However, at high concentrations, NO is not a messenger but a toxic molecule, capable of diffusing across biological membranes and cytoplasm and reacting with diverse biomolecules such as iron–sulfur clusters (Cruz-Ramos et al. 2002), *S*-nitrosothiols and nitrosyl haems (Angelo, Singel and Stamler 2006). These reactions underpin the use of NO as a powerful weapon in the armoury of mammalian (Shiloh and Nathan 2000) and plant (Boccara et al. 2005; Mur, Carver and Prats 2006) cells to combat microbial infections. The behaviour of NO in biological systems is complicated by the ability of NO to be oxidised to the nitrosonium cation ($NO^+$), to be reduced to the nitroxyl anion ($NO^-$) and to react with oxygen to give nitrite ($NO_2^-$). The reactions of NO in biology lead to the production of other reactive nitrogen species (RNS), briefly reviewed before (Poole and Hughes 2000), such as peroxynitrite, which is formed during the oxidative burst of macrophages by the reaction of NO with superoxide anion. Indeed, the view of NO as a freely diffusible and promiscuous second messenger or toxin has been fundamentally altered by the realisation that NO targets cysteine thiols and transition metal centres of proteins and that these modifications dramatically modulate protein function.

The reactions of NO with biological targets give new species with modified activity and functions (Fig. 2). Transfer of the NO group intracellularly is exemplified by mammalian haemoglobin, in which NO can be bound to the haem as a ferrous nitrosyl adduct (Fe(II)NO) or to b-chain cysteine (CysbSNO), the NO retaining bioactivity and capable of transferring to the Cys residues in other peptides/proteins (Stamler 2003). In plants too, S-nitrosothiol formation and turnover regulate multiple processes (Feechan et al. 2005). The covalent attachment of NO groups to protein sulfydryls and transition metals is now regarded as a precisely regulated post-translational modification (Hess et al. 2005), in some respects analogous to protein phosphorylation. Recently the term 'nitrosoproteome' has been used (Rhee et al. 2005) to describe the cellular content of *S*-nitrosoproteins in *M. tuberculosis*. An exciting corollary to the substantial body of work by the Wittenbergs and others on the role of *M. tuberculosis* truncated haemoglobins in resisting nitrosative stress will be to explore the efficacy of these proteins in limiting or specifying the components of the nitrosoproteome. It is already known that deletion of Hmp (see below) in *S. typhimurium* increases the cellular content of *S*-nitrosothiols (Bang et al. 2006).

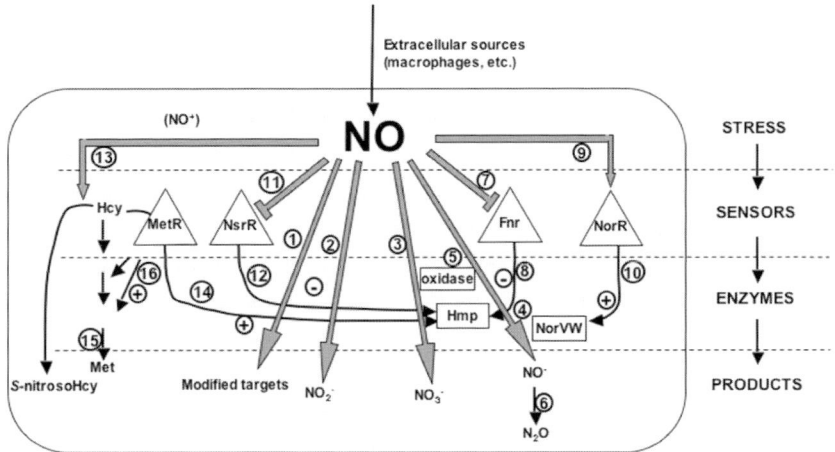

FIG. 2. Targets of NO and its fates in *E. coli*. NO, to which bacteria are exposed in natural bio-environments, or in the laboratory by administering the gas or NO-releasing compounds, has potent biological activity. It targets a variety of cellular molecules particularly thiols, haems and other metal centres (reaction 1). S-nitrosothiols are significant products. Unreacted NO has other fates including oxidation to nitrite (2). The detoxification mechanisms employed by enterobacteria involve primarily the aerobic conversion to nitrate ion (3) by flavohaemoglobin (Hmp) and the one-electron reduction to nitroxyl anion ($NO^-$) catalysed by flavorubredoxin (NorVW) (4) or possibly terminal respiratory oxidases (5) or other detoxifying enzymes. It is probable that nitroxyl dimerises in aqueous media to give nitrous oxide ($N_2O$) (6). Several NO-sensing and regulatory pathways have been identified. First, NO interacts with the [4Fe-4S] cluster of the transcriptional regulator FNR (7); since FNR represses hmp gene transcription, Hmp is synthesised (8). NorR senses NO (9) and activates transcription of NorVW (10). NsrR is a newly discovered regulator of *hmp* and a small number of other genes that may be involved in NO detoxification via unknown pathways (11, 12). NO (strictly $NO^+$, nitrosonium) reacts with homocysteine (Hcy, 13) forming S-nitrosoHcy. In the absence of Hcy, MetR activates *hmp* transcription (14). A further consequence of depletion of Hcy (an intermediate in methionine biosynthesis) is up-regulation (15) of many genes involved in the pathway of biosynthesis of this amino acid. Some of these regulatory events are directly mediated by MetR (16). The responses of the various sensors and targets to different RNS are largely unknown

# Flavohaemoglobins and Their Roles in Resisting NO and Nitrosative Stress

Unsurprisingly, due to the cytotoxic effects of NO and the presence of NO in diverse microbial habitats, bacterial cells have developed mechanisms for NO detoxification (Poole and Hughes 2000; Poole 2005). It is in this arena that interest in microbial haemoglobins has increased most dramatically. To my knowledge, the first evidence that any microbial haemoglobin might function in NO detoxification or metabolism came from our experiments in which exposure of *E. coli* to NO (or nitrite) elicited dramatic up-regulation of a gene fusion [Φ(*hmp-lacZ*)] in which the promoter of the *hmp* gene was fused to the gene encoding the easily assayed β-galactosidase (Poole et al. 1996). Prior to this, we entertained the idea that Hmp might function predominantly in oxygen metabolism, perhaps as an oxygen sensor or oxygen-activated switch (Poole, Ioannidis and Orii 1994), or perhaps in oxidative stress tolerance or respiration, as once proposed for the yeast flavohaemoglobin and *Vitreoscilla* haemoglobin, and critically reviewed by the Wittenbergs (Wittenberg and Wittenberg 1990). However, it is now well established that the *hmp* gene is subject to complex control (Poole and Hughes 2000; Spiro 2007), being up-regulated by NO, and that mutation of *hmp* alone is sufficient to render bacteria hypersensitive to NO and other nitrosative stresses (Membrillo-Hernández et al. 1999). A breakthrough in this field was the demonstration that flavohaemoglobin Hmp, in the presence of oxygen, acts as a nitric oxide denitrosylase (dioxygenase) (Gardner et al. 1998), converting NO to the relatively innocuous nitrate ion ($NO_3^-$) (Poole and Hughes 2000; Gardner 2005). The protective benefits of Hmp have been demonstrated by assays of viability after RNS treatment (Membrillo-Hernández et al. 1999), cell respiration (Stevanin et al. 2000) and intracellular (macrophage) killing assays (Stevanin et al. 2002; Stevanin, Read and Poole 2007). Both haem and flavin domains of the protein are required for protection from nitrosative stress (Hernandez-Urzua et al. 2003). The haem domain has a peroxidase-like active site (Mukai et al. 2001) appropriate for the 'push–pull' type of oxygen activation observed in the peroxidases.

Underlying the increase in Hmp synthesis that follows nitrosative stress is a complex regulatory network. First, transcription of the *hmp* gene is repressed by the 'oxygen-sensing' transcriptional regulator Fnr (Poole et al. 1996) but, in the presence of NO, the DNA-binding activity of FNR is diminished by formation of dinitrosyl–iron complexes by reaction of the iron–sulfur cluster with NO (Cruz-Ramos et al. 2002), so that NO drepresses *hmp* transcription. Second, MetR activates *hmp* transcription. Nitrosation of homocysteine (Hcy) withdraws this key intermediate from the biosynthetic pathway leading to methionine. In the absence of its cofactor (Hcy), MetR is proposed to bind to the *hmp* promoter and activate transcription (Membrillo-Hernández et al. 1998). Depletion of Hcy by nitrosation triggers enhanced expression of genes required for methionine biosynthesis (see below) (Flatley et al. 2005) in an "attempt" (teleologically) to recover pools of this essential amino acid. Third, *hmp* transcription is repressed by NsrR, an effect reversed by nitrite or, perhaps intracellularly, by NO (Bodenmiller and Spiro 2006; Spiro 2007). These mechanisms for flavohaemoglobin up-regulation by NO are probably common to many bacteria; unsurprisingly, a major transcription factor that controls Hmp synthesis in *S. typhimurium* is again NsrR (Gilberthorpe et al. 2007). There may be minor regulatory roles in the enterobacteria for Fur (D'Autreaux et al. 2002; Hernandez-Urzua et al. 2007) and SoxRS (Nunoshiba et al. 1993) in NO-regulated gene expression.

We suggest that the multilevel repression of Hmp synthesis in the absence of NO is necessitated by the propensity of the protein to act as a broad-specificity reductase and to reduce oxygen. However, unlike the proposals for the *Vitreoscilla* haemoglobin, which has been suggested to act as a terminal respiratory oxidase (Dikshit et al. 1992), and which are discussed below, we have demonstrated that Hmp reduces oxygen to superoxide and peroxide. Experimental evidence from this laboratory for superoxide generation is now compelling: (i) overexpression of Hmp in *E. coli* activates transcription of the superoxide-responsive gene *sodA*, encoding superoxide dismutase (Membrillo-Hernandez, Ioannidis and Poole 1996); (ii) *in vitro*, Hmp catalyses aerobically the oxidation of NADH, leading to superoxide and peroxide accumulation and ultimately protein destruction (Orii, Ioannidis and Poole 1992; Wu et al. 2004); (iii) in *S. typhimurium*, overexpression of Hmp (achieved experimentally by mutation of the repressor NsrR) results *in vitro* in hypersensitivity to peroxide and superoxide, and *in vivo* (within macrophages) to diminished bacterial survival (Gilberthorpe et al. 2007).

Recently, several attempts have been made to define the entire complement of genes up-regulated by NO using transcriptional profiling (gene arrays). The genome-wide responses of *E. coli* to acidified nitrite (Mukhopadhyay et al. 2004), *S*-nitrosoglutathione (GSNO) (Flatley et al. 2005) and NO (Justino et al. 2005; Hyduke et al. 2007; Pullan et al. 2007) have all been reported, with only partial agreement. One explanation of the diverse results reported is that many, various forms of nitrosative stress have been applied to bacterial cultures, despite clear evidence that GSNO and NO elicit quite different effects, for example in caspase activation (Borutaite, Morkuniene and Brown 2000). Comparison between datasets has been frustrated by (i) the use of different media – defined media and complex broths, (ii) poorly controlled batch growth conditions and, most worryingly, (iii) attributing an effect to NO when the effective stress may be, for example, a nitrosating agent, $NO_2^-$ or even iron from SNP (Schroder 2006). Indeed, only *hmp* (flavohaemoglobin), *norVW* (flavorubredoxin and its reductase) and *nrdH* – a glutaredoxin-like protein with an unknown role in RNS tolerance (Jordan et al. 1997) – have been consistently observed as up-regulated (Spiro 2006).

We have made serious efforts to resolve these issues by introducing three critical aspects of experimental design – (i) the use of a chemically defined growth medium in which all ingredients are known, (ii) the exploitation of chemostats to eliminate undesired changes in culture physiology, particularly growth rate, and (iii) a comparison under otherwise identical conditions of different RNS agents. Such chemostat studies of the GSNO stimulon in *E. coli* (Flatley et al. 2005) not only confirmed the roles of Hmp and NorVW, but also revealed that genes required for methionine biosynthesis are critical for GSNO tolerance, in accord with pre-array molecular genetic data (Membrillo-Hernández et al. 1998), in which we postulated that GSNO nitrosates intracellular thiols, including homocysteine *in vivo*. A recent study (Pullan et al. 2007) under identical conditions with NO has allowed a direct comparison of the effects of NO and GSNO in well controlled continuous culture conditions. On adding NO, *hmp* and *norVW* are again the prominent up-regulated genes, but we do not see up-regulation of *met* genes, indicating that nitrosation of homocysteine does not occur under these conditions. Furthermore, we see striking changes in gene expression controlled by Fnr (which is inactivated by NOCs (Cruz-Ramos et al. 2002)) and Fur (which is also inactivated by NO (D'Autreaux et al. 2002)). These data provide the clearest evidence yet in any microbial system that different forms of RNS have discrete effects on gene expression. The findings from several laboratories that not only NO, but also GSNO and SNP, up-regulate Hmp expression, are attributable to the release of NO from S-nitrosothiol sites *in vivo*.

Very recently, these findings have been largely confirmed and extended by network component analysis, which was elegantly used to identify transcription factors that are perturbed by NO. Such information was screened with potential NO reaction mechanisms and phenotypic data from genetic knockouts to identify active chemistry and direct NO targets in *E. coli* (Hyduke et al. 2007). This approach identified the comprehensive NO response network of this bacterium and demonstrated that NO halts bacterial growth via inhibition of the branched-chain amino acid biosynthesis enzyme, dihydroxyacid dehydratase. Clearly, Hmp plays a critical role in bacterial resistance to NO and related agents.

There are now many lines of evidence that this protein is important in the pathogenic lifestyle of several bacteria. First, mutants of *E. coli* and *S. typhimurium* are compromised in their ability to survive within macrophages (Stevanin et al. 2002; Bang et al. 2006; Stevanin, Read and Poole 2007, Gilbertherpe et al. 2007) and cause disease in experimental animals (Bang et al. 2006). Second, transcription of *hmp* and other genes revealed in recent microarray studies is evident in disease models *in situ*. These include the apparently nitrate-rich environment of the urine in patients with urinary tract infections (Snyder et al. 2004; Roos and Klemm 2006) and of experimental models of bubonic plague. *Yersinia pestis*, the causative agent of bubonic plague, has been isolated from experimental buboes in the rat, to study by microarray analysis gene expression *in situ* (Sebbane et al. 2006). These studies reveal an adaptive response to NO-derived species and to iron limitation in the extracellular environment of the bubo. Polymorphonuclear neutrophils recruited to the infected lymph node expressed abundant inducible NO synthase, and several *Y. pestis* homologues of genes involved in the protective response to reactive nitrogen species were consequently up-regulated in the bubo. Significantly, mutation of one of these genes, *hmp*, attenuated virulence (Sebbane et al. 2006).

# Bacterial NO Detoxification by Non-flavohaemoglobins

Intriguingly, the genome sequence of *C. jejuni* does not contain genes for the major NO detoxification pathways of enterobacteria (flavohaemoglobin and flavorubredoxin) (Parkhill et al. 2000). However, *C. jejuni* gene Cj1586 (*cgb*) encodes a single-domain haemoglobin, mutation of which leads to hypersensitivity to GSNO and the NO-releasing compound, spermine NONOate (Elvers et al. 2004). Consistent with the protective role of Cgb against RNS, *cgb* expression is minimal in laboratory media but is strongly and specifically induced after exposure to nitrosative stress. Cells expressing up-regulated levels of Cgb as a result of GSNO treatment exhibit enhanced resistance to the inhibitory effects of NO on cell respiration (Elvers et al. 2004). Expression of Cgb in response to NO is mediated via NssR (Cj0466), a member of the Crp-Fnr superfamily, which acts as a positive regulator of a small regulon that includes both *cgb* and *ctb* (Cj0465c), the latter encoding the *C. jejuni* truncated haemoglobin (Elvers et al. 2005). Since an *nssR* mutant shows increased hypersensitivity to RNS when compared to a *cgb* mutant, there appears to be some role for members of the regulon other than *cgb* in the nitrosative stress response (Elvers et al. 2005). Although the haemoglobin Cgb is perceived to be the major NO-detoxifying protein, the nitrite reductase NrfA appears to play a supporting role (Pittman et al. 2007).

The purified Cgb protein has been studied with resonance Raman scattering, revealing the iron-histidine stretching mode at 251 cm$^{-1}$ (Lu et al. 2007b). This frequency is unusually high, suggesting an imidazolate character of the proximal histidine as a result of the H-bonding network linking the catalytic triad involving the F8His, H23Glu and G5Tyr residues. Lowering the pH caused the shift of the iron-histidine stretching frequency from 251 to 229 cm$^{-1}$, presumably due to the protonation of the H23Glu residue. Furthermore, the mutation of the G5Tyr residue to Phe caused the protein to convert to a six-coordinate low-spin configuration. These data demonstrate the important structural and functional role of the proximal catalytic triad in Cgb. In the CO complex, two conformers were identified with the $v_{C-O}/v_{Fe-CO}$ at 529/1914 cm$^{-1}$ and 492/1963 cm$^{-1}$. The former is assigned to a "closed" conformation, in which the haem-bound CO is stabilised by the H-bond(s) donated from the B10Tyr-E7Gln residues, whereas the latter are assigned to an "open" conformer, in which the H-bonding interaction is absent. The presence of the two alternative conformations demonstrates the plasticity of the protein matrix of Cgb. In the O$_2$ complex, the iron-O$_2$ stretching frequency was identified at 554 cm$^{-1}$, which is unusually low, indicating that the haem-bound O$_2$ is stabilised by strong H-bond(s) donated by the B10Tyr-E7Gln residues. This scenario is consistent with its low O$_2$ off-rate of 0.87 s$^{-1}$. Taken together, the data suggest that the NO-detoxifying activity of Cgb is facilitated by the imidazolate character of the proximal F8His and the distal positive polar environment provided by the B10Tyr-E7Gln. They may offer electronic "push" and "pull", respectively, for the O–O bond cleavage reaction required for the isomerisation of the presumed peroxynitrite intermediate to the product, nitrate.

A major unresolved issue in our understanding of Cgb function is the nature of the reducing mechanism *in vivo*. Without an 'on-board' reductase module, as in the flavohaemoglobins, which demonstrate robust oxygen-reducing and NO-detoxifying activities in the absence of other proteins, we must presumably look to cognate partner proteins to provide the necessary electron flux for the NO-consuming reaction. Such a protein has been proposed to partner Cgb as its NADH-dependent methaemoglobin reductase (Jakob, Webster and Kroneck 1992). In *C. jejuni*, the partner protein – if it exists – is unknown, but may lie among the proteins whose synthesis is regulated by NssR (Elvers et al. 2005).

# The Oxygen-avid Truncated Haemoglobin of *C. jejuni*

There seems little doubt that Cgb performs a direct detoxification reaction to protect *C. jejuni* against NO attack. On the other hand, the physiological function of Ctb, a class III truncated haemoglobin, remains unclear. (Note that conventions in bacterial genetic nomenclature require that the protein name reflects the gene name; thus, the most appropriate name for this protein is Ctb, encoded by the *ctb* gene, annotated as Cj0465c in the *C. jejuni* genome sequence) (Parkhill et al. 2000). The function of Ctb is unclear; although its expression is elevated by nitrosative stress, a *ctb* mutant is not compromised in its tolerance of nitrosative stress-generating agents (Wainwright et al. 2005). Instead, and in view of its propensity to form an oxygenated species, it has been implicated in cellular oxygen management, perhaps in delivering oxygen to a terminal oxidase (see below) (Wainwright et al. 2005) or to Cgb itself (Wainwright et al. 2006).

By using CO as a structural probe, resonance Raman data show that the distal haem pocket of Ctb exhibits a positive electrostatic potential (Lu et al. 2007a). In addition, two ligand-related vibrational modes, $v(Fe–O_2)$ and $v(O–O)$, were identified in the oxy-derivative, with frequencies at 542 and 1132 cm$^{-1}$, respectively, suggesting the presence of an intertwined H-bonding network surrounding the haem-bound ligand, which accounts for its unusually high oxygen affinity. Mutagenesis studies of various distal mutants suggest that the haem-bound dioxygen is stabilised by H-bonds donated from the Tyr(B10) and Trp(G8) residues, which are highly conserved in the class III truncated haemoglobins. Furthermore, an additional H-bond donated from the His(E7) to the Tyr(B10) further regulates these H-bonding interactions by restricting the conformational freedom of the phenolic side chain of the Tyr(B10). Taken together, the data suggest that it is the intricate balance of the H-bonding interactions that determines the unique ligand-binding properties of Ctb (Lu et al. 2007a). The extremely high oxygen affinity of Ctb makes it unlikely to function as an oxygen transporter; on the other hand, the distal haem environment of Ctb is surprisingly similar to that of cytochrome c peroxidase, suggesting a role for Ctb in performing a peroxidase or P450-type of oxygen chemistry.

Until recently, no structural data were reported for group III trHbs. However, elegant crystallographic studies have now solved the three-dimensional structure of Ctb, named trHbP in the paper from the Bolognesi laboratory (Nardini et al. 2006). The 2.15-Å resolution structure of the cyano-met form shows that the 2-on-2 trHb fold is substantially conserved in this group III trHb, despite the absence of the Gly-based sequence motifs that were considered necessary for the attainment of the trHb specific fold. The haem crevice, however, reveals important structural modifications in the C-E region and in the FG helical hinge, with novel surface clefts at the proximal haem site. Contrary to earlier studies of group I and II trHbs, no protein matrix tunnel/cavity system is evident in *C. jejuni* Ctb. Instead, a gating movement of the His(E7) side chain (found in two alternate conformations in the crystal structure) may be the key to ligand entry to the haem distal site. Sequence conservation within the group III trHbs suggests that these structural findings may be relevant to the entire group (Nardini et al. 2006).

# Do Bacterial Haemoglobins Have Direct Roles in Oxygen Management and Cellular Respiration?

It might seem perverse to pose this question so late. From the time of Keilin in the 1950s to 1990, when the Wittenbergs reviewed 'cytoplasmic haemoglobins', it has seemed second nature to propose that microbial globins primarily have physiological roles in oxygen metabolism, perhaps as terminal oxidases (Wittenberg and Wittenberg 1990). This view has been overturned in recent years, especially by genetic approaches, which, as described above, have established clear functional roles for flavohaemoglobins in NO metabolism. However, the jury is still out on the functions of other globin classes and an open mind is needed. Certainly, in some respects, bacterial globins do "share the chemical reactivity of terminal oxidases" (Wittenberg and Wittenberg 1990), in that, in *E. coli*, expression of the *Vitreoscilla* globin can restore the aerobic growth capacity of a mutant defective in both established oxidases (cytochromes *bo'* and *bd*) (Dikshit et al. 1992). Surprisingly, there has been little published work since on this observation. *Vitreoscilla* is not a tractable experimental organism and this has frustrated efforts to establish whether such an oxidase activity occurs in this organism and represents the normal physiological function of the globin. However, in a striking demonstration, Park et al. (2002) used a two-hybrid approach to demonstrate that *Vitreoscilla* haemoglobin binds specifically to subunit I of cytochrome *bo'* ubiquinol oxidases. This is precisely the type of interaction cleverly anticipated by the Wittenbergs (Wittenberg and Wittenberg, 1990): "we suggest that cytoplasmic *Vitreoscilla* haemoglobin, in temporary association with a transmembrane protein, could bind oxygen and transfer electrons to ligated oxygen". The 1990 review

(Wittenberg and Wittenberg, 1990) also "casts doubt upon the attractive idea that periplasmic haemoglobin could facilitate diffusion of oxygen to plasma membrane terminal oxidases" based on the finding that periplasmic proteins exhibit slow rates of diffusion; indeed, to my knowledge, it has not been demonstrated that the bacterial globins sometimes found in the periplasmic compartment (Khosla and Bailey 1989; Vasudevan et al. 1995) interact with membrane-spanning oxidases.

# The Future of Haemoglobin Research

In my view, one of the most exciting and important directions in haemoglobin research will be *functional* studies. In the haemoglobin family, we have beautiful examples of functional diversity and adaptation of an apparently simple protein motif to serve diverse biology functions: oxygen carrier, oxygen store, sensor and detoxification machine. It seems to me that recognition of this plurality of function has been central to the Wittenberg's view of haemoglobins and will inspire many others to answer the questions raised in this volume.

However, the answers may be elusive. I am reminded of the address given by the *Rektor magnificus* of the University of Vienna in 1892 at a gala reception at which the composer Anton Bruckner was the guest of honour. Dr Adolf Exner concluded his address thus (Schonzeler, 1978): "Where science must come to a halt, where its progress is barred by unsurmountable barriers, there begins the realm of art which knows how to express that which will ever remain a closed book to scientific knowledge." There is much to cherish near the end of a scientific career.

*Acknowledgements*

I am grateful to the Editors for giving me this opportunity to record my appreciation of the work of Beatrice and Jonathan. Jonathan was also honoured at the 2006 International Meeting on Dioxygen-Binding Proteins in Naples, where he most modestly accepted the appreciation of the conference participants. Jonathan might have used similar words (but did not) as heard at that occasion in Vienna when Bruckner – the virtuoso improviser on the organ – was so overcome with emotion that he lost the thread of his speech and finished in confusion: "I cannot find the words to thank you as I would wish, but if there were an organ here, I could tell you".

Finally, I would like to record my deep appreciation of the work and support of another of the contributors to this volume, Cyril Appleby. Cyril ignited my interest in non-oxidase haemoproteins through his passionate interest in plant haemoglobins (when my horizon was oxidases!). At that time – before the discovery of the flavohaemoglobin gene – I had no idea how involved I would become in the haemoglobin field, or how much his passion was transferrable.

I thank Dr Syun-Ru Yeh for continued collaboration on these fascinating proteins and some of the data presented here. All work in my laboratory was supported by the Biotechnology and Biological Sciences Research Council (BBSRC, UK).

# *References*

Angelo, M., Singel, D. J., and Stamler, J. S. 2006. An S-nitrosothiol (SNO) synthase function of hemoglobin that utilizes nitrite as a substrate. Proc. Natl. Acad. Sci. U.S.A. 103:8366–8371.

Bang, I. S., Liu, L. M., Vazquez-Torres, A., Crouch, M. L., Stamler, J. S., and Fang, F. C. 2006. Maintenance of nitric oxide and redox homeostasis by the *Salmonella* flavohemoglobin Hmp. J. Biol. Chem. 281:28039–28047.

Boccara, M., Mills, C. E., Zeier, J., Anzi, C., Lamb, C., Poole, R. K., and Delledonne, M. 2005. Flavohaemoglobin HmpX from *Erwinia chrysanthemi* confers nitrosative stress tolerance and affects the plant hypersensitive reaction by intercepting nitric oxide produced by the host. Plant J. 43:226–237.

Bodenmiller, D. M., and Spiro, S. 2006. The *yjeB* (*nsrR*) gene of *Escherichia coli* encodes a nitric oxide-sensitive transcriptional regulator. J. Bacteriol. 188:874–881.

Borutaite, V., Morkuniene, R., and Brown, G. C. 2000. Nitric oxide donors, nitrosothiols and mitochondrial respiration inhibitors induce caspase activation by different mechanisms. FEBS Lett. 467:155–159.

Couture, M., Das, T. K., Lee, H. C., Peisach, J., Rousseau, D. L., Wittenberg, B. A., Wittenberg, J. B., and Guertin, M. 1999. *Chlamydomonas* chloroplast ferrous hemoglobin-Heme pocket structure and reactions with ligands. J. Biol. Chem. 274:6898–6910.

Couture, M., Das, T. K., Savard, P. Y., Ouellet, Y., Wittenberg, J. B., Wittenberg, B. A., Rousseau, D. L., and Guertin, M. 2000. Structural investigations of the hemoglobin of the cyanobacterium *Synechocystis* PCC6803 reveal a unique distal heme pocket. Eur. J. Biochem. 267:4770–4780.

Cruz-Ramos, H., Crack, J., Wu, G., Hughes, M. N., Scott, C., Thomson, A. J., Green, J., and Poole, R. K. 2002. NO sensing by FNR: regulation of the *Escherichia coli* NO-detoxifying flavohaemoglobin, Hmp. EMBO J. 21:3235–3244.

Dantsker, D., Samuni, U., Ouellet, Y., Wittenberg, B. A., Wittenberg, J. B., Milani, M., Bolognesi, M., Guertin, M., and Friedman, J. M. 2004. Viscosity-dependent relaxation significantly modulates the kinetics of CO recombination in the truncated hemoglobin TrHbN from *Mycobacterium tuberculosis*. J. Biol. Chem. 279:38844–38853.

Das, T. K., Couture, M., Lee, H. C., Peisach, J., Rousseau, D. L., Wittenberg, B. A., Wittenberg, J. B., and Guertin, M. 1999. Identification of the ligands to the ferric heme of *Chlamydomonas* chloroplast hemoglobin: Evidence for ligation of tyrosine-63 (B10) to the heme. Biochemistry 38:15360–15368.

Das, T. K., Weber, R. E., Dewilde, S., Wittenberg, J. B., Wittenberg, B. A., Yamauchi, K., VanHauwaert, M. L., Moens, L., and Rousseau, D. L. 2000. Ligand binding in the ferric and ferrous states of *Paramecium* hemoglobin. Biochemistry 39:14330–14340.

D'Autreaux, B., Touati, D., Bersch, B., Latour, J. M., and Michaud-Soret, I. 2002. Direct inhibition by nitric oxide of the transcriptional ferric uptake regulation protein via nitrosylation of the iron. Proc. Natl. Acad. Sci. U.S.A. 99:16619–16624.

Dikshit, R. P., Dikshit, K. L., Liu, Y. X., and Webster, D. A. 1992. The bacterial hemoglobin from *Vitreoscilla* can support the aerobic growth of *Escherichia coli* lacking terminal oxidases. Arch. Biochem. Biophys. 293:241–245.

Elvers, K. T., Wu, G., Gilberthorpe, N. J., Poole, R. K., and Park, S. F. 2004. Role of an inducible single-domain hemoglobin in mediating resistance to nitric oxide and nitrosative stress in *Campylobacter jejuni* and *Campylobacter coli*. J. Bacteriol. 186:5332–5341.

Elvers, K. T., Turner, S. M., Wainwright, L. M., Marsden, G., Hinds, J., Cole, J. A., Poole, R. K., Penn, C. W., and Park, S. F. 2005. NssR, a member of the Crp-Fnr superfamily from *Campylobacter jejuni*, regulates a nitrosative stress-responsive regulon that includes both a single-domain and a truncated haemoglobin. Mol. Microbiol. 57:735–750.

Feechan, A., Kwon, E., Yun, B.-W., Wang, Y., Pallas, J. A., and Loake, G. J. 2005. A central role for S-nitrosothiols in plant disease resistance. Proc. Natl. Acad. Sci. U.S.A. 102:8054–8059.

Flatley, J., Barrett, J., Pullan, S. T., Hughes, M. N., Green, J., and Poole, R. K. 2005. Transcriptional responses of *Escherichia coli* to S-nitrosoglutathione under defined chemostat conditions reveal major changes in methionine biosynthesis. J. Biol. Chem. 280:10065–10072.

Gardner, P. R. 2005. Nitric oxide dioxygenase function and mechanism of flavohemoglobin, hemoglobin, myoglobin and their associated reductases. J. Inorg. Biochem. 99:247–266.

Gardner, P. R., Gardner, A. M., Martin, L. A., and Salzman, A. L. 1998. Nitric oxide dioxygenase: an enzymic function for flavohemoglobin. Proc. Natl. Acad. Sci. U.S.A. 95:10378–10383.

Gilberthorpe, N. J., Lee, M. E., Stevanin, T. M., Read, R. C., and Poole, R. K. 2007. NsrR: a key regulator circumventing *Salmonella enterica* serovar Typhimurium oxidative and nitrosative stress in vitro and in IFN-gamma-stimulated J774.2 macrophages. Microbiology 153:1756–1771.

Hernandez-Urzua, E., Zamorano-Sanchez, D. S., Ponce-Coria, J., Morett, E., Grogan, S., Poole, R. K., and Membrillo-Hernandez, J. 2007. Multiple regulators of the Flavohaemoglobin (hmp) gene of *Salmonella enterica* serovar Typhimurium include RamA, a transcriptional regulator conferring the multidrug resistance phenotype. Arch. Microbiol. 187:67–77.

Hernandez-Urzua, E., Mills, C. E., White, G. P., Contreras-Zentella, M. L., Escamilla, E., Vasudevan, S. G., Membrillo-Hernandez, J., and Poole, R. K. 2003. Flavohemoglobin Hmp, but not its individual domains, confers protection from respiratory inhibition by nitric oxide in *Escherichia coli*. J. Biol. Chem. 278:34975–34982.

Hess, D. T., Matsumoto, A., Kim, S.-O., Marshall, H. E., and Stamler, J. S. 2005. Protein S-nitrosylation: purview and parameters. Nat. Rev. Mol. Cell. Biol. 6:150–166.

Hyduke, D. R., Jarboe, L. R., Tran, L. M., Chou, K. J. Y., and Liao, J. C. 2007. Integrated network analysis identifies nitric oxide response networks and dihydroxyacid dehydratase as a crucial target in *Escherichia coli*. Proc. Natl. Acad. Sci. U.S.A. 104:8484–8489.

Jakob, W., Webster, D. A., and Kroneck, P. M. H. 1992. NADH-dependent methemoglobin reductase from the obligate aerobe *Vitreoscilla*. Improved method of purification and reexamination of prosthetic groups. Arch. Biochem. Biophys. 292:29–33.

Jordan, A., Aslund, F., Pontis, E., Reichard, P., and Holmgren, A. 1997. Characterization of *Escherichia coli* NrdH-A glutaredoxin-like protein with a thioredoxin-like activity profile. J. Biol. Chem. 272:18044–18050.

Justino, M. C., Vicente, J. B., Teixeira, M., and Saraiva, L. M. 2005. New genes implicated in the protection of anaerobically grown *Escherichia coli* against nitric oxide. J. Biol. Chem. 280:2636–2643.

Keilin, D. 1953. Haemoglobin in fungi. Occurrence of haemoglobin in yeast and the supposed stabilization of the oxygenated cytochrome oxidase. Nature 172:390–393.

Keilin, D., and Ryley, J. F. 1953. Haemoglobin in protozoa. Nature 172:451.

Keilin, D., and Tissieres, A. 1953. Haemoglobin in moulds: *Neurospora crassa* and *Penicillium notatum*. Nature 172:393–394.

Khosla, C., and Bailey, J. E. 1989. Evidence for partial export of Vitreoscilla hemoglobin into the periplasmic space in *Escherichia coli*. Implications for protein function. J. Mol. Biol. 210:79–89.

Lu, C. Y., Egawa, T., Wainwright, L. M., Poole, R. K., and Yeh, S.-R. 2007a. Structural and functional properties of a truncated hemoglobin from a food-borne pathogen *Campylobacter jejuni*. J. Biol. Chem. 282:13627–13636.

Lu, C. Y., Mukai, M., Lin, Y., Wu, G. H., Poole, R. K., and Yeh, S. R. 2007b. Structural and functional properties of a single domain hemoglobin from the food-borne pathogen Campylobacter jejuni. J. Biol. Chem. 282:25917–25928.

Membrillo-Hernandez, J., Ioannidis, N., and Poole, R. K. 1996. The flavohaemoglobin (HMP) of *Escherichia coli* generates superoxide in vitro and causes oxidative stress in vivo. FEBS Lett. 382:141–144.

Membrillo-Hernández, J., Coopamah, M. D., Channa, A., Hughes, M. N., and Poole, R. K. 1998. A novel mechanism for upregulation of the Escherichia coli K-12 hmp (flavohaemoglobin) gene by the 'NO releaser', S-nitrosoglutathione: nitrosation of homocysteine and modulation of MetR binding to the *glyA-hmp* intergenic region. Mol. Microbiol. 29:1101–1112.

Membrillo-Hernández, J., Coopamah, M. D., Anjum, M. F., Stevanin, T. M., Kelly, A., Hughes, M. N., and Poole, R. K. 1999. The flavohemoglobin of *Escherichia coli* confers resistance to a nitrosating agent, a ''nitric oxide releaser,'' and paraquat and is essential for transcriptional responses to oxidative stress. J. Biol. Chem. 274:748–754.

Mukai, M., Mills, C. E., Poole, R. K., and Yeh, S. R. 2001. Flavohemoglobin, a globin with a peroxidase-like catalytic site. J. Biol. Chem. 276:7272–7277.

Mukhopadhyay, P., Zheng, M., Bedzyk, L. A., LaRossa, R. A., and Storz, G. 2004. Prominent roles of the NorR and Fur regulators in the *Escherichia coli* transcriptional response to reactive nitrogen species. Proc. Natl. Acad. Sci. U.S.A. 101:745–750.

Mur, L. A. J., Carver, T. L. W., and Prats, E. 2006. NO way to live; the various roles of nitric oxide in plant-pathogen interactions. J. Exp. Bot. 57:489–505.

Nardini, M., Pesce, A., Labarre, M., Richard, C., Bolli, A., Ascenzi, P., Guertin, M., and Bolognesi, M. 2006. Structural determinants in the group III truncated hemoglobin from *Campylobacter jejuni*. J. Biol. Chem. 281:37803–37812.

Nunoshiba, T., DeRojas, T., Wishnok, J. S., Tannenbaum, S. R., and Demple, B. 1993. Activation by nitric oxide of an oxidative-stress response that defends *Escherichia coli* against activated macrophages. Proc. Natl. Acad. Sci. U.S.A. 90:9993–9997.

Orii, Y., Ioannidis, N., and Poole, R. K. 1992. The oxygenated flavohaemoglobin from *Escherichia coli*: Evidence from photodissociation and rapid-scan studies for two kinetic and spectral forms. Biochem. Biophys. Res. Commun. 187:94–100.

Ouellet, H., Juszczak, L., Dantsker, D., Samuni, U., Ouellet, Y. H., Savard, P. Y., Wittenberg, J. B., Wittenberg, B. A., Friedman, J. M., and Guertin, M. 2003. Reactions of *Mycobacterium tuberculosis* truncated hemoglobin O with ligands reveal a novel ligand-inclusive hydrogen bond network. Biochemistry 42:5764–5774.

Ouellett, H., Ouellett, Y., Richard, C., Labarre, M., Wittenberg, B., Wittenberg, J., and Guertin, M. 2002. Truncated hemoglobin HbN protects *Mycobacterium bovis* from nitric oxide. Proc. Natl. Acad. Sci. U.S.A. 99:5902–5907.

Park, K. W., Kim, K. J., Howard, A. J., Stark, B. C., and Webster, D. A. 2002. *Vitreoscilla* hemoglobin binds to subunit I of cytochrome bo ubiquinol oxidases. J. Biol. Chem. 277:33334–33337.

Parkhill, J., Wren, B. W., Mungall, K., Ketley, J. M., Churcher, C., Basham, D., Chillingworth, T., Davies, R. M., Feltwell, T., Holroyd, S., Jagels, K., Karlyshev, A. V., Moule, S., Pallen, M. J., Penn, C. W., Quail, M. A., Rajandream, M. A., Rutherford, K. M., vanVliet, A. H. M., Whitehead, S., and Barrell, B. G. 2000. The genome sequence of the food-borne pathogen *Campylobacter jejuni* reveals hypervariable sequences. Nature 403:665–668.

Pittman, M. S., Elvers, K. T., Lee, L., Jones, M. A., Poole, R. K., Park, S. F., and Kelly, D. J. 2007. Growth of *Campylobacter jejuni* on nitrate and nitrite: electron transport to NapA and NrfA via NrfH and distinct roles for NrfA and the globin Cgb in protection against nitrosative stress. Mol. Microbiol. 63:575–590.

Poole, R. K. 2005. Nitric oxide and nitrosative stress tolerance in bacteria. Biochem. Soc. Trans. 33:176–180.

Poole, R. K., and Hughes, M. N. 2000. New functions for the ancient globin family: bacterial responses to nitric oxide and nitrosative stress. Mol. Microbiol. 36:775–783.

Poole, R. K., Anjum, M. F., Membrillo-Hernández, J., Kim, S. O., Hughes, M. N., and Stewart, V. 1996. Nitric oxide, nitrite, and Fnr regulation of *hmp* (flavohemoglobin) gene expression in *Escherichia coli* K-12. J Bacteriol. 178:5487–5492.

Poole, R. K., Ioannidis, N., and Orii, Y. 1994. Reactions of the *Escherichia coli* flavohaemoglobin (Hmp) with oxygen and reduced nicotinamide adenine dinucleotide: evidence for oxygen switching of flavin oxidoreduction and a mechanism for oxygen sensing. Proc. R. Soc. Lond. Series B Biol. Sci. 255:251–258.

Pullan, S. T., Gidley, M. D., Jones, R. A., Barrett, J., Stevanin, T. A., Read, R. C., Green, J., and Poole, R. K. 2007. Nitric oxide in chemostat-cultured *Escherichia coli* is sensed by Fnr and other global regulators: unaltered methionine biosynthesis indicates lack of S-nitrosation. J. Bacteriol. 189:1845–1855.

Rhee, K. Y., Erdjument-Bromage, H., Tempst, P., and Nathan, C. F. 2005. S-nitroso proteome of *Mycobacterium tuberculosis*: enzymes of intermediary metabolism and antioxidant defense. Proc. Natl. Acad. Sci. U.S.A. 102:467–472.

Roos, V., and Klemm, P. 2006. Global gene expression profiling of the asymptomatic bacteriuria *Escherichia coli* strain 83972 in the human urinary tract. Infect. Immun. 74:3565–3575.

Schonzeler, H. H. 1978. Bruckner. London, Boston: Marion Boyars Publishers Ltd.

Schroder, H. 2006. No nitric oxide for HO-1 from sodium nitroprusside. Mol. Pharmacol. 69:1507–1509.

Sebbane, F., Lemaitre, N., Sturdevant, D. E., Rebeil, R., Virtaneva, K., Porcella, S. F., and Hinnebusch, B. J. 2006. Adaptive response of *Yersinia pestis* to extracellular effectors of innate immunity during bubonic plague. Proc. Natl. Acad. Sci. U.S.A. 103:11766–11771.

Shiloh, M. U., and Nathan, C. F. 2000. Reactive nitrogen intermediates and the pathogenesis of *Salmonella* and mycobacteria. Curr. Opin. Microbiol. 3:35–42.

Snyder, J. A., Haugen, B. J., Buckles, E. L., Lockatell, C. V., Johnson, D. E., Donnenberg, M. S., Welch, R. A., and Mobley, H. L. T. 2004. Transcriptome of uropathogenic *Escherichia coli* during urinary tract infection. Infect. Immun. 72:6373–6381.

Spiro, S. 2006. Nitric oxide-sensing mechanisms in *Escherichia coli*. Biochem. Soc. Trans. 34:200–202.

Spiro, S. 2007. Regulators of bacterial responses to nitric oxide. FEMS Microbiol. Rev. 31:193–211.

Stamler, J.S. 2003. Hemoglobin and nitric oxide. N. Engl. J. Med. 349:402.

Stevanin, T. M., Ioannidis, N., Mills, C. E., Kim, S. O., Hughes, M. N., and Poole, R. K. 2000. Flavohemoglobin Hmp affords inducible protection for *Escherichia coli* respiration, catalyzed by cytochromes *bo'* or *bd*, from nitric oxide. J. Biol. Chem. 275:35868–35875.

Stevanin, T. M., Poole, R. K., Demoncheaux, E. A. G., and Read, R. C. 2002. Flavohemoglobin Hmp protects *Salmonella enterica* serovar Typhimurium from nitric oxide-related killing by human macrophages. Infect Immun. 70:4399–4405.

Stevanin, T. A., Read, R. C., and Poole, R. K. 2007. The hmp gene encoding the NO-inducible flavohaemoglobin in *Escherichia coli* confers a protective advantage in resisting killing within macrophages, but not in vitro: links with swarming motility. Gene 398:62–68.

Vasudevan, S. G., Tang, P., Dixon, N. E., and Poole, R. K. 1995. Distribution of the flavohaemoglobin, HMP, between periplasm and cytoplasm in *Escherichia coli*. FEMS Microbiol. Lett. 125:219–224.

Wainwright, L. M., Elvers, K. T., Park, S. F., and Poole, R. K. 2005. A truncated haemoglobin implicated in oxygen metabolism by the microaerophilic food-borne pathogen *Campylobacter jejuni*. Microbiology 151:4079–4091.

Wainwright, L. M., Wang, Y. H., Park, S. F., Yeh, S. R., and Poole, R. K. 2006. Purification and spectroscopic characterization of ctb, a group III truncated hemoglobin implicated in oxygen metabolism in the food-borne pathogen *Campylobacter jejuni*. Biochemistry. 45:6003–6011.

Wittenberg, J. B., and Wittenberg, B. A. 1990. Mechanisms of cytoplasmic hemoglobin and myoglobin function. Annu. Rev. Biophys. Biophys. Chem. 19:217–241.

Wittenberg, J. B., Bolognesi, M., Wittenberg, B. A., and Guertin, M. 2002. Truncated hemoglobins: A new family of hemoglobins widely distributed in bacteria, unicellular eukaryotes, and plants. J. Biol. Chem. 277:871–874.

Wu, G., Corker, H., Orii, Y., and Poole, R. K. 2004. *Escherichia coli* Hmp, an "oxygen-binding flavohaemoprotein", produces superoxide anion and self-destructs. Arch. Microbiol. 182:193–203.

Wu, G., Wainwright, L. M., and Poole, R. K. 2003. Microbial globins. Adv. Microb. Physiol. 47:255–310

# Subject Index

Printed in June 2008